# COMPUTATIONAL ACOUSTICS

**Wiley Series in Acoustics, Noise and Vibration**

# COMPUTATIONAL ACOUSTICS
## THEORY AND IMPLEMENTATION

**David R. Bergman**
*Exact Solution Scientific Consulting LLC, Morristown, NJ, USA*

*Registered Office(s)*
John Wiley & Sons, Inc., 111 River Street, Hoboken, NJ 07030, USA
John Wiley & Sons Ltd, The Atrium, Southern Gate, Chichester, West Sussex, PO19 8SQ, UK

*Editorial Office*
The Atrium, Southern Gate, Chichester, West Sussex, PO19 8SQ, UK

For details of our global editorial offices, customer services, and more information about Wiley products visit us at www.wiley.com.

Wiley also publishes its books in a variety of electronic formats and by print-on-demand. Some content that appears in standard print versions of this book may not be available in other formats.

MATLAB® is a trademark of The MathWorks, Inc. and is used with permission. The MathWorks does not warrant the accuracy of the text or exercises in this book. This work's use or discussion of MATLAB® software or related products does not constitute endorsement or sponsorship by The MathWorks of a particular pedagogical approach or particular use of the MATLAB® software.

Library of Congress Cataloging-in-Publication Data

Names: Bergman, David R., author.
Title: Computational acoustics : theory and implementation / David R. Bergman.
Description: Hoboken, NJ : John Wiley & Sons, 2018. | Includes
    bibliographical references and index. |
Identifiers: LCCN 2017036469 (print) | LCCN 2017046745 (ebook) |
    ISBN 9781119277330 (pdf) | ISBN 9781119277279 (epub) |
    ISBN 9781119277286 (cloth)
Subjects: LCSH: Sound-waves–Measurement. | Sound-waves–Computer simulation. |
    Sound-waves–Mathematical models.
Classification: LCC QC243 (ebook) | LCC QC243 .B38 2018 (print) |
    DDC 534.0285–dc23
LC record available at https://lccn.loc.gov/2017036469

Cover Design: Wiley
Cover Image: © Vik_Y/Gettyimages

Set in 10/12.5pt Times by SPi Global, Pondicherry, India
Printed and bound by CPI Group (UK) Ltd, Croydon, CR0 4YY

10  9  8  7  6  5  4  3  2  1

# Contents

# Series Preface

This book series will embrace a wide spectrum of acoustics, noise, and vibration topics from theoretical foundations to real-world applications. Individual volumes will range from specialist works of science to advanced undergraduate and graduate student texts. Books in the series will review scientific principles of acoustics, describe special research studies, and discuss solutions for noise and vibration problems in communities, industry, and transportation.

The first books in the series include those on *Biomedical Ultrasound*; *Effects of Sound on People, Engineering Acoustics, Noise and Vibration Control, Environmental Noise Management*; and *Sound Intensity and Windfarm Noise*. Books on a wide variety of related topics.

The books I edited for Wiley—*Encyclopedia of Acoustics* (1997), *The Handbook of Acoustics* (1998), and *Handbook of Noise and Vibration Control* (2007)—included over 400 chapters written by different authors. Each author had to restrict their chapter length on their special topics to no more than about 10 pages. The books in the current series will allow authors to provide much more in-depth coverage of their topic.

The series will be of interest to senior undergraduate and graduate students, consultants, and researchers in acoustics, noise, and vibration and in particular those involved in engineering and scientific fields, including aerospace, automotive, biomedical, civil/structural, electrical, environmental, industrial, materials, naval architecture, and mechanical systems. In addition the books will be of interest to practitioners and researchers in fields such as audiology, architecture, the environment, physics, signal processing, and speech.

*Malcolm J. Crocker*
Series Editor

# 1

# Introduction

Computers have become an invaluable tool in science and engineering. Over time their use has evolved from a device to aid in complex lengthy calculations to a self-contained discipline or field of study. Coursework in science and engineering often involves learning analytic techniques and exact solutions to a sizable collection of problems. Although educational, these are rarely useful beyond the classroom. On the other side of the spectrum is the practical experimental approach to investigating nature and developing engineering solutions to practical everyday problems. Most readers are familiar with the nonideal nature of things that, in many cases, prevents one from seeing the utility of the theoretical approach. In science theorist and experimentalist see nature from a different perspective in a quest for understanding its laws but agree that the facts can only be found through observation, as patterns in data acquired via well-planned and executed experiments designed to isolate certain degrees of freedom, to replicate an ideal circumstance to the best of our ability. Experiments can be costly but are the only mechanism for determining scientific truth. While the scientist works to create ideal circumstances to verify a fundamental law or hypothesis, an engineer must design and build with nonideal conditions in mind. In many cases the only approach available is trial and error. This requires the resources to build new versions of a device or invention every time it fails, each prototype being built and used to see what will happen and how it will fail and to learn from the experience. In this regard, computational science offers a path toward testing prototypes in a virtual environment. When executed carefully this approach can save time and money, prevent human injury or loss of life, and reduce impact to the environment.

As computers became larger, faster, and more efficient, the size of the tasks that could be performed also became larger, evolving from modeling the stress on a single beam to that found throughout the structure of a building, ship, or aircraft under dynamic loading. In recent times the use of computer-based modeling and simulation has gained a certain credibility in fields where there is no possible experimental method available and theory does not offer a suitable

*Computational Acoustics: Theory and Implementation*, First Edition. David R. Bergman.
© 2018 John Wiley & Sons Ltd. Published 2018 by John Wiley & Sons Ltd.

path forward in exploring the consequences of natural law, in particular the fields of numerical relativity (NR) and computational fluid mechanics (CFM). In these fields of study, the computer has become the laboratory, offering us the ability to experiment on systems we cannot build in the physical world.

Over the past decade or so, we are perhaps seeing the emergence of a new class of scientist or scientific specialist, along with the experimentalist and theorist, the numericist. Just as an experimentalist needs to be aware of the science behind the inner working of their probes and detectors, the numericist must understand the limitations imposed by working with finite precision, or a discrete representation of the real number system. The computer is the device we used to probe our virtual world, and discrete mathematics imposes constraints on the precision of the probe. In moving from the world of smooth operations on the continuum to discrete representations of the same, we lose some basic kernels of truth we rely on as common sense. Namely, certain operations are abelian. More precisely there are certain procedures that when carried out by hand produce the same results regardless of the order in which the steps are performed but when executed on a computer could lead to different results for different implementations. The development of computational procedures requires attention to this fact, a new burden for the numericist, and understanding of the impact of this behavior on expected results.

This text focuses on the application of computational methods to the fields of linear acoustics. Acoustics is broadly defined as the propagation of mechanical vibration in a medium. Several aspects of this make acoustics an interesting field of study. First is the need for a medium to support the acoustic phenomenon, which unlike light propagates in free space at constant speed relative to all inertial observers. Another point of interest is that there are as many types of acoustic phenomena as there are media, from longitudinal pressure waves in a fluid to S and P waves in seismology. The material properties of the medium determine the number and type of acoustic waves that may be created and observed. We typically think of acoustics as a macro phenomenon, the result of bulk movement of the medium. However, as we probe nature at smaller scales, this type of phenomenon is precisely what is creating the acoustic phenomenon in solids and similarly particle collisions in fluids. The acoustic phenomenon is seen at small scales in lattice vibrations in crystals. Here the acoustic field is quantized and the quanta are referred to as phonons. This model is the result of an attempt to understand a phenomenon that exists at scales too large to be described by the fundamental process and too small to be a purely classical phenomenon.

The goal of this text is to introduce to the reader those numerical methods associated with the development of computational procedures for solving problems in acoustics and understanding linear acoustic propagation and scattering. The intended audience are students and professionals who are interested in the ingredients needed for the development of these procedures. The presentation of the material in this text is unique in the sense that it focuses on modeling paradigms first and introduce the numerical methods appropriate to that modeling paradigm rather than offer them in a preliminary chapter or appendix. Along the way, implementation issues that readers should be aware of are discussed. Examples are provided along with suggested exercises and references. The intent is to be pedagogical in the approach to presenting information so that readers who are new to the subject can begin experimenting. Classic methods and approaches are featured throughout the text while additional comments are included that highlight modern advances and novel modeling approaches that have appeared in the literature.

Since the intended audience consists of upper-level undergraduate students, graduate students, or professionals interested in this discipline, expected prerequisites to this material are:

- An introductory course that covers acoustics or fluid dynamics
- Familiarity with ordinary differential and partial differential equations, perhaps a course in mathematical methods for scientists and engineers
- Some exposure to programming in a high-level language such as Maple, Mathematica, and MATLAB or its open-source counterparts, SCILAB and Octave

The key feature of the presentation contained in this text is that it serves to bridge the gap between theory and implementation. The main focus is on techniques for solving the linear wave equation in homogeneous medium as well as inhomogeneous and anisotropic fluid medium for modeling wave propagation from a source and scattering from objects. Therefore, the starting point for much of this text will be the standard wave equation or the Helmholtz equation.

The transition from equations to computer procedures is not always a straightforward path. High-level programming languages come with easy-to-use interfaces for solving differential equations, matrix equations, and performing signal processing. Beyond these are professional software packages designed to allow users to build and run specific types of simulations using common modeling paradigms. Examples include ANSYS, FEMLAB, and FEKO, just to name a few. An understanding of the math, physics, and numerics is required to evaluate and interpret the results, but low-level programming is not necessary. Why learn these techniques? Specialized software can be very expensive, in fact cost prohibitive for students or those engaging in self-study. Many software companies offer personal or student versions of their software at a severely discounted price and with a restricted user license. If the reader is using this text for coursework in computational acoustics at a college or university, chances are student licenses for some professional software packages are made available through the campus bookstore. If not, it is easy to find this information online. Open-source versions of professional software exist and are worth trying. The downside to this is that bugs exist and due to certain constraints a fix may not be available in a hurry. Also, some open-source tools are not compatible with all operating systems. Readers who like programming and are amenable to the open-source philosophy can always contribute their fixes and upgrades (read the license). Pure curiosity drives most scientists and engineers to want to know what's going on in any system, and this is a driver for developing homegrown algorithms even when libraries are available.

A brief description of each chapter is provided. Chapter 2 introduces topics related to numerics, computers, and algorithm development. These topics include binary representation of numbers, floating-point numbers, and $O(N)$ analysis, to name a few. Chapter 3 contains a survey of the linear wave equation and its connection to the supporting medium, from elastic bodies to fluids. In this chapter the linear wave equation for acoustics in a moving medium is introduced and discussed in detail. Chapter 4 introduces a variety of mathematical techniques and methods for solving the wave equation and describing the general behavior of the acoustic field. Chapter 5 discusses a variety of topics related to the analysis of acoustic waves: dispersion, refraction, attenuation, and Fourier analysis. After these chapters the structure of the text focuses on specific modeling techniques. In Chapter 6 normal modes are discussed. The wave equation is solved for a variety of 1-dimensional (1-dim) refractive profiles using exact methods, perturbation theory, and the numerical technique of relaxation. The chapter closes with a brief description of coupled modes and their use in modeling acoustics in realistic environments. Chapter 7 provides an introduction to ray theory and ray tracing techniques. Exact solutions to 1-dim problems are discussed along with methods of developing ray trace procedures that account for 3-dim propagation without simplifying assumptions. Numerical techniques are also discussed and the Runge–Kutta method is introduced. In Chapter 8 the finite difference (FD) and finite difference time domain (FDTD) technique are discussed in theory and applied to the wave equation in the frequency and time domains. Following the FD method, Chapter 9 discusses the parabolic equation and its application to modeling sound in ducted environments. Chapter 10 provides an introduction to the finite element method (FEM), introducing numerical techniques required for building an FEM model of the acoustic field in the frequency domain. The last chapter, Chapter 11, is dedicated to the boundary element method (BEM). This chapter discusses the integral equation form of the Helmholtz equation and its discretization into a matrix equation. The exterior and interior problems are discussed, but attention is spent on developing models of the scattering cross section of hard bodies. This chapter introduces techniques for dealing with singular integrals.

# 2

# Computation and Related Topics

This chapter introduces a collection of topics related to computation, model and simulation development, and code writing, starting with an introduction to floating-point numbers that introduces representations of numbers in bases other than 10 and floating-point representations of numbers. Following this is an introduction to estimating computational cost using $O(N)$ analysis. The next section provides a discussion on simulation fidelity and complexity followed by a simple example of converting an equation to pseudo code. The last section provides a compiled list of open-source alternative to professional software and open-source numerical libraries for C/C++.

## 2.1  Floating-Point Numbers

### 2.1.1  Representations of Numbers

A number, $x$, is represented by a power series in powers of a fixed number $b$ called the base. The power series may be finite or infinite depending on the number:

$$x = \sum_{n=-N_1}^{N_2} a_n b^n \qquad (2.1)$$

The coefficients in the expansion are given by $\{a_n\}$, and $N_1$ and $N_2$ are the limits of the expansion. Coefficients obey the inequality $0 \le a_n < b$, and count how many of that power are present in the number. For irrational and rational numbers with infinitely repeating patterns, $N_1 = \infty$. For all other rational numbers, $N_1$ is finite. One typically denotes the number by

*Computational Acoustics: Theory and Implementation*, First Edition. David R. Bergman.
© 2018 John Wiley & Sons Ltd. Published 2018 by John Wiley & Sons Ltd.

writing the coefficients in a sequence without the base explicitly present. A decimal notation is used to separate positive powers from negative powers:

$$x = a_{N_2} a_{(N_2-1)} a_{(N_2-2)} \cdots a_1 a_0 . a_{-1} a_{-2} \cdots a_{(1-N_1)} a_{-N_1} \tag{2.2}$$

Reading the digit sequence from left to right gives the number of each power of the base contained in the series expansion. We grow up learning base 10 and most readers have likely encountered base 2, or binary, representation of numbers. The following notation is used to keep tabs on which base is being used:

$$x = \left( a_{N_2} a_{(N_2-1)} a_{(N_2-2)} \cdots a_1 a_0 . a_{-1} a_{-2} \cdots a_{(1-N_1)} a_{-N_1} \right)_b \tag{2.3}$$

In some cases the parentheses are omitted. As an example, the number $(237.4631)_{10}$ is represented as a series expansion:

$$1 \times 10^{-4} + 3 \times 10^{-3} + 6 \times 10^{-2} + 4 \times 10^{-1} + 7 \times 10^0 + 3 \times 10^1 + 2 \times 10^2$$

Notice the reversed order of appearance of the coefficients. Now consider the base, $b = 2$. In terms of a power series expansion, numbers are represented in terms of "ones place," "twos place," "fours place," and so on. Coefficients in the expansion are bound by the inequality $0 \le a_n < 2$; hence the coefficients can only be 0 or 1. Table 2.1 provides a list of the integers from 0 to 10 in binary representation.

This example illustrates the value in using notation that references the base. The third row of the right column contains 10, which is not the integer 10 but the binary representation of 2, one in the twos place and zero in the ones place. Using the base notation in (2.3), $7_{10} = 111_2$, both are representations of the number 7. Numbers between 0 and 1 are represented in terms of negative powers of base 2, that is, a halves place, a quarters place, and so on. A few examples are presented in Table 2.2.

**Table 2.1** Binary representation of integers

| Integer (base 10) | Expansion in base 2 | Base 2 representation |
|---|---|---|
| 0 | $0 \times 2^0$ | 0 |
| 1 | $1 \times 2^0$ | 1 |
| 2 | $0 \times 2^0 + 1 \times 2^1$ | 10 |
| 3 | $1 \times 2^0 + 1 \times 2^1$ | 11 |
| 4 | $0 \times 2^0 + 0 \times 2^1 + 1 \times 2^2$ | 100 |
| 5 | $1 \times 2^0 + 0 \times 2^1 + 1 \times 2^2$ | 101 |
| 6 | $0 \times 2^0 + 1 \times 2^1 + 1 \times 2^2$ | 110 |
| 7 | $1 \times 2^0 + 1 \times 2^1 + 1 \times 2^2$ | 111 |
| 8 | $0 \times 2^0 + 0 \times 2^1 + 0 \times 2^2 + 1 \times 2^3$ | 1000 |
| 9 | $1 \times 2^0 + 0 \times 2^1 + 0 \times 2^2 + 1 \times 2^3$ | 1001 |
| 10 | $0 \times 2^0 + 1 \times 2^1 + 0 \times 2^2 + 1 \times 2^3$ | 1010 |

**Table 2.2**  Binary representations of fractions

| Fraction (base 10) | Expansion in base 2 | Base 2 representation |
|---|---|---|
| 0.5 | $1 \times 2^{-1}$ | 0.1 |
| 0.25 | $0 \times 2^{-1} + 1 \times 2^{-2}$ | 0.01 |
| 0.75 | $1 \times 2^{-1} + 1 \times 2^{-2}$ | 0.11 |
| 0.125 | $0 \times 2^{-1} + 0 \times 2^{-2} + 1 \times 2^{-3}$ | 0.001 |
| 0.375 | $0 \times 2^{-1} + 1 \times 2^{-2} + 1 \times 2^{-3}$ | 0.011 |
| 0.625 | $1 \times 2^{-1} + 0 \times 2^{-2} + 1 \times 2^{-3}$ | 0.101 |
| 0.875 | $1 \times 2^{-1} + 1 \times 2^{-2} + 1 \times 2^{-3}$ | 0.111 |

For the example above, fractions that can be expressed with three or fewer coefficients down to 1/8th are presented. Some of the numbers in the table contain the same number of significant figures in both bases. This is serendipitous.

As one final example, consider $5.625_{10} = 101.101_2$. This example illustrates the fact that a different number of significant figures is required to express a number in different bases. This is an important fact whose consequences cannot be overlooked, especially in the world of finite precision arithmetic [1]. One consequence is that some fractions may have a finite number of coefficients in one representation while producing an infinite repeating sequence in another representation. The maximum number of coefficients for representing a number that can be stored in memory is restricted, which means that error will necessarily exist when approximating such numbers. Recall how the fraction 1/3 is dealt with in base 10, 0.3333..., or $0.\bar{3}$ to be exact. The bar notation indicates that the sequence repeats an infinite number of times. When using the number in a calculation, it would be truncated, keeping as many places as necessary to maintain the proper number of significant figures in the final answer, for example, $1/3 \approx 0.3333$. Now consider what would happen if only four significant figures were allowed for a number in any representation. This limitation imposes a new constraint called precision. The last example now reads $5.625_{10} \approx 101.1_2$. Starting with this number in base 10 representation, converting to base 2, truncating to four significant figures, and then converting back to base 10 gives $5.500_{10}$, or $5.625_{10} \approx 5.500_{10}$. This is not a horrible approximation, but can we do better? Not to this level of precision.

Three more bases commonly used in computer science are septal (base 7), octal (base 8), and hexadecimal (base 16). Coefficients in septal and octal can be represented by their integer values in base 10, 0–6, and 0–7, respectively. For hexadecimal, 16 characters are needed for each coefficient. The convention used is that each $a_n$ takes a value in the set $\{0, 1, 2, \ldots 9, A, B, \ldots, F\}$. Table 2.3 lists the first 16 whole numbers in all representations introduced in this section.

## 2.1.2  Floating-Point Numbers

The IEEE Std 754-1985 defines a standard for representing various numbers as a sequence of bits called a bit string [2]. To represent arbitrary numbers, a form of base 2 scientific notation is used:

$$N = (-1)^S m \times 2^E \tag{2.4}$$

**Table 2.3**  Four alternate representations of the first 16 (base 10) integers

| Integer (base 10) | Binary | Septal | Octal | Hexadecimal |
|---|---|---|---|---|
| 1 | 1 | 1 | 1 | 1 |
| 2 | 10 | 2 | 2 | 2 |
| 3 | 11 | 3 | 3 | 3 |
| 4 | 100 | 4 | 4 | 4 |
| 5 | 101 | 5 | 5 | 5 |
| 6 | 110 | 6 | 6 | 6 |
| 7 | 111 | 10 | 7 | 7 |
| 8 | 1000 | 11 | 10 | 8 |
| 9 | 1001 | 12 | 11 | 9 |
| 10 | 1010 | 13 | 12 | A |
| 11 | 1011 | 14 | 13 | B |
| 12 | 1100 | 15 | 14 | C |
| 13 | 1101 | 16 | 15 | D |
| 14 | 1110 | 20 | 16 | E |
| 15 | 1111 | 21 | 17 | F |
| 16 | 10000 | 22 | 20 | 10 |

**Table 2.4**  Exponent and mantissa widths and exponent bias for floating-point numbers

| Type | $N$ | $M$ | Bias |
|---|---|---|---|
| Single | 8 | 23 | 127 |
| Double | 11 | 52 | 1023 |
| Extended | 15 | 64 | 16383 |
| Quad | 15 | 112 | 16383 |

Three quantities specify the number $N$. $S$ is the sign and can be either $(0, 1)$ for + or –, respectively. The number $m$ is called the mantissa and is a binary fraction of the form 1.F. The exponent is given by $E$. Normalized numbers have the exponent biased depending on the precision. When represented as a bit string, the binary representation of these numbers is placed in the order $(S, E, m)$. In binary notation, the mantissa digits are denoted $b_i$, and the exponent is denoted $a_j$, where the limits on $i, j$ are related to the type, that is, single precision or double precision. A visual representation of the bit string is given as follows:

$$\pm \quad a_1 a_2 \cdots a_N \quad b_1 b_2 \cdots b_M$$

In normalized format the leading bit of the mantissa is 1 and the exponent has a bias of $2^{N-1} - 1$. The width of the exponent and mantissa along with the bias is listed in for single-, double-, extended-, and quad-precision floating-point number (Table 2.4).

As an example, the single-precision floating-point representation of 3.7 is given as follows:

$$0 \quad 10000000 \quad 11011001100110011001100$$

**Table 2.5**  Machine epsilon for floating-point types

|         | Machine epsilon | |
| --- | --- | --- |
| Type | Epsilon (binary) | Epsilon (decimal) |
| Single | $2^{-23}$ | $\sim 1.19 \times 10^{-7}$ |
| Double | $2^{-52}$ | $\sim 2.22 \times 10^{-16}$ |
| Extended | $2^{-63}$ | $\sim 1.08 \times 10^{-19}$ |
| Quad | $2^{-112}$ | $\sim 1.93 \times 10^{-112}$ |

The binary representation of 0.7 produces an infinite repeated sequence, $0.10\overline{1100}$, which is truncated. There are a few special numbers in the IEEE standard. When the exponent reaches its maximum value and the mantissa is 0, this is defined as infinity in the floating-point system. Another floating-point number is NaN, or not a number. This arises in situations such as 0/0, $\infty/\infty$, $\infty - \infty$, $-\infty + \infty$, and $0 \times \infty$. There are two types of NaNs, quiet and signaling. A signaling NaN raises an exception, whereas the quiet NaN will propagate through a routine without any problem, producing NaN for the output. Associated with floating-point arithmetic is the notion of machine epsilon. Given a base and precision, the width of the mantissa including the implicit bit and the machine epsilon is defined as

$$\varepsilon = b^{-(p-1)} \tag{2.5}$$

Machine epsilon in binary and decimal representation is listed in Table 2.5 for the four floating-point types in Table 2.4.

## 2.2  Computational Cost

There are two types of cost to consider in simulation development: processing time and memory. One is typically concerned with estimating how these quantities scale with the size of the simulation input, for example, number of degrees of freedom. For calculations involving a large number of operations or a set of operations acting on a large number of degrees of freedom, the "Big $O$" notion is useful for estimating the processing time and memory. If an algorithm operates on a large number of array elements, $N$, Big $O$ refers to the largest power of $N$ obtained in counting the operations. It represents the limiting behavior of the algorithm. The purpose of the notation is to provide insight into how processes scale. Overall constants are dropped and the order of the process is quoted as $O(N^{\alpha})$, where $\alpha$ is a real number, usually but not necessarily an integer. Leading order estimates do not have to be a power of $N$. Examples include $O(N \log N)$, $O(2^N)$, and $O(N!)$. To estimate the cost of an algorithm, values for the cost of various processes are needed. A rough estimate could assume that all calculations are equal in cost for a single execution, but this is not very accurate. The fastest operation on a computer is addition (and subtraction). On current processors, multiplication of floating-point numbers is about the same cost as addition. Division of two numbers requires more than one addition and is a more expensive operation. Denoting the cost of an addition by $c_a$ for a floating-point operation, division can cost between 3 and 10 times $c_a$, the square root function between 1 and 18 times, and transcendental functions, trigonometric, and so on up to 15–50 times the cost of an addition. These ranges were estimated using information from Ref. [3], scaling by the cost of a floating-point addition. The range of values is due to differences in architecture location of data in memory and other factors. The take-away is that more complex operations have a larger individual unit of cost. As an example, consider the cost of evaluating the dot product between two vectors of size $N$. The operation will require $N$ multiplications and $N-1$ additions. The total cost of these operations is estimated to be $C_{Dot} = 2c_aN - c_a$. Taking the largest power of $N$ and dropping constant, the dot product is an $O(N)$ algorithm. Doubling the size of an array would double the processing time for this calculation. Matrix multiplication can be thought of as taking a dot product for every row of one matrix with every column of the other matrix. For two $N \times N$ matrices, this is $N^2$ dot products so the process is $O(N^3)$. Inverting a matrix by Gauss–Jordan elimination is also $O(N^3)$. Consider the process of solving the linear system, $A \cdot x = b$, for $M$ distinct r.h.s. inputs, $b$. Once the matrix is inverted, there is no need to invert it again to solve the system with different input vectors. Each multiplication of a matrix times a vector consists of $N$ dot products, each row of $A^{-1}$ with $b$. Solving the system is an $O(N^2)$. Given $M$ r.h.s. vectors, the total cost is $\sim N^3 + MN^2$. Clearly the inverting is the most expensive part of the procedure. Even for a large number of inputs, say, $M \sim N$, the entire process is still only $O(N^3)$. Of course, it would be wrong to conclude that there is no extra cost to adding more instances of $b$ on the pile. That would be comparing apples to oranges. If the inversion is a one-time task followed by a series of applications to an input, then the cost to solve for multiple inputs scales as $MN^2$. Lastly, if one were to run a routine for solving a linear system $M$ times that only accepted a single $b$ and did not save the inverse, the cost would be $O(MN^3)$. This is clearly never a wise move.

The use of Big $O$ provides an estimate of how algorithms scale, but it is equally important to consider the low-level unit cost of operations. The following example is a modified version of that found in Ref. [3]. Consider the cross term of an interference pattern. Two $N$-dimensional ($N$-dim) arrays of data, $A$ and $B$, are given, and the calculation being done is $\cos(A_i - B_j)$ for all pairs $i \neq j$. Since the operation is symmetric, there is no need to calculate this for each pair. The total number of calculations is $N(N-1)/2$:

```
for i = 1 to N
    for j = i + 1 to N
        Zij = cos (Ai - Bj)
```

The cost of this algorithm is $c_T N(N-1)/2$, where $c_T$ is the cost of a trigonometric function evaluation, the cost of subtraction being ignored relative to transcendental function evaluation. The same calculation can be done by evaluating the trigonometric functions in a single loop and evaluating the quantity $Z_{ij}$ using the trig identity $\cos(a-b) = \cos a \cos b + \sin a \sin b$:

```
for i = 1 to N
    X1 = cos Ai
    X2 = cos Bi
    Y1 = sin Ai
    Y2 = sin Bi

for i = 1 to N
    for j = i + 1 to N
        Zij = X1iX2j + Y1iY2j
```

**Table 2.6**   Size of various numbers in bytes

| Number type | Size in bytes | Symbol |
|---|---|---|
| Short integer | 2 | s |
| Integer | 4 | i |
| Long integer | 8 | l |
| Single | 4 | f |
| Double | 8 | d |
| Quad | 16 | q |

The cost for this process is $4c_T N$ for the first loop, evaluation of the trigonometric functions, and $3c_a N(N-1)/2$ for the double loop, for a total of $4c_T N + 3c_a N(N-1)/2$. The leading orders of these algorithms are $O(c_T N^2)$ for the first approach and $O(3c_a N^2)$ for the second. For large $N$ the ratio of these is $3c_a/c_T$. Based on the values for evaluating transcendental functions provided earlier, the second approach can result in significant time savings.

Another cost estimate is memory, either the amount of memory that is required in RAM for a given function or the amount of memory that will be produced as output and written to a file. The same estimation process used for estimating process time carries over. The difference is in the unit cost, in this case bytes. To get an accurate estimate, one needs to know the cost of various data types. This boils down to knowing the number of bytes that a single type of data will occupy. Table 2.6 lists some common numeric data types, their size in bytes, and the symbol used to refer to them in this text. The size of a double, *sizeof (double)* in C, is denoted by d.

Consider a memory estimate for an $N \times N$ matrix. Clearly there are $N^2$ numbers to store, and the total size will depend on the type of data in the matrix. The memory required for this is sizeof(*type*)$N^2$. A square matrix double-precision numbers with 10,000 rows and columns would occupy $8 \times (10^4)^2$ bytes $= 0.8$GB, almost one gigabyte of memory. If a single-precision result is sufficient for later use, this can be cut in half. It is helpful to estimate the memory that will be stored by temporary variables in a software program to ensure that it will not grow too large for the available RAM. Avoiding unnecessary copies of data or the use of additional variables and steps can streamline the performance of algorithms and simulation code. Similarly, it is always a good idea to get an estimate of the amount of output data that will be produced by a simulation and written to file to ensure there is enough storage.

## 2.3   Fidelity

The purpose of a simulation is to represent the behavior of something in the real world. Fidelity is related to the ability of a simulation to accurately represent the state or behavior of a system. When developing a simulation, it is important to decide ahead of time what is really important. Trying to model a complex system with unknown variables using detailed physics-based simulations may likely not succeed in reproducing data gathered in the field. Under ideal circumstances, carefully measured data under controlled conditions should match results from a well-designed model of the system. Common adjectives like "high," "medium," and "low" are frequently used in describing the fidelity of simulations. There is nothing wrong with these descriptions as long as they are meaningful to the person using them. That is to say, when someone with expertise in a field describes a simulation as being high fidelity, most likely they are applying a series of technical comparisons to the results of the simulation in their mind, possibly recalling past experience using the simulation and a lifetime of knowledge regarding what to expect. Defining a fidelity requirement to drive the scope of a simulation development effort requires a detailed description of what the simulation is meant to model, specific descriptions of allowed errors in outputs, and conditions under which the simulation and its results will be used. Using a term like "high fidelity" or even "low fidelity" to set expectations for a simulation will lead to an endless spiral of rework to meet an undefined, and hence unrealizable, expectation. This topic has been the subject of efforts to develop a quantitative, technically meaningful definition of fidelity. The reader is encouraged to review Ref. [4]. Another factor related to fidelity decisions is complexity. Simulations that are very complex can be difficult to test, debug, and maintain. They can also become unstable, being well behaved and producing reliable results only in a limited range of cases.

For the purposes of this text, the system under study is an equation, the wave equation. True its value only resides in the fact that it can describe physical phenomena like acoustics, optics, and quantum mechanics, but it is this equation and the behavior of its solutions under mathematically ideal circumstances that is of interest. A particular numerical implementation of the wave equation can be considered to have "good" fidelity if it reproduces the expected results to within some tolerance when compared with known exact solutions. This is really a validation criterion. Ensuring a good match between two models does not ensure matching with nature, and it does not mean that the implementation will continue to perform as well when it is used outside the set of validation test cases. Validating a procedure against other known results is used to build confidence in the new implementation and, if they fail to match, uncover unintentional errors. Again, the focus of the text is on methods for modeling the wave equation. How the outputs of this equation are used to model music on a windy day, sonar pings in a noisy environment, and so on is a different matter. The person developing models for these purposes has to take into account the unknowns and their potential impact on fidelity. Through this process reasonable compromises can be made, leading to a reduction in complexity while producing useful results.

## 2.4   Code Development

This section deals with the process of converting a mathematical expression into some type of computer program. This is done by example to illustrate the kind of issues that can arise. Given some proficiency in at least one language, even a high-level language, the real task is deciding how to approach the problem of turning equations into code. This assumes that a modeling paradigm has been chosen.

Sometimes a set of equations written on paper already looks like pseudo code but this is typically not the case. One equation does not easily translate to one line of code. Points that are often overlooked are; (1) that there are preconditions on data, and (2) constraints required for the answers to make sense. Sometimes simple math equations come with lists of caveats, and these caveats are not usually considered as part of the "equation," but in fact they are and need to be implemented for things to work. Whether or not a particular condition or constraint requires additional processing steps or data types depends on the language used.

A simple example is provided to illustrate this point. The task is to write a function that calculates the two roots of a quadratic expression given the coefficients in standard form:

$$r_{1,2} = \frac{-b \pm \sqrt{b^2 - 4ac}}{2a} \tag{2.6}$$

There are a few things hidden in this equation. The first is that it is two equations, one for each of two roots. This isn't hard to handle but may look different in MATLAB than in C [5]. Sample code for both is provided as follows:

```
%
% Example of quadratic formula in MatLab
%
function [roots] = quadraticRoots(a,b,c)

    inv_2a = 1.0/(2*a);
    R = sqrt(b^2 - 4*a*c);
    Roots = inv_2a*[-b + R; -b - R];

% end example
```

Next here is the same example in C:

```
/*
Example of quadratic formula in C
*/

double quadraticRoots(double a, double b, double c, double *r){

    double R, inv_2a;

    inv_2a = 1.0/(2.0*a);
    R = sqrt(b*b - 4*a*c);

    r[0] = inv_2a*(R-b);
    r[1] = -inv_2a*(R+b);

}
```

An obvious difference is the occurrence of "double." This is a key word in C and many other languages to indicate what type of number is being used to approximate the variable and how much memory to allocate. In C the type of data each variable represents needs to be specified. In MATLAB and other high-level languages, all variables are assumed to be double unless specified otherwise. In defining a function in C, the output, a pointer to an array, is specified as an input, indicated by the "double *r". Finally notice that the index for $r$ starts at 0, whereas in MATLAB it would start at 1. The next thing to notice in the quadratic formula is the possibility of a negative radicand and hence a complex result for the root. A human can understand all the possible cases and what they mean geometrically, but a computer can't. It needs to be told how to give the user what they want, and the user needs to interpret the results based on an understanding of the original problem. In MATLAB, this is not an issue since MATLAB is capable of handling complex numbers without special instructions. One could say that all numbers in MATLAB are double, complex, 1 by 1 matrices. The number 1 would be $1.000... + 1i\,0.000...$, $1i$ is the syntax for $\sqrt{-1}$. In C/C++ one would need to either make each root a 2-dim array containing the real and imaginary parts or declare a complex double data type that requires an additional library. This is the least of our trouble since C will not understand what it means to calculate sqrt(−1) under normal circumstances. To continue this example, assume the use of a programming language that does not understand complex numbers but avoid C/C++ specific syntax [6]. The goal is to write a function that is capable of producing both roots and expressing all the information required to interpret them. Start by writing down exactly what will be done and what is needed in a detailed set of instructions. In this example as many as four output variables are needed; self-explanatory names are provided as follows:

```
realRoot1, imagRoot1, realRoot2, imagRoot2
```

Given the coefficients $a$, $b$, and $c$, calculate the radicand $b*b - 4*a*c$ and call it rdcnd. We now have a variable $rdcnd = b*b - 4*a*c$.

Check the sign on the radicand, $b*b - 4*a*c$, since that determines the type of roots. There are three cases:

Case 1: rdcnd $= 0$, there is one real root not two.
Case 2: rdncd $> 0$, there are two real roots.
Case 3: rdcnd $< 0$, there are two complex roots.

The third case is the cause of possible issues. Here is a strategy for dealing with case 3. If rdcnd $< 0$, then change the sign, calculate the radical term in the quadratic formula, and store the result as the imaginary part of the output.
  Calculate the two parts of the root and give them names:

```
x = -b/(2*a), y = sqrt(abs(rdcnd))/(2*a)
```

Now store the results depending on which of the three cases is realized.
  If case 1 is true, then let

```
realRoot1 = x, imagRoot1 = 0, realRoot2 = x, imagRoot2 = 0
```

  If case 2 is true, then let

```
realRoot1 = x + y, imagRoot1 = 0, realRoot2 = x - y, imagRoot2 = 0
```

Finally, if case 3 is true, then let

```
realRoot1 = x, imagRoot1 = y, realRoot2 = x, imagRoot2 = -y
```

The previous dialog goes through all the logic behind the three cases. A cleaner version of the pseudo code is provided as follows:

```
Input variables
a, b, c

Output variables
    realRoot1, imagRoot1, realRoot2, imagRoot2

Processing instructions
    rdcnd = b*b - 4*a*c
    x = -b/(2*a)
    y = sqrt(abs(rdcnd))/(2*a)

    If (rdcnd = 0)
        realRoot1 = x
        imagRoot1 = 0
        realRoot2 = x
        imagRoot2 = 0

    If (rdcnd > 0)
        realRoot1 = x + y
        imagRoot1 = 0
        realRoot2 = x - y
        imagRoot2 = 0

    If (rdcnd < 0)
        realRoot1 = x
        imagRoot1 = y
        realRoot2 = x
        imagRoot2 = -y
```

Writing things out this way makes the logic flow and the steps clear. And it looks like code in many languages. The virtue of writing pseudo code is that it serves as an organizing tool revealing redundant steps and potential optimizations. It serves as a shorthand notation for a script, and that is exactly what code is, a script. This example was meant to illustrate the steps involved in writing a simple procedure. Since all languages are meant to allow scripts to be turned into binary instructions, it makes sense to start organizing things in terms of this approach. Identify what the inputs and outputs are and if possible what data type is needed. Then write instructions as if you were teaching someone how to perform the calculation. Once the instructions are complete, start organizing those instructions in shorthand. At this point the rest depends on the choice of programming language.

## 2.5   List of Open-Source Tools

This section offers a small collection of open-source tools that falls into one of three categories: libraries, compilers, and other tools for use with C/C++, open-source equivalents of MATLAB, Maple or Mathematica, and miscellaneous software that is relevant to the subject matter of this text (Tables 2.7, 2.8, and 2.9).

*Disclaimer(s)*

The author is not endorsing any one package over another or the use of open source as an alternative to professional software packages. These lists are provided as a courtesy for readers interested in using and contributing to open source.

It is the responsibility of the reader to read and understand the license agreement associated with these software libraries and alternatives to professional software. What you can or cannot do is expressed in the license. Read the license.

**Table 2.7**   Open-source alternatives to MATLAB, MAPLE, or Mathematica

| List of software tools and libraries | | |
|---|---|---|
| Name | Where to find | Brief description |
| SCILAB | http://www.scilab.org/ | Open-source version of MATLAB |
| Octave | https://www.gnu.org/software/octave/ | Open-source version of MATLAB |
| Maxima | http://maxima.sourceforge.net/ | Open-source symbolic math software, similar to Maple or Mathematica |
| Euler Math Toolbox | http://euler.rene-grothmann.de/ | Open-source version of MATLAB with symbolic math |
| FreeMat | http://freemat.sourceforge.net/ | Open-source version of MATLAB |
| SageMath | http://www.sagemath.org/ | Open-source alternative to Maple, Mathematica, and MATLAB |

**Table 2.8**   List of C/C++ numeric libraries and other items

| List of useful libraries for C/C++ | | |
|---|---|---|
| GNU Science Library | https://www.gnu.org/software/gsl/ | Numerical library for C/C++ |
| Boost | http://www.boost.org/ | Numerical library for C/C++ |
| Eigen | http://eigen.tuxfamily.org/ | Linear algebra library for C/C++ |
| Armadillo | http://arma.sourceforge.net/ | Open-source linear algebra library for C++ |
| Visual Studio Express | https://www.visualstudio.com/vs/visual-studio-express/ | Free C/C++ compiler and IDE |
| GCC | https://gcc.gnu.org/ | GNU open-source C/C++ compiler |
| FFTW | http://www.fftw.org/ | Free discrete FFT library for C |
| Odeint | http://headmyshoulder.github.io/odeint-v2/ | Free C++ library of numerical ordinary differential equation solvers |

**Table 2.9** List of tools related to finite and boundary element analysis

| Additional software for scientific computing | | |
|---|---|---|
| Python | https://www.python.org/ | Interpreted high-level programming language |
| Advanced Simulation Library | http://asl.org.il/ | Open-source tool for solving partial differential equation |
| BEM++ | http://www.bempp.org/ | Open-source boundary element method library for Python |
| GetFEM++ | http://download.gna.org/getfem/html/homepage/ | Open-source FEM library for Python, MATLAB, and SCILAB |
| Gmsh | http://gmsh.info/ | Open-source 3D finite element mesh generator |
| enGrid | http://www.engits.com/engrid---mesher.html?lang=en | Open-source mesh generation |

## 2.6 Exercises

1. For the $\cos(a - b)$ evaluation, find the value of $N$ when the two approaches are equal as a function of $r = c_T/c_a$.
2. Write the binary representation of 123.45.
3. For the result of Exercise 2, write the single-precision floating-point representation of this number.
4. Write pseudo code for evaluating Green's function:

$$\frac{\exp(ikx)}{x}$$

where, $x, k$ are real number and $x \geq 0$, $k$ can be either sign. Assume that only real number is possible in the evaluation of the function. You will need to decide what the behavior should be if $x = 0$ is an input.

5. The function in Exercise 4 is used to estimate the interaction among $N$ point line objects.
   (a) How many times will the equation need to be used to calculate all interactions?
   (b) Estimate the computational cost of this calculation including the unit cost of the operations expressed in the equation.
   (c) Estimate the memory required to store all results
   (d) What is the memory requirement for 100,000 objects assuming double precision?

## References

[1] Rump, S. M., Ogita, T., and Oishi, S., Accurate floating-point summation, part I: Faithful rounding, SIAM J. Sci. Comput., Vol. 31, No. 1, pp. 189–224, 2008.

[2] IEEE, ANSI/IEEE Std 754-1985: IEEE Standard for Binary Floating-Point Arithmetic, IEEE, New York, 1985.

[3] Pancratov, C., Kurzer, J. M., Shaw, K. A., and Trawick, M. L., Why computer architecture matters, Comput. Sci. Eng., Vol. 10, No. 3, pp. 59–63, 2008.

[4] Gross, D. C., Report from the Fidelity Implementation Study Group, paper no. 99S-SIW-167, Proceedings of 1999 Spring Simulation Interoperability Workshop, Simulation Interoperability Standards Organization, Orlando, FL, 1999.

[5] Press, W. H., Teukolsky, S. A., Vetterling, W. T., and Flannery, B. P., Numerical Recipes in C, The Art of Scientific Computing, Second Edition, Cambridge University Press, Cambridge, 1999.

[6] Deitel, H. M. and Deitel, P. J., C++ How to Program, Fifth Edition, Pearson, Prentice Hall, Upper Saddle River, NJ, 2005.

# 3

# Derivation of the Wave Equation

## 3.1 Introduction

The linear wave equation can be derived from several different starting points depending on whether the vibrations are traveling in a solid or fluid. In the case of vibrations in a solid, the form of the wave equation will depend on the symmetries of the solid structure and material properties [1]. Since the subject matter of this text deals with methods for modeling wave propagation in fluids, this will be dealt with in detail. Derivations of acoustic vibrations in solids for various models are presented briefly in the interest of demonstrating the connection of the wave equation, and generalizations of the wave equation, to the equations of the supporting medium. Reviewing the connection between the medium and the wave equation will serve two purposes: one is to provide a tangible connection between two paradigms, a micro description of the system and the macroscopic phenomena supported by the system, and the other is to show a direct connection between material properties, such as density, bulk modulus, and so on, and the observed properties of the wave as a separate entity, *e.g.* velocity.

Sections appear as follows. The first section describes wave phenomenon without any reference to the medium. This is to establish basic vocabulary. The next section is a review of the coupled mass–spring system from classical mechanics along with a description of normal modes of vibration. The limit to a continuum is taken to arrive at an equation for wave propagation in one spatial dimension. Following the treatment of one-dimensional waves is a presentation of the equations for wave propagation in three-dimensional elastic solids. This presentation starts with the stress–strain relation, Hook's law, and continuum mechanics and passes to a second-order wave equation. The next section presents a detailed derivation of the linear first- and second-order wave equations derived from the equations of fluid mechanics. This will be the foundation for the discussion of modeling methods for describing the acoustic field. This section introduces a novel way of looking at the differential operator in the wave equation, appearing in the acoustics literature in the late 1970s, which uses concepts

*Computational Acoustics: Theory and Implementation*, First Edition. David R. Bergman.
© 2018 John Wiley & Sons Ltd. Published 2018 by John Wiley & Sons Ltd.

from differential geometry and tensor analysis. Following waves in fluids is a very brief section on thin plates and rods, followed by a description of phonons from solid-state physics. After a survey of the various manifestations of the wave equation, this chapter concludes with a brief section on tensors and manifolds to serve as preparation for later chapters.

## 3.2    General Properties of Waves

This section introduces some basic concepts used to describe wave phenomena without reference to a specific type of physical wave. The two most common wave phenomena are optics and acoustics; hence some jargon related to waves is derived from these subjects. Waves share much in common with the phenomenon of oscillation, or periodic motion. A perfect example is waves traveling on the surface of a body of water. Looking out at an ocean or lake, under the right conditions, if one were to sketch a picture of the water surface, it might look like a set of long parallel crests separated by valleys. If pressed to give a description of the terrain of the water surface, one might say it rises and falls with a fixed height and at fixed intervals in a specific direction on the surface of the water. The apparent fixed height is related to the amplitude of the wave, while the fixed horizontal distance between crests or valleys is called the wavelength. Looking at the waves as they travel across the water surface, one would likely see a fixed spatial profile move in a well-defined direction at a constant speed. In short, based on observation, it would appear that this collection of crests has a structure and moves as an entity on the water. This structure is not eternal, as it will eventually crash into the shore or encounter obstacles, changing direction and shape. However, for the time that it appears to exist, the parameters amplitude and wavelength describe this shape, and "moving with constant velocity" describes its behavior. A reasonable mathematical description of this profile is

$$A_0 \cos\left(2\pi \frac{\hat{n} \cdot \vec{x} - vt}{\lambda}\right) \tag{3.1}$$

All the components of the previous description appear in this expression. The amplitude $A_0$, from which the distance from crest to valley is measured is $2A_0$. The distance between crests is referred to as wavelength, $\lambda$. The velocity of the traveling wave has a speed, $v$, and direction of travel, $\hat{n}$. Points in space are defined in Cartesian coordinates by $\vec{x}$, and time is denoted in the expression by $t$. The function $\cos\theta$ was chosen since it has the periodic structure observed in the wave (this could have just as easily been $\sin\theta$). The factor of $2\pi$ is required to ensure that for a fixed time our crests fall at exactly $\hat{n} \cdot \vec{x} = m\lambda$, $m = 1, 2, \ldots$, from some initial crest. The valleys will occur in between the crests, at $m = 1/2, 3/2, \ldots$. The term $2\pi vt/\lambda$ will shift the entire picture in the direction $\hat{n}$ as time increases. The frequency is defined by

$$f = \frac{v}{\lambda} \tag{3.2}$$

and is a measure of the number of times per second a fixed point of the surface of the water completes one oscillation. Based on this description the wave has a single wavelength and constant speed, and therefore, it has a single frequency. It is more convenient to absorb the factor of $2\pi$ into the expression in parentheses and define the angular frequency and wavenumber:

$$\omega = 2\pi f \tag{3.3}$$

$$k = \frac{2\pi}{\lambda} \tag{3.4}$$

It is also customary to define the wavevector:

$$\vec{k} = k\hat{n} \tag{3.5}$$

In terms of these quantities, (3.1) becomes

$$A_0 \cos\left(\vec{k} \cdot \vec{x} - \omega t\right) \tag{3.6}$$

A lot of physics has been overlooked and in fact surface waves on water are quite complicated, their motion being a combination of vertical and horizontal motions. But the exercise has served its purpose, namely, to impress upon the reader that waves have an anatomy, that there is a language for describing this anatomy, and that this language will carry over into many other areas of physics.

So far nothing has been said about what laws of nature govern these waves, how they are created, or what makes them propagate at constant speed. One might naturally ask what makes these waves have the shape parameters they do. In other words, will all waves have the same distance between crests and the same amplitude? Are these a property of the water? Do all waves travel at the same speed, or can they accelerate like particles under the influence of a force? Based on experience water and other media can appear to be free of waves, calm, under the right conditions. Waves require a source of mechanical vibration. Once created, they will travel away from the source through the medium, with the anatomy of the wave depending to some degree on the source. Placing one's hand in water and shaking it at a certain rate creates waves of a particular wavelength. If the rate of vibration is increased (decreased), the wave crests will become closer together (farther apart). The water will vibrate at the same rate as the hand, and the wavelength of the wave will change according to (3.2). The amplitude of the wave is also dependent on the source: shaking one's hand gently creates a small amplitude wave, while shaking more forcefully creates a large amplitude wave. The speed of the traveling waves depends on the nature of the medium. Waves of all frequencies may appear to travel at the same speed, each with a different anatomy, but all moving together reaching the same place at the same time if initiated from the same source.

One can also describe the reaction of water to say the impulsive act of a stone, or other object, being tossed upon its surface. In this case one would likely see a single circular crest propagate away from the point of impact. Unlike the previous description, this type of disturbance, and that produced by shaking one's hand in the water, will create a wave that travels in all directions at once, away from the source. As these waves or pulses travel farther and farther away, the radius of the circle will become large enough that someone observing a portion of the wave would likely use (3.6) as a description. Yet close to the source this description will fail to be accurate. Careful observation would suggest that the amplitude of the crests get smaller as they get farther away from the source. One might appeal to (3.6) and attempt to make a description of these waves by placing the origin of coordinates at the source and replacing $\hat{n}$ with $\hat{r}$. Then the periodic nature of the wave would look like

$$\cos\left(k\hat{r}\cdot\vec{x}-\omega t\right) = \cos\left(kr-\omega t\right) \tag{3.7}$$

where the replacement $\hat{r}\cdot\vec{x} = r$ has been made. Without anything else to serve as a guide regarding the amplitude variation, one might assume a power law or exponential law and use observed data to determine free parameters:

$$\frac{A_0 \cos\left(kr-\omega t\right)}{r^{\alpha}} \tag{3.8}$$

In Equation (3.8) $\alpha$ is a free parameter.

Wave phenomenon can be identified in many physical systems, and a large class of phenomenon are now understood to be wave phenomenon. Optics, for example, was studied for many centuries before it was understood that light was an electromagnetic vector wave, and later a photon. It has a set of laws that have been determined by observation and can be used to make predictions regarding the formation of images by mirrors and lenses and the deformation of images by changes in refractive index of materials. It wasn't until a complete theory of electromagnetism had developed that a derivation of such laws mathematically from first principles was possible. Sound, as experienced by us, is the result of mechanical waves traveling through air from a source to our bodies. Impinging on our bodies, these waves induce mechanical vibrations that in turn stimulate signals in nerves that are sent to the brain for processing. The frequency of a wave is for the most part determined by the source. Different media will produce waves of different wavelength for the same source, but all will register the same frequency. In optics wavelength is associated with color. It is understood that when light is passed through different materials, its wavelength will change, but its frequency will not. Perhaps for historical reasons color is labeled by wavelength, in vacuum, rather than frequency. In music, the pitch of a note is determined by frequency. When a source generates waves of a single frequency, it is referred to as monochromatic, borrowing jargon from optics.

A description of waves as an independent entity with certain properties related to a source and the medium of propagation has been presented here. All acoustics is mechanical vibration, so one should expect that a theory of how connected particles respond to each other's movement will also describe acoustic phenomenon. Motivated by this several descriptions of waves and their underlying medium will be presented. The reader should keep in mind that the intent is not to build a particle model of acoustics. The last thing anyone wants to do is model the movement of every water molecule in the ocean to arrive at a solution for a plane wave. The virtue of studying the wave as an entity is that the tiny details of the particle movement that help produce and sustain the wave become ignorable. Before delving into the mathematical theory of waves, a few topics that have not been covered in the previous description are mentioned. The first is attenuation. Any wave traveling through a material is going to experience some form of damping, much like a simple harmonic oscillator with air friction added. Damping can be included in the description of a wave by letting the wavenumber acquire an imaginary component. Note that (3.6) can be simplified using

$$\cos\theta = \mathrm{Re}\left(\exp i\theta\right) \tag{3.9}$$

The wave traveling in one direction is now expressed by the complex exponential

$$\exp\left(i\left(\vec{k}\cdot\vec{x}-\omega t+\theta_0\right)\right) \tag{3.10}$$

An additional parameter, $\theta_0$, called the phase of the wave has been added. This is determined by the initial conditions and represents the fact that one might not have placed their coordinates exactly on a crest at $t=0$. Attenuation is an exponential damping of the amplitude that usually occurs in the direction of propagation:

$$\exp\left(-\gamma\hat{n}\cdot\vec{x}\right)\exp\left(i\left(\vec{k}\cdot\vec{x}-\omega t+\theta_0\right)\right) \tag{3.11}$$

where it is assumed that the source is at $\hat{n}\cdot\vec{x}=0$. Combining the exponents gives

$$\exp\left(i\left(\kappa\hat{n}\cdot\vec{x}-\omega t+\theta_0\right)\right) \tag{3.12}$$

with $\kappa\equiv k+i\gamma$. For the most part the basic laws of optics hold for acoustics as well. This includes the law of reflection from a smooth hard barrier, that is, the angle of reflection equals the angle of incidence, and the law of refraction at a flat interface between two media, Snell's law. Up to this point the presentation of wave phenomenon has been phenomenological, based on observation (pseudo-observation) and an attempt to categorize the basic properties observed. Now that a clear vocabulary for the description of wave phenomenon is in place, and some guidance as to how the observed parameters are related to a source versus the medium, the presentation will be less descriptive and more mathematical.

## 3.3   One-Dimensional Waves on a String

To derive a linear wave equation for one-dimensional waves, consider a line of identical coupled masses, $m_i=m$ connected to nearest neighbors by identical springs, with spring constant $k_i=k$. The number of masses is finite and the first and last springs are connected to rigid boundaries. Each mass has a coordinate position, $x_i$ measured relative to its equilibrium position, and velocity, $\dot{x}_i=dx_i/dt$. The kinetic energy of the system may be written as

$$K=\sum_{i=1}^{N}\frac{m}{2}\dot{x}_i^2 \tag{3.13}$$

The total potential energy of the springs is a function of the position of each mass. Except for the end springs, each is stretched or compressed relative to its original length by an amount $x_i-x_{i-1}$:

$$V=\frac{k}{2}x_1^2+\sum_{i=2}^{N}\frac{k}{2}(x_i-x_{i-1})^2+\frac{k}{2}x_N^2 \tag{3.14}$$

The first and last terms account for the end springs that are attached to large infinitely massive walls. The equations of motion for the masses are derived from the Lagrangian $L=K-V$:

$$\frac{d}{dt}\frac{\partial L}{\partial\dot{x}_i}-\frac{\partial L}{\partial x_i}=0 \tag{3.15}$$

Applying (3.15) to the Lagrangian associated with (3.13) and (3.14) gives a system of coupled equations for the motion of the masses:

$$m\ddot{x}_1 = -k(x_1 - (x_2 - x_1)) \tag{3.16}$$

$$m\ddot{x}_i = -k((x_i - x_{i-1}) - (x_{i+1} - x_i)), \quad i = 2, \ldots, N-1 \tag{3.17}$$

$$m\ddot{x}_N = -k(x_N + (x_N - x_{N-1})) \tag{3.18}$$

A solution of the form $x_i(t) = u_i \cos(\omega t)$ is assumed, which leads to the characteristic equation

$$
\begin{pmatrix}
2k - m\omega^2 & -k & 0 & \cdots & 0 \\
-k & 2k - m\omega^2 & -k & \cdots & 0 \\
0 & -k & 2k - m\omega^2 & \cdots & 0 \\
\vdots & \vdots & \vdots & \ddots & \vdots \\
0 & 0 & 0 & \cdots & 2k - m\omega^2
\end{pmatrix}
\begin{pmatrix}
u_1 \\ u_2 \\ u_3 \\ \vdots \\ u_N
\end{pmatrix}
= 0 \tag{3.19}
$$

This is an eigenvalue equation. For nontrivial solutions to exist, the determinant of the matrix must be equal zero, leading to an eigenvalue equation for $\omega^2$. For each eigenvalue $\omega_n$, there is a corresponding eigenvector $\vec{u}_n$. A system of $N$-coupled masses will yield $N$ eigenvalues:

$$\omega_n = 2\omega_0 \sin\left(\frac{n\pi/2}{N+1}\right) \tag{3.20}$$

$$u_{i(n)} = U \sin\left(\frac{n\pi i}{N+1}\right) \tag{3.21}$$

The somewhat awkward notation for the eigenvector references the normal mode via $n$ and the particle by $i$. The natural frequency of the single mass–spring system is introduced in (3.20):

$$\omega_0 = \sqrt{\frac{k}{m}} \tag{3.22}$$

The aforementioned treatment provides a complete description of the discrete mass–spring system that is used as a starting point for a continuum equation. Passing to the continuum involves letting the number of masses go to infinity as the distance between them goes to zero under constraints that the mass density and other physical quantities remain finite. To develop a suitable equation in the continuum limit, attention should be focused on an arbitrary mass, not coupled to either boundary, for example, the $i$th mass. The equilibrium distance between two adjacent masses, $L_0$, is the quantity that will go to zero in the limiting process. As the number of masses goes to infinity, the individual mass will tend to zero and the local mass density,

$\rho = m/L_0$, will be held constant. Young's modulus is defined via $Y = kL_0$ in this limit. Rearranging terms in the equation of motion and using these definitions gives the following equation for the $i$-th mass:

$$\rho \ddot{x}_i = YL_0^{-2}((x_{i+1} - x_i) - (x_i - x_{i-1}))$$

(3.23)

A spatial parameter, denoted $\sigma$, is defined and used to locate a position on the $x$-axis independent of the mass locations, labeled $x_i$. The location of the mass element, as the limit is taken, is a function of time and this new spatial parameter, $x(t, \sigma)$. The continuous parameter $\sigma$ acts as, or replaces, the discrete index $i$, for example, $x(t, iL_0) = x_i(t)$. The following approximations to the first and second partial derivative of, $x(t, \sigma)$ with respect to $\sigma$ are made:

$$\left.\frac{\partial x}{\partial \sigma}\right|_{\sigma = iL_0 \pm L_0/2} \approx \pm \frac{(x_{i\pm1} - x_i)}{L_0} \equiv x'_{\pm}$$

(3.24)

$$\left.\frac{\partial^2 x}{\partial \sigma^2}\right|_{y = iL_0} \approx \frac{x'_+ - x'_-}{L_0}$$

(3.25)

With this, the discrete equation for the motion of the $i$-th mass becomes

$$\frac{\partial^2 x}{\partial t^2} = \frac{Y}{\rho}\frac{\partial^2 x}{\partial \sigma^2}$$

(3.26)

The end points, infinite massive walls, become boundary conditions on the field $x(t, \sigma)$. Defining the walls to be at $\sigma = 0$, and $\sigma = (N+1)L_0 \approx NL_0 = L$, leads to the conditions $x(t,0) = x(t,L_0) = 0$. The discrete system of $N$ masses has $N$ normal modes of vibration. Passing the limit to a continuum model, the spectrum becomes

$$\omega_n = n\frac{\pi}{L}\sqrt{\frac{Y}{\rho}}$$

(3.27)

which can be derived by taking the small argument approximation of (3.20) and using the definitions of the macroscopic quantities previously defined. Similarly, for the amplitude vector:

$$u_{i(n)} = U\sin\left(\frac{\pi\sigma}{L}\right)$$

(3.28)

The discrete set of normal modes becomes a field:

$$x_n(t,\sigma) = U\sin\left(\frac{\pi\sigma}{L}\right)\cos(\omega_n t)$$

(3.29)

The eigenfrequencies can be written in terms of the lowest frequency, $n = 1$, $\omega_n = n\omega_1$, with $\omega_1$ derived from (3.27). This process is meant to demonstrate how the two paradigms, discrete versus continuous, are related and impress upon the reader the value of reductionism. Building models of bulk behavior from particle models is a common approach to understanding how discrete properties of the micro system lead to properties of the macro system [2].

## 3.4   Waves in Elastic Solids

In the study of elasticity, the medium is assumed to be a continuum. The laws governing the deformation of the continuum under application of a force are derived by applying Newton's laws to an infinitesimal volume of the solid. The stress tensor is defined as the force per unit area acting in each direction for any planar slice through the medium. Conversely, projecting the stress tensor onto different directions will give the internal forces acting at any point in the solid medium. For any plane through the material, the projection of the stress tensor in the direction normal to the plane is the normal stress or tensile stress, while the projection along the direction tangent to the plane defines the shear stress. Also defined is a strain tensor, which is related to the differential deformation of the material relative to its original state. In three dimensions, each may be expressed as 3-by-3 matrix:

$$\sigma = \begin{pmatrix} \sigma_{xx} & \sigma_{xy} & \sigma_{xz} \\ \sigma_{yx} & \sigma_{yy} & \sigma_{yz} \\ \sigma_{zx} & \sigma_{zy} & \sigma_{zz} \end{pmatrix} \tag{3.30}$$

$$\varepsilon = \begin{pmatrix} \varepsilon_{xx} & \varepsilon_{xy} & \varepsilon_{xz} \\ \varepsilon_{yx} & \varepsilon_{yy} & \varepsilon_{yz} \\ \varepsilon_{zx} & \varepsilon_{zy} & \varepsilon_{zz} \end{pmatrix} \tag{3.31}$$

The individual components of the strain are defined as follows:

$$\varepsilon_{ij} = \frac{1}{2}\left(\frac{\partial u_i}{\partial x_j} + \frac{\partial u_j}{\partial x_i}\right), \quad i,j = 1,2,3 \tag{3.32}$$

The set of variables $\{u_k\}$ are the infinitesimal displacements along the axes defined by $\{x_k\}$, where the following notation has been adopted for position and displacement of an element of the solid, $\{x_1, x_2, x_3\} = \{x, y, z\}$. It is worth noting that the stress and strain tensor components are defined relative to different systems. Stress is defined by looking at the equilibrium conditions applied to a deformed body, referred to as Eulerian coordinates. Strain is defined relative to the undeformed set of axes, a Lagrangian coordinate system. For small deformations, it can be assumed that these are approximately the same coordinates. Based on symmetry and physical arguments, one can prove that each of these is a symmetric matrix, $\sigma_{ij} = \sigma_{ji}$. In the study of elasticity, there is a generalization of Hooke's law that relates local stress and strain of a body at each point. This can be expressed easily as a 6-by-6 matrix:

$$
\begin{pmatrix} \sigma_{xx} \\ \sigma_{yy} \\ \sigma_{zz} \\ \sigma_{xy} \\ \sigma_{yz} \\ \sigma_{zx} \end{pmatrix} = \begin{pmatrix} c_{11} & c_{12} & c_{13} & c_{14} & c_{15} & c_{16} \\ c_{21} & c_{22} & c_{23} & c_{24} & c_{25} & c_{26} \\ c_{31} & c_{32} & c_{33} & c_{34} & c_{35} & c_{36} \\ c_{41} & c_{42} & c_{43} & c_{44} & c_{45} & c_{46} \\ c_{51} & c_{52} & c_{53} & c_{54} & c_{55} & c_{56} \\ c_{61} & c_{62} & c_{63} & c_{64} & c_{65} & c_{66} \end{pmatrix} \begin{pmatrix} \varepsilon_{xx} \\ \varepsilon_{yy} \\ \varepsilon_{zz} \\ 2\varepsilon_{xy} \\ 2\varepsilon_{yz} \\ 2\varepsilon_{zx} \end{pmatrix}
\tag{3.33}
$$

Equation (3.33) can be written more succinctly as $\Sigma = C \cdot E$. This relationship is a three-dimensional generalization of the law used to describe the mass–spring system, $F = -kx$. It is worth noting that (3.33) could be written in terms of a fourth rank tensor:

$$
\sigma_{ij} = c_{ijmn} \varepsilon_{mn}
\tag{3.34}
$$

The four index tensors are related to the matrix, for example, $c_{1111} = C_{11}$, $c_{1122} = C_{12}$, $c_{1212} = C_{44}$, and so on. The 36 values of the matrix $C$ are called the elastic constants of the material. For homogeneous materials, these are constant, that is, the same value at every point inside the material. In general, a solid material will exhibit anisotropic behavior. There exist a wide variety of elastic matrices depending on the symmetries of the material structure. For an isotropic material, there are only two independent elastic constants. The isotropic elastic tensor is given as follows:

$$
\begin{pmatrix} c_1 & c_2 & c_2 & 0 & 0 & 0 \\ c_2 & c_1 & c_2 & 0 & 0 & 0 \\ c_2 & c_2 & c_1 & 0 & 0 & 0 \\ 0 & 0 & 0 & c_3 & 0 & 0 \\ 0 & 0 & 0 & 0 & c_3 & 0 \\ 0 & 0 & 0 & 0 & 0 & c_3 \end{pmatrix}
\tag{3.35}
$$

The three parameters are related to each other:

$$
c_3 = \frac{c_1 - c_2}{2}
\tag{3.36}
$$

These two independent parameters can be related to either Lamé's parameters or Young's modulus and Poisson's ratio. Another type of material is one with cubic symmetry. For such materials, the stiffness matrix is identical in form to that of a purely isotropic material but without (3.36) relating the three variables. Cubic materials have three independent parameters.

The free dynamic behavior of the material, assuming small deformations, is governed by the equation

$$\frac{\partial \sigma_{ij}}{\partial x_j} = \rho \frac{\partial^2 u_i}{\partial t^2} \tag{3.37}$$

Using the generalized Hooke's law to replace $\sigma_{ij}$, and the definition of the strain in terms of the displacement field leads to a second-order linear partial differential equation (PDE) for the material displacement field:

$$c_{ijmn} \frac{\partial^2 u_n}{\partial x_j \partial x_m} = \rho \frac{\partial^2 u_i}{\partial t^2} \tag{3.38}$$

The presence of the material tensor will mix the various spatial derivatives, but (3.37) is similar in form to the second-order wave equation. For constant elastic matrix values and constant density, the equation can be solved by assuming a monochromatic plane wave:

$$u_n = U_n e^{i\left(\vec{k}\cdot\vec{x} - \omega t\right)} \tag{3.39}$$

The derivatives of this function are

$$\frac{\partial^2 u_n}{\partial t^2} = -\omega^2 u_n \tag{3.40}$$

$$\frac{\partial^2 u_n}{\partial x_j \partial x_m} = -k_j k_m u_n \tag{3.41}$$

Inserting (3.40) and (3.41) into (3.38) yields the following:

$$c_{ijmn} k_j k_m U_n = \rho \omega^2 U_i \tag{3.42}$$

The wavevector is expressed in terms of its amplitude and direction of propagation $k_j = k n_j$, and the velocity parameter is defined via $v = \omega/k$. These definitions are used to define a second rank tensor to simplify notation:

$$D_{in} \equiv \frac{c_{ijmn} n_j n_m}{\rho} \tag{3.43}$$

Using (3.43), (3.42) can be written as

$$\left(D_{ij} - v^2 \delta_{ij}\right) U_j = 0 \tag{3.44}$$

Taking the determinant of (3.44) gives the corresponding eigenvalue equation for this system:

$$\det\left(D_{ij} - v^2 \delta_{ij}\right) = 0 \tag{3.45}$$

Equation (3.45) determines the allowed values of wave speed in terms of the material properties. For each eigenvalue, there is an associated eigenvector that determines the direction of the displacement perturbation within the material. While this is a straightforward system to solve, the interpretation can get tricky. The solution is a plane wave with a constant amplitude vector traveling in a fixed direction. It is useful to define the polarization modes of the wave relative to the propagation direction, $n_j$. Displacements parallel to the direction of motion are longitudinal wave, similar to those in a fluid. These are referred to as $P$-waves in seismology. This leaves two transverse modes, referred to as $S$-waves in seismology. For each fixed direction of propagation $n_j$, the eigenvalues and eigenvectors may be different. In other words, a pure $P$-wave may propagate with different speeds along the $x$-axis as compared with the $y$- or $z$-axes.

## 3.5   Waves in Ideal Fluids

Wave phenomena in fluids can be derived from the equations of fluid dynamics. These equations are nonlinear; therefore, a linear wave equation is an approximation that can be expected to fail in some cases. Based on experience there seem to be as many derivations of a linear wave equation in moving fluids as there are acousticians, each containing specific approximations and assumptions suited for a particular application. This is not a bad thing. Part of the art of modeling and simulation is having a good feel for what can and should be neglected in developing a model of a system. Another driver in special cases is the desire to reuse existing software to investigate new phenomena. The derivations presented in this section will remain as general as possible. Rather than gloss over details, the goal is to arrive at the most general linear equation possible then start imposing assumption to see what results are obtained. The reader should keep this in mind while reading this section as it is lengthy. In the process three distinct approaches will be presented in subsections and the results compared in the discussion. Regardless, the reader will be able to default to the linearized equation as a starting point for deriving their own approximate wave equation. This section presents (1) a derivation of a linear equation for the acoustics perturbation propagating in a moving fluid, (2) a derivation of a second-order wave equation for the acoustic perturbations, (3) a discussion of the various assumptions made in such derivations, and (4) the introduction of a special form for the second-order linear differential operator that will be used in later sections.

### 3.5.1   Setting Up the Derivation

The equations describing the behavior of a classical Newtonian fluid are the equation of continuity, Newton's second law applied to a volume element of the fluid, and an equation of state relating the thermodynamic variables of the fluid. For the purpose of deriving the equation of linear acoustics in a fluid medium, it is assumed that the fluid is isentropic (constant entropy) and inviscid (no viscosity). The fluid state is described locally by its density, $\rho$, pressue, $p$, and velocity, $\vec{v}$. The equation of continuity states that matter is neither created nor destroyed:

$$\frac{\partial \rho}{\partial t} + \vec{\nabla} \cdot \left( \rho \vec{v} \right) = 0 \qquad (3.46)$$

The continuity equation should look familiar as it is the same regardless of where it is encountered. It states that the time rate of change of material density at a point is equal to the negative divergence of the current at that point. A positive divergence at a point means that matter is leaving or moving away from that point and hence a reduction in density in time. Applying Newton's second law to a volume element of the fluid leads to Euler's equation:

$$\rho\left(\frac{\partial \vec{v}}{\partial t} + \vec{v}\cdot\vec{\nabla}\vec{v}\right) = -\vec{\nabla}p + \vec{F} \tag{3.47}$$

One frequently sees a material derivative defined, identical to the total time derivative, $d/dt = \partial/\partial t + \vec{v}\cdot\vec{\nabla}$; for example, see Bergmann [3]. The local pressure gradient is always present, acting as a force on the fluid elements. The second term on the right-hand side (r.h.s.) represents other external forces, expressed as force densities, acting on all fluid elements. It is assumed that these are conservative forces and can be expressed in terms of a potential $\vec{F} = -\vec{\nabla}\Phi$. Lastly, there is an equation of state that connects the thermodynamic state variables of the fluid. The assumptions regarding the fluid allow one to write a state equation relating pressure and density, $p(\rho)$.

The variables in (3.46) and (3.47) are $\vec{v}, p, \rho$, with the last two constrained by an equation of state. These variables are written in terms of background terms and a small perturbation:

$$\vec{v} = \vec{v}_0 + \vec{v}_1 \tag{3.48}$$

$$p = p_0 + p_1 \tag{3.49}$$

$$\rho = \rho_0 + \rho_1 \tag{3.50}$$

where $\vec{v}_0, p_0$ and $\rho_0$ are the velocity, pressure, and density of the fluid medium and $\vec{v}_1, p_1$, and $\rho_1$ are contributions due to acoustic phenomenon. In addition to the current list of assumptions, it is required that the background variables obey the equations of fluid mechanics independent of the perturbations and that the perturbations are small compared to the background fields.

## 3.5.2   A Simple Example

To give the reader some background on the linear wave equation in fluids and a view of the end game, a specialized approach is presented here, following closely the assumption made in Landau and Lifshitz [4]. Namely, the velocity field is so small that the nonlinear term in (3.47) can be completely neglected, there is no background flow, and the velocity is determined by a scalar potential, $\vec{v} = \vec{\nabla}\psi$. The pressure and density fields are expanded using (3.49) and (3.50). A consequence of the last assumption is that the fluid motion is irrotational, $\vec{\nabla} \times \vec{v} = 0$. For the acoustic perturbations, these assumptions lead to a fairly simple version of the fluid equations:

$$\frac{\partial p_1}{\partial t} + \rho_0 c^2 \nabla^2 \psi = 0 \tag{3.51}$$

$$\frac{\partial \psi}{\partial t} + \frac{1}{\rho_0} p_1 = 0 \tag{3.52}$$

The perturbation $\rho_1$ would only show up in nonlinear terms and is therefore ignored. The sound speed is defined by the equation of state:

$$c^2 = \left(\frac{\partial p}{\partial \rho}\right)_S \equiv p' \tag{3.53}$$

The subscript $S$ indicates that the partial derivative is taken at constant entropy and is defined simply as $p'$. Lastly, the density, $\rho_0$, is assumed constant and the sound speed time independent. Taking the time derivative of (3.52) and using (3.51) to replace $\partial p_1/\partial t$,

$$-\frac{\partial^2 \psi}{\partial t^2} + c^2 \nabla^2 \psi = 0 \tag{3.54}$$

Equation (3.54) is the wave equation most commonly presented in elementary texts. It was not necessary to explicitly require the sound speed to be constant, only time independent. It is easy to show that the pressure field also obeys the wave equation. Equation (3.54) is sometimes the final word on the wave equation, and for many applications it is good enough. Since the sound speed can be a function of position, (3.54) can be used as the fundamental equation for many problems in underwater acoustics and aeroacoustics to account for the refractive properties caused by temperature and pressure variations. The reader has likely seen (3.54) and a complete description of its solutions in several coordinate systems for $c = $ constant.

### 3.5.3 Linearized Equations

In this section a complete linearized version of the fluid equation is derived. The only alteration is the use of the state equation to eliminate derivatives in density. Assuming an equation of state, $p(\rho)$, the local sound speed is defined:

$$c^2 \equiv \frac{dp}{d\rho} = \left(\frac{d\rho}{dp}\right)^{-1} \tag{3.55}$$

Any derivative of $\rho$ can be replaced via

$$\delta\rho = \frac{d\rho}{dp}\delta p = \frac{\delta p}{c^2} \tag{3.56}$$

where $\delta$ represents any derivative. With (3.55) and (3.56) the equations of fluid dynamics can be put in the following form:

$$\frac{\partial p}{\partial t} + \vec{v}\cdot\vec{\nabla}p + \rho c^2 \vec{\nabla}\cdot\vec{v} = 0 \tag{3.57}$$

$$\rho\left(\frac{\partial\vec{v}}{\partial t}+\vec{v}\cdot\vec{\nabla}\vec{v}\right)+\vec{\nabla}p+\vec{\nabla}\Phi=0 \tag{3.58}$$

The pressure and density perturbations are related by

$$p_1=\frac{dp}{d\rho}\rho_1\equiv c^2\rho_1 \tag{3.59}$$

Applying (3.48)–(3.50) and (3.59) to (3.57) gives the linearized continuity equation

$$\frac{\partial p_0}{\partial t}+\frac{\partial p_1}{\partial t}+\vec{v}_0\cdot\vec{\nabla}p_0+\vec{v}_0\cdot\vec{\nabla}p_1+\vec{v}_1\cdot\vec{\nabla}p_0+c^2\left(\rho_0\vec{\nabla}\cdot\vec{v}_0+\rho_0\vec{\nabla}\cdot\vec{v}_1+\rho_1\vec{\nabla}\cdot\vec{v}_0\right)\approx0 \tag{3.60}$$

After some rearrangement (3.60) is massaged into the following form:

$$\left[\frac{\partial p_0}{\partial t}+\vec{v}_0\cdot\vec{\nabla}p_0+c^2\rho_0\vec{\nabla}\cdot\vec{v}_0\right]+\frac{\partial p_1}{\partial t}+\vec{v}_0\cdot\vec{\nabla}p_1+\vec{v}_1\cdot\vec{\nabla}p_0+c^2\left(\rho_0\vec{\nabla}\cdot\vec{v}_1+\rho_1\vec{\nabla}\cdot\vec{v}_0\right)\approx0 \tag{3.61}$$

The same steps applied to Euler's equation give us the following linearized version:

$$\left[\rho_0\left(\frac{\partial\vec{v}_0}{\partial t}+\vec{v}_0\cdot\vec{\nabla}\vec{v}_0\right)+\vec{\nabla}p_0+\vec{\nabla}\Phi\right]+\rho_1\left(\frac{\partial\vec{v}_0}{\partial t}+\vec{v}_0\cdot\vec{\nabla}\vec{v}_0\right)+\rho_0\left(\frac{\partial\vec{v}_1}{\partial t}+\vec{v}_0\cdot\vec{\nabla}\vec{v}_1+\vec{v}_1\cdot\vec{\nabla}\vec{v}_0\right)$$
$$+\vec{\nabla}p_1\approx0 \tag{3.62}$$

All terms that are nonlinear in the perturbations have been neglected. Equations (3.61) and (3.62) make explicit the original form of the fluid equations in square brackets. The background fields are constrained to obey the original fluid equations independent of the perturbations. Setting the terms in square brackets to zero leads to a set of linear first-order PDE for the perturbations:

$$\frac{\partial p_1}{\partial t}+\vec{v}_0\cdot\vec{\nabla}p_1+\vec{v}_1\cdot\vec{\nabla}p_0+c^2\left(\rho_0\vec{\nabla}\cdot\vec{v}_1+\rho_1\vec{\nabla}\cdot\vec{v}_0\right)\approx0 \tag{3.63}$$

$$\rho_1\left(\frac{\partial\vec{v}_0}{\partial t}+\vec{v}_0\cdot\vec{\nabla}\vec{v}_0\right)+\rho_0\left(\frac{\partial\vec{v}_1}{\partial t}+\vec{v}_0\cdot\vec{\nabla}\vec{v}_1+\vec{v}_1\cdot\vec{\nabla}\vec{v}_0\right)+\vec{\nabla}p_1\approx0 \tag{3.64}$$

Defining the derivative operator, $D_0\equiv\partial/\partial t+\vec{v}_0\cdot\vec{\nabla}$, and rearranging terms again produces the following:

$$\left(D_0p_1+\rho_0c^2\vec{\nabla}\cdot\vec{v}_1\right)+\left(\vec{v}_1\cdot\vec{\nabla}p_0+p_1\vec{\nabla}\cdot\vec{v}_0\right)=0 \tag{3.65}$$

$$\left( D_0 \vec{v}_1 + \frac{1}{\rho_0} \vec{\nabla} p_1 \right) + \left( \vec{v}_1 \cdot \vec{\nabla} \vec{v}_0 + \frac{p_1}{\rho_0 c^2} D_0 \vec{v}_0 \right) = 0 \tag{3.66}$$

The equations have been arranged so that the first term in parenthesis in each contains only first-order partials of the fields while the second parenthesis contains terms linear in the fields. The set of equations is a linear homogeneous PDE for the four fields, $\{p_1, \vec{v}_1\}$. As it stands this set of equations is completely general, the only neglected terms being nonlinear terms in the perturbations. To simplify these expressions, the following are defined:

$$F_1 \equiv \left( \vec{v}_1 \cdot \vec{\nabla} p_0 + p_1 \vec{\nabla} \cdot \vec{v}_0 \right) \tag{3.67}$$

$$\vec{F}_2 = \left( \vec{v}_1 \cdot \vec{\nabla} \vec{v}_0 + \frac{p_1}{\rho_0 c^2} D_0 \vec{v}_0 \right) \tag{3.68}$$

The final form of the linear equations is given in (3.69) and (3.70):

$$\left( D_0 p_1 + \rho_0 c^2 \vec{\nabla} \cdot \vec{v}_1 \right) + F_1 = 0 \tag{3.69}$$

$$\left( D_0 \vec{v}_1 + \frac{1}{\rho_0} \vec{\nabla} p_1 \right) + \vec{F}_2 = 0 \tag{3.70}$$

### 3.5.4 A Second-Order Equation from Differentiation

Equations (3.69) and (3.70) can be used as a starting point for modeling acoustics. Following the first example a second-order equation is derived from the final first-order equations. Equation (3.69) is divided through by $\rho_0 c^2$ and differentiated with respect to time. Then the same equation is multiplied by $\vec{v}_0$, and after division by $\rho_0 c^2$ the divergence is taken. The results are combined to give the following:

$$-\frac{\partial}{\partial t}\left( \frac{1}{\rho_0 c^2} D_0 p_1 \right) - \frac{\partial}{\partial t}\left( \vec{\nabla} \cdot \vec{v}_1 \right) - \frac{\partial}{\partial t}\left( \frac{F_1}{\rho_0 c^2} \right) - \vec{\nabla} \cdot \left( \frac{\vec{v}_0}{\rho_0 c^2} D_0 p_1 + \vec{v}_0 \vec{\nabla} \cdot \vec{v}_1 + \frac{\vec{v}_0 F_1}{\rho_0 c^2} \right) = 0 \tag{3.71}$$

Equation (3.70) is used to eliminate the time derivative of $\vec{\nabla} \cdot \vec{v}_1$. Taking the divergence of (3.70) and commuting time and space derivatives,

$$\frac{\partial}{\partial t}\left( \vec{\nabla} \cdot \vec{v}_1 \right) = -\vec{\nabla} \cdot \left( \vec{v}_0 \cdot \vec{\nabla} \vec{v}_1 \right) - \vec{\nabla} \cdot \left( \frac{1}{\rho_0} \vec{\nabla} p_1 \right) - \vec{\nabla} \cdot \vec{F}_2 \tag{3.72}$$

Inserting (3.72) into (3.71) gives the following:

$$-\frac{\partial}{\partial t}\left(\frac{1}{\rho_0 c^2} D_0 p_1\right) + \vec{\nabla} \cdot \left(\vec{v}_0 \cdot \vec{\nabla} \vec{v}_1\right) + \vec{\nabla} \cdot \left(\frac{1}{\rho_0}\vec{\nabla} p_1\right) + \vec{\nabla} \cdot \vec{F}_2 - \frac{\partial}{\partial t}\left(\frac{F_1}{\rho_0 c^2}\right)$$

$$-\vec{\nabla} \cdot \left(\frac{\vec{v}_0}{\rho_0 c^2} D_0 p_1 + \vec{v}_0 \vec{\nabla} \cdot \vec{v}_1 + \frac{\vec{v}_0 F_1}{\rho_0 c^2}\right) = 0 \tag{3.73}$$

The terms containing divergence of the velocity fields are combined to yield

$$\vec{\nabla} \cdot \left(\vec{v}_0 \cdot \vec{\nabla} \vec{v}_1\right) - \vec{\nabla} \cdot \left(\vec{v}_0 \vec{\nabla} \cdot \vec{v}_1\right) = \frac{\partial v_{0k}}{\partial x_i}\frac{\partial v_{1i}}{\partial x_k} - \frac{\partial v_{0k}}{\partial x_k}\frac{\partial v_{1i}}{\partial x_i} \equiv F_3 \tag{3.74}$$

Equation (3.73) may be written using (3.74):

$$-\frac{\partial}{\partial t}\left(\frac{1}{\rho_0 c^2} D_0 p_1\right) - \vec{\nabla} \cdot \left(\frac{\vec{v}_0}{\rho_0 c^2} D_0 p_1 - \frac{1}{\rho_0}\vec{\nabla} p_1\right) = \frac{\partial}{\partial t}\left(\frac{F_1}{\rho_0 c^2}\right) + \vec{\nabla} \cdot \left(\frac{\vec{v}_0 F_1}{\rho_0 c^2}\right) - \vec{\nabla} \cdot \vec{F}_2 - F_3 \tag{3.75}$$

The r.h.s. still contains the field variables but an important point is that there are no second-order derivatives in these variables. This point will come up again in discussions on the method of characteristics and rays. The significance of the operator appearing on the left-hand side (l.h.s.) will be addressed in the discussion. A similar second-order equation may be derived for the velocity perturbations.

### 3.5.5   A Second-Order Equation from a Velocity Potential

If it is assumed that the fluid vector is irrotational, a second-order wave equation can be derived for perturbations in a velocity potential. The irrotational constraint is $\vec{\nabla} \times \vec{v} = 0$. This approach and the results are due to Unruh [5]. Under such conditions the velocity field can be expressed in terms of a scalar potential, $\psi$:

$$\vec{v} = \vec{\nabla} \psi \tag{3.76}$$

The steps in the derivation are slightly different and worth working through. Referring back to the original fluid equations, and neglecting external forces

$$\frac{1}{\rho}\frac{\partial \rho}{\partial t} + \vec{\nabla} \cdot \vec{v} + \frac{1}{\rho}\vec{v} \cdot \vec{\nabla} \rho = 0 \tag{3.77}$$

$$\frac{\partial \vec{v}}{\partial t} + \frac{1}{2}\vec{\nabla}\left(\vec{v} \cdot \vec{v}\right) + \frac{1}{\rho}\vec{\nabla} p = 0 \tag{3.78}$$

The identity $\vec{\nabla}\left(\vec{v} \cdot \vec{v}\right) = 2\vec{v} \cdot \vec{\nabla}\vec{v}$ follows from the irrotational condition. A new variable, $\sigma = \ln\rho$, and the potential function

$$\varphi(\sigma) \equiv \int^{\rho(\sigma)} \left(\frac{dp}{d\rho}\right)' \frac{d\rho'}{\rho'} \tag{3.79}$$

are defined. Equation (3.79) along with the velocity potential allows the l.h.s. of (3.78) to be written as the gradient of a scalar function:

$$\vec{\nabla} \left( \frac{\partial \psi}{\partial t} + \frac{1}{2} \vec{v} \cdot \vec{v} + \varphi(\sigma) \right) = 0 \qquad (3.80)$$

This gives a set of fluid equations in the variables $\psi$ and $\sigma$:

$$\frac{\partial \sigma}{\partial t} + \vec{\nabla} \cdot \vec{v} + \vec{v} \cdot \vec{\nabla} \sigma = 0 \qquad (3.81)$$

$$\frac{\partial \psi}{\partial t} + \frac{1}{2} \vec{v} \cdot \vec{v} + \varphi(\sigma) = 0 \qquad (3.82)$$

The linearization procedure takes a slightly different form with $\psi = \psi_0 + \psi_1$ or $\vec{v} = \vec{v}_0 + \vec{\nabla} \psi_1$ and $\sigma = \sigma_0 + \sigma_1 \approx \ln \rho_0 + \rho_1 / \rho_0$ to first order. The potential is also expanded to first order using the definition of the sound speed:

$$\varphi(\sigma) \approx \varphi(\sigma_0) + c^2 \sigma_1 \qquad (3.83)$$

As in the previous linearization, (3.81) and (3.82) are expanded and the zeroth-order terms constrained to independently obey the fluid equations. The operator $D_0$ is also used to simplify the final expressions:

$$D_0 \sigma_1 + \frac{1}{\rho_0} \vec{\nabla} \cdot \left( \rho_0 \vec{\nabla} \psi_1 \right) = 0 \qquad (3.84)$$

$$D_0 \psi_1 + c^2 \sigma_1 = 0 \qquad (3.85)$$

The second term in (3.84) results from using $\sigma_0 = \ln \rho_0$. Equation (3.84) is divided through by $\rho_0$ and (3.85) by $\rho_0 c^{-2}$. Afterwards, the operator $D_0$ is applied to (3.85), and a factor of $\vec{\nabla} \cdot \vec{v}$ is applied to (3.85) in a separate operation. The two resulting equations are combined with (3.84) to produce the following:

$$-\left[ \frac{\partial}{\partial t} + \vec{\nabla} \cdot \left( \vec{v}_0(\cdot) \right) \right] \left( \frac{\rho_0}{c^2} D_0 \psi_1 + \rho_0 \sigma_1 \right) + \left( \rho_0 D_0 \sigma_1 + \vec{\nabla} \cdot \left( \rho_0 \vec{\nabla} \psi_1 \right) \right) = 0 \qquad (3.86)$$

The quantity in parenthesis in the first term is multiplied by $\vec{v}_0$ before the divergence is taken, hence the notation $(\cdot)$. Expanding and grouping terms together,

$$-\frac{\partial}{\partial t} \left( \frac{\rho_0}{c^2} D_0 \psi_1 \right) - \vec{\nabla} \cdot \left( \vec{v}_0 \frac{\rho_0}{c^2} D_0 \psi_1 - \rho_0 \vec{\nabla} \psi_1 \right) + \rho_0 D_0 \sigma_1 - \frac{\partial(\rho_0 \sigma_1)}{\partial t} - \vec{\nabla} \cdot \left( \vec{v}_0 \rho_0 \sigma_1 \right) = 0 \quad (3.87)$$

Notice that the first two terms have the same form as that in (3.75) of the last section. Consider the last three terms in (3.87). Applying the Leibnitz rule to the last two terms and rearranging,

$$\rho_0 D_0 \sigma_1 - \sigma_1 \left( \frac{\partial \rho_0}{\partial t} + \vec{\nabla} \cdot \left( \vec{v}_0 \rho_0 \right) \right) - \rho_0 D_0 \sigma_1 = 0 \tag{3.88}$$

In (3.88) the first and last terms cancel and the middle term is zero since the background fields are constrained to obey the fluid equations. The final result for a second-order linear PDE for perturbations and irrotational background flow is (3.89):

$$-\frac{\partial}{\partial t} \left( \frac{\rho_0}{c^2} D_0 \psi_1 \right) - \vec{\nabla} \cdot \left( \vec{v}_0 \frac{\rho_0}{c^2} D_0 \psi_1 - \rho_0 \vec{\nabla} \psi_1 \right) = 0 \tag{3.89}$$

The approach to the derivation presented in this section follows closely that of Unruh (Ref. [5]).

### 3.5.6   Second-Order Equation without Perturbations

In 1973 R. White published a derivation of a second-order wave equation without assuming perturbations, that is, a nonlinear second-order PDE [6]. The motive of White's work is centered on the geometry of the propagation paths determined by the method of characteristics. The main points of White's work are presented here for later comparison with the other approaches taken in this section. First, a new variable is defined:

$$q = \int_{p_0}^{p} \frac{dp}{c\rho} \tag{3.90}$$

In terms of this variable the fluid equations transform into the following:

$$\frac{1}{c} Dq + \vec{\nabla} \cdot \vec{v} = 0 \tag{3.91}$$

$$D\vec{v} + c \vec{\nabla} q = 0 \tag{3.92}$$

The steps to the desired second-order equation are the same as in the first original derivation:

$$-\frac{\partial}{\partial t} \left( \frac{1}{c} Dq \right) - \vec{\nabla} \cdot \left( \frac{\vec{v}}{c} Dq - c \vec{\nabla} q \right) = \vec{\nabla} \cdot \left( \vec{v} \vec{\nabla} \cdot \vec{v} - \vec{v} \cdot \vec{\nabla} \vec{v} \right) \tag{3.93}$$

A salient feature of White's approach is that the equations have not been linearized.

### 3.5.7   Special Form of the Operator

Regardless of the approach taken or assumptions made in the derivation, the second-order operator on the l.h.s. could always be arranged in such a way that one scalar field, $f$, appeared in the following form:

$$-\frac{\partial}{\partial t} \left( \frac{a}{c^2} D_0 f \right) - \vec{\nabla} \cdot \left( \vec{v}_0 \frac{a}{c^2} D_0 f - a \vec{\nabla} f \right) \tag{3.94}$$

The factor $a$ is an arbitrary function of coordinates and time. In all the derivations no assumption was made regarding the form of the background fields other than they obey the fluid equations independent of the acoustic perturbation, except for the case that assumed irrotational flow. Hence, the local sound speed, $c$, fluid flow, $\vec{v}_0$, and density, $\rho_0$, can all be functions of position and time. The scalar function $a$ depends on the particular variables chosen to express the fluid equations. Equation (3.94) can be written in a compact form by introducing a generalization of the gradient operator to include time:

$$\vec{\partial} = \hat{e}_t \frac{\partial}{\partial t} + \vec{\nabla} \tag{3.95}$$

A unit vector in the time direction, $\hat{e}_t$, is introduced. In array notation,

$$\left[ \frac{\partial}{\partial t} \ \frac{\partial}{\partial x} \ \frac{\partial}{\partial y} \ \frac{\partial}{\partial z} \right]^T \tag{3.96}$$

It is customary to use a numerical index for the four coordinates, one time and three spatial. A lowercase Greek letter is used to represent the coordinate index, $\nu = 0, 1, 2, 3$, allowing the four coordinates to be expressed as $\{x^\nu\}$, with the identification $\{x^0, x^1, x^2, x^3\} = \{t, x, y, z\}$. Similarly, a shorthand for the four-dimensional gradient operator is introduced:

$$\partial_\nu = \begin{cases} \partial/\partial t & \nu = 0 \\ \partial/\partial x & \nu = 1 \\ \partial/\partial y & \nu = 2 \\ \partial/\partial z & \nu = 3 \end{cases} \tag{3.97}$$

Using this notation, the 4-dimensional gradient of a scalar function can be written as follows:

$$\vec{\partial} f = \left[ \frac{\partial f}{\partial t} \ \frac{\partial f}{\partial x} \ \frac{\partial f}{\partial y} \ \frac{\partial f}{\partial z} \right]^T \tag{3.98}$$

The ordinary wave equation, (3.54), can then be written in matrix form:

$$-\frac{1}{c^2} \frac{\partial^2 f}{\partial t^2} + \nabla^2 f = \left[ \frac{\partial}{\partial t} \ \frac{\partial}{\partial x} \ \frac{\partial}{\partial y} \ \frac{\partial}{\partial z} \right] \begin{bmatrix} -c^{-2} & 0 & 0 & 0 \\ 0 & 1 & 0 & 0 \\ 0 & 0 & 1 & 0 \\ 0 & 0 & 0 & 1 \end{bmatrix} \begin{bmatrix} \partial f/\partial t \\ \partial f/\partial x \\ \partial f/\partial y \\ \partial f/\partial z \end{bmatrix} \tag{3.99}$$

As it stands this may not seem like an improvement. Many popular approaches to acoustic modeling and solving the wave equation involve transforming away the time variable and working in the frequency domain. The form of (3.99) is similar to an inner product on a vector space with the matrix acting as a metric tensor. For this example, the sound speed is constant and therefore can be moved inside the derivative operator. It is also customary when dealing with constant wave speeds to redefine the zeroth coordinate, $x_0 \equiv ct$. In this manner, all four coordinates have units of length, and the upper left entry of the metric is $-1$.

Returning to the general case, one can arrange the factors of (3.94) so that it may be written in matrix form as well:

$$
\begin{bmatrix} \dfrac{\partial}{\partial t} & \dfrac{\partial}{\partial x} & \dfrac{\partial}{\partial y} & \dfrac{\partial}{\partial z} \end{bmatrix} \dfrac{a}{c^2}
\begin{bmatrix}
-1 & -v_{0x} & -v_{0y} & -v_{0z} \\
-v_{0x} & c^2 - v_{0x}^2 & -v_{0x}v_{0y} & -v_{0x}v_{0z} \\
-v_{0y} & -v_{0x}v_{0y} & c^2 - v_{0y}^2 & -v_{0y}v_{0z} \\
-v_{0z} & -v_{0x}v_{0z} & -v_{0y}v_{0z} & c^2 - v_{0z}^2
\end{bmatrix}
\begin{bmatrix}
\partial f/\partial t \\
\partial f/\partial x \\
\partial f/\partial y \\
\partial f/\partial z
\end{bmatrix}
\tag{3.100}
$$

In a more succinct notation,

$$
\begin{bmatrix} \dfrac{\partial}{\partial t} & \vec{\nabla}^T \end{bmatrix} \dfrac{a}{c^2}
\begin{bmatrix}
-1 & -\vec{v}_0^{\,T} \\
-\vec{v}_0 & c^2 \mathbf{I} - \vec{v}_0\,\vec{v}_0^{\,T}
\end{bmatrix}
\begin{bmatrix}
\partial f/\partial t \\
\vec{\nabla} f
\end{bmatrix}
\tag{3.101}
$$

The matrix $\mathbf{I}$ is the 3-by-3 identity. The scalar factor to the left of the matrix is operated on by the derivative operator to its left. The form of the operator is reminiscent of the Laplacian in curvilinear coordinates, where a metric is introduced to account for the curvature of the coordinate surfaces in Euclidean space. Note that (3.100) can be written in the same notation as Ref. [6]. Following this example, if (3.100) can be massaged into the same form, then the wave equation can be interpreted as a scalar field propagation in a generalized curved manifold where the background geometric structure is determined by the background fluid variables. The goal is to cast the operator in the following form:

$$
\dfrac{1}{\sqrt{|g|}} \partial_\nu \sqrt{|g|} g^{\nu\mu} \partial_\mu f
\tag{3.102}
$$

The notation is as follows. The factors with upper indices, $g^{\nu\mu}$, are elements of the inverse metric. The metric tensor itself is denoted by $g_{\alpha\beta}$. The metric is a covariant tensor, while its inverse, with upper indices, is a contravariant tensor [7]. Both are symmetric tensors, $g_{\alpha\beta} = g_{\beta\alpha}$. The Einstein summation convention is assumed. By definition,

$$
g_{\alpha\beta} g^{\beta\sigma} = g^{\beta\sigma} g_{\alpha\beta} = \delta_\alpha^\sigma
\tag{3.103}
$$

The mixed tensor $\delta_\alpha^\sigma$ is a generalization of the Kronecker delta. Equation (3.102) also contains the determinant of the metric tensor:

$$
g = \det g_{\alpha\beta}
\tag{3.104}
$$

With these definitions in place, it is customary to define a Laplacian operator, $\mathbf{D}^2$:

$$
\mathbf{D}^2 f \equiv D^\mu \left( \partial_\mu f \right) = \dfrac{1}{\sqrt{|g|}} \partial_\nu \sqrt{|g|} g^{\nu\mu} \partial_\mu f
\tag{3.105}
$$

It is left to the reader to explore the references for more details on tensor analysis on curved manifolds [7, 8]. Both the metric and its inverse may be cast in matrix form. It is stated without proof that the inverse of the matrix appearing in (3.101) is

$$\frac{1}{a}\begin{bmatrix} -c^2 + \vec{v}_0^T\vec{v}_0 & -\vec{v}_0^T \\ -\vec{v}_0 & \mathbf{I} \end{bmatrix} \tag{3.106}$$

where the notation $\vec{v}_0^T\vec{v}_0$ has been used instead of the dot product. The reader can verify as an exercise that the product of (3.106) with the matrix appearing in (3.101) is the 4-by-4 identity. From (3.103) the following is derived:

$$\det g^{\alpha\beta} = \frac{1}{\det g_{\alpha\beta}} \tag{3.107}$$

This can be used to identify the overall factor $\sqrt{|g|}$ appearing in (3.102) and with that identify the correct form of the metric. It is left as an exercise to the reader to verify the following form the definitions provided in this section:

$$g_{\alpha\beta} = \frac{a}{c}\begin{bmatrix} -c^2 + \vec{v}_0^T\vec{v}_0 & -\vec{v}_0^T \\ -\vec{v}_0 & \mathbf{I} \end{bmatrix} \tag{3.108}$$

$$g^{\alpha\beta} = \frac{1}{ac}\begin{bmatrix} -1 & -\vec{v}_0^T \\ -\vec{v}_0 & c^2\mathbf{I} - \vec{v}_0\vec{v}_0^T \end{bmatrix} \tag{3.109}$$

$$\det g_{\alpha\beta} = -\frac{a^4}{c^2} \tag{3.110}$$

As a final point the factor $a$ is identified for the three cases: $a = \rho_0$ for the irrotational case, $a = \rho_0^{-1}$ in the more general linear case, and $a = c$ in the unperturbed treatment. One can transform from one version to another by an appropriate scale factor. These results are summarized in Table 3.1, noting that the metric appearing in White [6], $a = c$, is taken to be the default.

**Table 3.1** Three representations of the second-order wave equation

| Equation | Metric | Assumptions/comments |
|---|---|---|
| $D^2 q = \frac{1}{c}\vec{\nabla}\cdot\left(\vec{v}\vec{\nabla}\cdot\vec{v} - \vec{v}\cdot\vec{\nabla}\vec{v}\right)$ | $g_{\alpha\beta}$ | Source is Ref. [6]. No extra assumptions are made; no perturbation performed. Primary scalar field, $q$, is defined in Equation (3.90) of this text |
| $D^2 p_1 = R_1$ | $(\rho_0 c)^{-1} g_{\alpha\beta}$ | Source is this text. A fully linearized version of the equations is performed, but no additional assumptions are made. The primary field is the pressure fluctuation, $p_1$ |
| $D^2 \psi_1 = 0$ | $\rho_0 c^{-1} g_{\alpha\beta}$ | Source is Ref. [5]. In addition to the standard assumptions, the fluid flow is assumed to be irrotational. The primary field is the velocity potential, $\psi_1$ |

The factor $R_1$ appearing in Table 3.1 is defined:

$$R_1 \equiv \rho_0^2 c \left( \frac{\partial}{\partial t} \left( \frac{F_1}{\rho_0 c^2} \right) + \vec{\nabla} \cdot \left( \frac{\vec{v}_0 F_1}{\rho_0 c^2} \right) - \vec{\nabla} \cdot \vec{F}_2 - F_3 \right) \tag{3.111}$$

### 3.5.8   Discussion Regarding Fluid Acoustics

In this subsection, several versions of a second-order equation were derived from first principles, two of which are generalizations of the linear wave equation. Which is correct? All
of them, depending on the situation. Effort was invested in massaging the format of the second-order operator into a common form, that of a covariant Laplacian operator in generalized
curvilinear coordinates in four-dimensional space with time as one of the coordinates. The
author has presented different versions of this derivation with slightly different scalar operators
(see Ref. [9]). Of the three presentations, the one following White's original work is the most
general in that it applies to the entire field and not just the small perturbations. The author's
approach to linearizing the original equations contains no approximations other than ignoring
all but linear terms in the perturbed fields but contains a lower-order term that depends on the
background fields and the perturbations. The version that follows Unruh's reference [5] is only
valid for irrotational flow, but otherwise completely general. The last form has been of particular interest in the field of general relativity owing to the fact that the acoustic equation is identical to that of a massless scalar field propagating on a curved background manifold. This is
precisely a scenario of interest to people doing research on the behavior of quantum fields
in the presence of strong gravitational fields, that is, curved space–time. The identification
is due to a serendipitous cancelation of all terms that would contribute to the r.h.s. of the equation, as in the other two cases, and to the fact that the perturbations in the different variables are
uncoupled. Looking at the other set of linearized equations, one can see that in general one has
to complete the equation for the velocity perturbations in order to develop a full solution for the
pressure perturbations. The factor of $R_1$ contains the velocity field perturbations as a source of
the pressure perturbations. This term prevents the identification made by Unruh holding in the
general case. For many real-world applications, ignoring this term may be justified, at least as a
further approximation. For example, one may consider special cases of background conditions
appropriate to underwater or aeroacoustics where the background flow is weak and the derivative of background fields is also weak. Then $R_1 \approx 0$ may be appropriate at high frequencies. In
other cases, it is common to have a background flow that is either incompressible, $\vec{\nabla} \cdot \vec{v}_0$, or
pseudo-incompressible, $\vec{\nabla} \cdot (\rho_0 \vec{v}_0)$. Realistic environments often have a background flow that
is purely horizontal in direction but depends only on depth or height. Again, for such special
cases, the form of $R_1$ may simplify or vanish.

   One of the virtues of the form of the operator appearing in all three forms of the second-order
equation is that it compacts the information related to the background field into a simple form and
identifies this with an effective manifold. This doesn't necessarily make the acoustic equations
easier to solve in general, but it does make explicit the nature of the waves and the PDE as being
hyperbolic in nature. It turns out that only the highest-order terms in the PDE govern the behavior
of the bicharacteristics, which in the high frequency limit are identical to the rays of the system.

## 3.6   Thin Rods and Plates

Thin plates and rods support bending waves that are governed by a PDE that is fourth order in the spatial variables. A detailed account of the derivation of these equations can be found in Rossing and Fletcher, *Principles of Vibration and Sound* [10]. A brief survey is included here for completeness. Consider a thin rod with longitudinal axis along the $x$-axis and 1-dimensional (1-dim) transverse displacements along the $y$-axis. The equation governing the transverse vibrations in the small amplitude limit is

$$\frac{\partial^2 y}{\partial t^2} = -\frac{Yk^2}{\rho}\frac{\partial^4 y}{\partial x^4} \tag{3.112}$$

The constants in (3.112) are Young's modulus, $Y$, material density, $\rho$, and the radius of gyration, $k$. A solution of the form $y = u(x)\exp(-i\omega t)$ reduces (3.112) to an ODE for $u(x)$, and the velocity parameter is defined:

$$v_t^2 = \omega k c_l \tag{3.113}$$

In (3.113) $c_l = \sqrt{Y/\rho}$ is the speed of longitudinal elastic vibrations. The speed of transverse bending modes depends on the frequency as well and experiences dispersion, and different frequencies will travel at different speeds. Since the equation is linear, a solution of the form $u(x) = A\exp(\alpha x)$ is assumed, leading to the general solution

$$u = \left(A_1\exp\left(\frac{\omega x}{v_t}\right) + A_2\exp\left(\frac{-\omega x}{v_t}\right) + A_3\exp\left(\frac{i\omega x}{v_t}\right) + A_4\exp\left(\frac{-i\omega x}{v_t}\right)\right) \tag{3.114}$$

The rod can be subjected to three types of boundary conditions at the ends. A free end has $\partial^2 y/\partial x^2 = 0$, and $\partial^3 y/\partial x^3 = 0$; for a supported end, $\partial^2 y/\partial x^2 = 0$, and $y = 0$; and a clamped end obeys $y = 0$ and $\partial y/\partial x = 0$. Application of boundary conditions leads to a discrete set of modes, as with the string, through a set of transcendental equations. For a long thin rod supported at both ends, for example, by simple ideal hinges, the frequencies are $\omega_n = \omega_1 n^2$, where $\omega_1 = \pi^2 k c_l/L^2$, and $n = 1, 2, 3, \ldots$ (Rossing and Fletcher [10]). In addition to the type of transverse vibrations discussed earlier, rods can exhibit torsional modes, where the material twists in the transverse plane of the rod.

Consider a thin plate in the $x-y$ plane, with $z$ identified with the transverse direction. The corresponding equation for transverse vibrations in a stiff plate is

$$\frac{\partial^2 z}{\partial t^2} = -\frac{Yh^2}{12\rho(1-\nu^2)}\nabla^4 z \tag{3.115}$$

The constants $Y$ and $\rho$ have the same meaning as for the rod, $\nu$ is Poisson's ratio, and $h$ is the thickness of the plate. The speed of longitudinal vibrations in an infinite solid is

$$c_l = \sqrt{\frac{Y}{\rho(1-\nu^2)}} \tag{3.116}$$

From this, the wave velocity is

$$v_t^2 = \frac{\omega h c_l}{\sqrt{12}} \qquad (3.117)$$

The specific normal modes of vibration depend on the type of boundary conditions placed on the edge of the plate, free, supported, or clamped.

## 3.7 Phonons

In the treatment of interactions between electrons and photons with solid matter, the elastic vibrations of the solid need to be taken into account [11]. Classical normal mode models, similar in form to the coupled mass–spring linear system discussed in (3.3), are developed by considering coupling between nearest-neighbor coupling between atomic lattice sites, Bravais lattice, in a crystal and expanding the potential energy function to second order. When evaluated at equilibrium, this process produces a set of equations identical in form to the mass–spring system. Plane wave solutions of the form $\exp(i(\alpha x - \omega t))$ are sought, where $x$ identifies a lattice site. Periodic boundary conditions are placed on the ends of the lattice chain, Born–von Karman boundary conditions. This leads to the following dispersion relation for a monoatomic 1-dim chain:

$$\omega = 2\sqrt{\frac{k}{m}} \left| \sin\left(\frac{\alpha a}{2}\right) \right| \qquad (3.118)$$

The constant $a$ is the equilibrium spacing of the nearest-neighbor lattice sites, $m$ and $k$ are the effective mass and "spring constant" for the system, and $\alpha$ is the wavenumber. Equation (3.118) can be compared with (3.20) of Section 3.3. Treatment of more complex systems, that is, two types of ions per primitive cell with alternating masses, coupling strengths, or both, leads to multiple dispersion relations. The classic case of a 1-dim lattice with alternating coupling strengths, $k_1$ and $k_2$, leads to the presence of two dispersion relations:

$$\omega^2 = \frac{k_1 + k_2}{m} \pm \sqrt{\frac{k_1^2 + k_2^2 + 2k_1 k_2 \cos(\alpha a)}{m}} \qquad (3.119)$$

The minus sign defines the acoustical branch, tending to zero as $\alpha \to 0$. The positive root defines the optical branch.

## 3.8 Tensors Lite

Since the concept of a metric has been introduced as a shorthand for encapsulating the environmental factors present in the wave equation, this seems like a good place to introduce some formalism related to tensor analysis on manifolds. The reader is directed to the references for additional information, Synge and Schild [7] and Lovelock and Rund [8]. It is assumed that the reader has been introduced to some of the basic concepts related to tensors and differential geometry of two-dimensional surfaces. Tensors are a generalization of vectors. They are entities that have multiple directional attributes associated with them. The components of a vector are denoted by indices:

$$\vec{V} = V_j \hat{e}_j \tag{3.120}$$

In (3.120) the vector nature is indicated by the arrow over the symbol. Also shown are the elements of an orthonormal basis, $\hat{e}_j$, while $j = 1, 2, 3, \ldots, N$, and the components of the vector expressed in that basis, $V_j$. A common example of a tensor is the stress tensor associated with solid materials in the study of elasticity. Each of these carries two indices and when projected in an orthonormal basis requires pairs of unit vectors:

$$\boldsymbol{\sigma} = \sigma_{jk} \hat{e}_j \hat{e}_k \tag{3.121}$$

The quantity $\hat{e}_j \hat{e}_k$ is a dyadic. Some texts indicate that a quantity is a tensor by placing multiple tildes above or below the symbol, for example, $\widetilde{\widetilde{\sigma}}$. In this text vectors are denoted by the traditional arrow, while tensors of higher rank are indicated by boldface lettering, as in (3.121). The use of special symbols is typically avoided by denoting a tensor by its "components," for example, $\sigma_{jk}$ is used to represent the stress tensor. This text makes use of that convention as well.

Vectors and other geometric quantities expressed in global Cartesian coordinates on an infinite plane do not require more than what has been presented to fully describe their properties. In particular, if one is dealing with a vector or tensor field, which changes value as a function of position, then the derivatives of the field can be expressed as

$$\frac{\partial \boldsymbol{\sigma}}{\partial x_i} = \frac{\partial \sigma_{jk}}{\partial x_i} \hat{e}_j \hat{e}_k \tag{3.122}$$

As it often happens, one finds a need to express vector and tensor quantities on curved surfaces, or spaces of higher dimension. The basic concepts introduced here will leverage those of two-dimensional surfaces embedded in three-dimensional Euclidian space, $\mathbf{E}^3$, for the purposes of establishing vocabulary. The surface in $\mathbf{E}^3$ can be described locally by two parameters, $(\xi_1, \xi_2)$:

$$\vec{x} = \vec{x}\,(\xi_1, \xi_2) \tag{3.123}$$

On a curved surface one does not have a single global set of unit vectors for expressing direction. Rather, at each point on the surface, there is a local tangent plane, or tangent space. The coordinate directions can be determined from the parameterized position vectors, if known, by taking partial derivatives with respect to the surface parameters:

$$\hat{e}_j = \frac{\partial \vec{x} / \partial \xi_j}{\left| \partial \vec{x} / \partial \xi_j \right|}, \quad j = 1, 2 \tag{3.124}$$

The movement of points constrained to lie within the surface, $S$, is described at each point by vectors expanded in the local basis of the tangent space. From this point of view, one should not think of the vector as pointing to another point within the surface, as it does not really live within the surface.

As a conceptual example consider a particle constrained to move in a curved surface. Its trajectory would be a curve constrained to the surface but not necessarily a coordinate curve. If the curve is parameterized by time, one can express the parameterized curve in $S$ as $(\xi_1(t), \xi_2(t))$. At each point along the trajectory, the velocity of the particle can be described by a vector in the tangent space at that specific point:

$$\vec{V} = \frac{d\xi_1}{dt}\hat{e}_1 + \frac{d\xi_2}{dt}\hat{e}_2 \tag{3.125}$$

In this case, not only is the velocity different at each point along the trajectory, but the vector space in which it lives is also different. The reader may wonder why this extra structure isn't required when describing vectors in ordinary space $\mathbf{E}^3$. It is, but the effect is trivial. One can think of $\mathbf{E}^3$, or the plane $\mathbf{E}^2$, as having a tangent space at every point. When they are built up from the standard Cartesian coordinate lines, they are all exact copies of each other, all pointing in the same direction. Hence there is really no danger in discarding this field of tangent planes. Introductions to vector algebra in the plane teach students to use an arrow to represent the vector and "slide" arrows around as long as they are not rotated or stretched and they will represent the same vector. From this the parallelogram rule for vector addition emerges. Thus it would seem that vectors that have different base points, that is, different tails, can be added. On a curved surface one cannot do vector algebra with vectors that have different base points. Only vectors in the same tangent space can be added, subtracted, or projected onto each other. To compare two vectors that reside in different tangent spaces requires the development of a procedure that brings them into one common space, a generalization of the "sliding" we are allowed to do in the plane. All of the quantities related to the previous discussion are depicted in Figure 3.1.

The components of a vector relative to the coordinate directions in the tangent space are an example of a contravariant quantity. The canonical example of a contravariant vectors is the velocity, or the differential of position, $dx$. In addition to the tangent space, there is a dual space

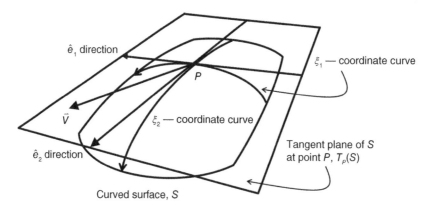

**Figure 3.1** Description of the tangent plane at a point on a curved manifold

at every point called the cotangent space. Its basis is determined by the normal to the coordinate planes and its elements called dual vectors. The canonical example of a dual vector is the gradient of a scalar function, $\partial f/\partial x$. The components of a contravariant vectors are denoted by a superscript, for example, $V^i$, while those of a covariant vector are denoted by a subscript, for example, $V_i$. So far, for coordinate labels, subscripts have been used. From this point on, superscripts will be used:

$$(x, y, z) = (x^1, x^2, x^3) \tag{3.126}$$

This convention is also adopted for the local coordinate curves in $S$, $(\xi^1, \xi^2)$. Covariance and contravariance are distinguished by their behavior under coordinate transformations. If a new set of coordinates, $(\bar{\xi}^1, \bar{\xi}^2)$, are defined, then differential elements and gradient components transform accordingly:

$$d\xi^j = \frac{\partial \xi^j}{\partial \bar{\xi}^k} d\bar{\xi}^k \tag{3.127}$$

$$\frac{\partial f}{\partial \xi^j} = \frac{\partial \bar{\xi}^k}{\partial \xi^j} \frac{\partial f}{\partial \bar{\xi}^k} \tag{3.128}$$

The factors $\partial \xi^j/\partial \bar{\xi}^k$ and $\partial \bar{\xi}^k/\partial \xi^j$ are general coordinate transformations and are, in general, nonlinear. Assuming that the coordinate transformation is invertible over the region where it is defined, these are inverses of each other:

$$\frac{\partial \bar{\xi}^k}{\partial \xi^j} \frac{\partial \xi^j}{\partial \bar{\xi}^i} = \bar{\delta}_i^k, \quad \frac{\partial \bar{\xi}^k}{\partial \xi^i} \frac{\partial \xi^j}{\partial \bar{\xi}^k} = \delta_i^j \tag{3.129}$$

The bar over the Kronecker delta serves as a reminder that the identity acts on vectors expanded in the new coordinate basis, whereas the second identity acts on the original coordinate basis. Finally, the product of (3.127) and (3.128) is the deferential of the scalar function $f$, $df$, which is a scalar under coordinate transformation:

$$df = \frac{\partial f}{\partial \xi^j} d\xi^j = \left( \frac{\partial \bar{\xi}^k}{\partial \xi^j} \frac{\partial f}{\partial \bar{\xi}^k} \right) \left( \frac{\partial \xi^j}{\partial \bar{\xi}^i} d\bar{\xi}^i \right) = \left( \frac{\partial \bar{\xi}^k}{\partial \xi^j} \frac{\partial \xi^j}{\partial \bar{\xi}^i} \right) \left( \frac{\partial f}{\partial \bar{\xi}^k} d\bar{\xi}^i \right) = \frac{\partial f}{\partial \bar{\xi}^k} d\bar{\xi}^k \tag{3.130}$$

The aforementioned is a generalization of rotation of vectors by a fixed angle using matrices. Based on these rules, tensors of higher rank are defined by the number of upper and lower indices they carry, and they transform properly under a general coordinate transformation. For example, consider the components of a tensor with three contravariant and two covariant indices, $T_{mn}^{ijk}$. Under a coordinate transform these components would transform according to the following rule:

$$\bar{T}_{pq}^{uvw} = \frac{\partial \bar{\xi}^u}{\partial \xi^i} \frac{\partial \bar{\xi}^v}{\partial \xi^j} \frac{\partial \bar{\xi}^w}{\partial \xi^k} \frac{\partial \xi^m}{\partial \bar{\xi}^p} \frac{\partial \xi^n}{\partial \bar{\xi}^q} T_{mn}^{ijk} \tag{3.131}$$

It is easy to see that one will use up the alphabet, both Latin and Greek, quickly dealing with tensors of high rank. Some authors decorate the index set with bars to indicate which set of coordinates they refer. The standard notation for indicating the rank of $T$ in (3.131) is (3, 2): "T is a rank three two tensor." In this new notation, the unit vectors that point along the coordinate directions, given in (3.124), are written as

$$\hat{e}_j = \frac{\partial \vec{x} / \partial \xi^j}{\left| \partial \vec{x} / \partial \xi^j \right|} \tag{3.132}$$

The unit vector retains a lower index and does transform as a covariant quantity. One defines a basis for the dual space, $\hat{e}^k$. Hence, dual vectors can be expressed in this basis:

$$A = A_k \hat{e}^k \tag{3.133}$$

The elements of the tangent space and the dual space are connected by a metric tensor. The metric, which measures infinitesimal lengths, can be thought of as a mapping from one space to the other. Given a vector, $\vec{V}$, with elements, $V^k$, the components of the corresponding dual vector are

$$V_k = g_{kj} V^j \tag{3.134}$$

In the language of tensors, the metric appearing in (3.134) represents the components of a second rank covariant tensor. The mapping of vectors to dual vectors, represented in (3.134), is often referred to as lowering indices. The inverse of the metric is a second rank contravariant tensor, that is, two upper indices, $g^{mn}$. The covariant and contravariant metrics obey the relation

$$g^{mn} g_{ml} = \delta_l^n \tag{3.135}$$

Just as the covariant metric lowers indices, the contravariant metric raises indices. The metric is a generalization of the dot product and provides the magnitude of vectors and the angle between pairs of vectors based at the same point on the surface:

$$|\vec{V}| = \sqrt{g_{ij} V^i V^j} \tag{3.136}$$

$$\cos \theta \left( \vec{V}, \vec{U} \right) = \frac{g_{ij} V^i U^j}{|\vec{V}||\vec{U}|} \tag{3.137}$$

Lastly, it is mentioned that for surfaces embedded in three-dimensional Euclidian space, one can build the metric of the surface from the unnormalized direction vectors:

$$g_{ij} = \frac{\partial \vec{x}}{\partial \xi^i} \cdot \frac{\partial \vec{x}}{\partial \xi^j} \tag{3.138}$$

The dot product in (3.138) is the equivalent of the Euclidian metric acting on the three-dimensional vectors constrained to the parameterized surface.

As an example, consider a sphere of radius $R$, and standard spherical coordinates, $(\xi^1, \xi^2) = (\theta, \varphi)$:

$$\vec{x} = R[\sin\theta \cos\varphi \quad \sin\theta\sin\varphi \quad \cos\theta]^T \tag{3.139}$$

The unnormalized basis vectors are

$$\frac{\partial \vec{x}}{\partial \theta} = R[\cos\theta \cos\varphi \quad \cos\theta\sin\varphi \quad -\sin\theta]^T \tag{3.140}$$

$$\frac{\partial \vec{x}}{\partial \varphi} = R[-\sin\theta\sin\varphi \quad \sin\theta \cos\varphi \quad 0]^T \tag{3.141}$$

From these one can verify the following:

$$\frac{\partial \vec{x}}{\partial \theta} \cdot \frac{\partial \vec{x}}{\partial \theta} = R^2 \tag{3.142}$$

$$\frac{\partial \vec{x}}{\partial \varphi} \cdot \frac{\partial \vec{x}}{\partial \varphi} = R^2 \sin^2\theta \tag{3.143}$$

$$\frac{\partial \vec{x}}{\partial \theta} \cdot \frac{\partial \vec{x}}{\partial \varphi} = 0 \tag{3.144}$$

In matrix form the metric tensor for the sphere is

$$R^2 \begin{bmatrix} 1 & 0 \\ 0 & \sin^2\theta \end{bmatrix} \tag{3.145}$$

If one considers a differential displacement in the coordinate directions, $(d\theta, d\varphi)$, then the differential element of arclength is determined by the following:

$$g_{ij}d\xi^i d\xi^j = R^2 \left(d\theta^2 + \sin^2\theta d\varphi^2\right) \tag{3.146}$$

In closing this section, it is emphasized that this introduction to the machinery of covariant and contravariant tensors on curved surfaces relied heavily on the example of two-dimensional surfaces embedded in three-dimensional space. This was done to provide a "lite" introduction to the vocabulary for reader less familiar with the subject. The formalism generalizes to $N$-dimensional spaces and does not require the embedding of the hypersurface in a higher-dimensional Euclidian space. This structure merely helps in visualizing the quantities defined on the surface and the action of the metric tensor.

## 3.9   Exercises

1.   Derive the second-order linear equation for the components of the velocity field.
2.   Verify that (3.109) is the inverse of (3.108), and verify the determinant (3.110).
3.   Derive an expression for the speed of sound in an ideal gas assuming an isentropic process.
4.   Parabolic coordinates $(u, v, w)$ are defined by the following relation to Cartesian coordinates:

$$x = uvw$$
$$y = uv\sqrt{1 - w^2}$$
$$z = \frac{1}{2}\left(u^2 - v^2\right)$$

From these relations determine the tangent vectors and the metric tensor for the coordinate surface defined by $u = $ constant and again for $w = $ constant.

## References

[1]  Ugural, A. C. and Fenster, S. K., Advanced Strength and Applied Elasticity, Prentice-Hall PTR, Upper Saddle River, NJ, 1995.

[2]  Shabana, A. A., Vibration of Discrete and Continuous Systems, Second Edition, Mechanical Engineering Series, F. F. Ling, Series Editor, Springer, New York, 1991.

[3]  Bergmann, P. G., The Wave Equation in a Medium with a Variable Index of Refraction, The Journal of the Acoustical Society of America, Volume 17, Number 4, pp. 329–333, 1946.

[4]  Landau, L. D. and Lifshitz, E. M., Fluid Mechanics, Second Edition, Butterworth & Heinemann, Oxford, 1987.

[5]  Unruh, W. G., Experimental black hole evaporation?, Phys. Rev. Lett., Vol. 46, pp. 1351–1353, 1981.

[6]  White, R. W., Acoustic ray tracing in moving inhomogeneous fluids, J. Acoust. Soc. Am., Vol. 53, No. 6, pp. 1700–1704, 1973.

[7]  Synge, J. L. and Schild, A., Tensor Calculus, University of Toronto Press, Toronto, ON, 1949.

[8]  Lovelock, D. and Rund, H., Tensors, Differential Forms, and Variational Principles, Dover Publications Inc., New York, 1989.

[9]  Bergman, D. R., Application of Differential Geometry to Acoustics: Development of a Generalized Paraxial Ray-Trace Procedure from Geodesic Deviation, Naval Research Laboratory Report, NRL/MR/ 7140–05-8835, Approved for public release: distribution unlimited, Naval Research Laboratory, Washington, DC, 2005.

[10] Rossing, T. D. and Fletcher, N. H., Principles of Vibration and Sound, Springer-Verlag, New York, 1995.

[11] Ashcroft, N. W. and Mermin, N. D., Solid State Physics, Saunders College Publishing, Fort Worth, TX, 1976.

# 4

# Methods for Solving the Wave Equation

## 4.1 Introduction

This chapter introduces well-known methods of describing the behavior of, and for solving, the wave equation. The first method introduced is the method of characteristics that when applied to the equations of fluid mechanics leads to its identification as a hyperbolic partial differential equation (PDE). The bicharacteristic curves are identified with the geodesics of a pseudo-Riemannian manifold, revisiting the discussion of the effective metric in Chapter 3. Through this discussion, the concept of Hamiltonian and Lagrangian representations of the characteristics is introduced. Following the method of characteristics will be a review on the separation of variables for the homogeneous equation. Solutions to the "free" wave equation in standard separable coordinates will be reviewed with attention to imposing ideal boundary conditions. An introduction to Green's function its construction using modes, and solutions of the in homogeneous equation are also presented. Lastly, the method of images is presented and examples applied to simple ideal boundaries.

## 4.2 Method of Characteristics

The equations of fluid dynamics and the wave equation are examples of PDE containing derivatives of the unknown fields with respect to spatial coordinates and time [1]. The highest derivative is referred to as the order, and sometimes space and time orders are quoted separately. A linear equation contains a sum of the fields and their derivatives up to and including the order of the equation. Nonlinear equations contain these terms raised to various powers. The ordinary wave equation is an example of a linear second-order PDE. When a source term is present, the equation is inhomogeneous; otherwise it is homogeneous. The equations of fluid dynamics are an example of what is called a quasi-linear first-order PDE. The term quasi-linear refers to the

*Computational Acoustics: Theory and Implementation*, First Edition. David R. Bergman.
© 2018 John Wiley & Sons Ltd. Published 2018 by John Wiley & Sons Ltd.

fact that the fields and their derivatives only appear to first order in power, but cross terms coupling the field to its derivative appear. The PDE commonly encountered in physics fall under three categories: elliptic, parabolic, and hyperbolic. An example of an elliptic equation is Laplace's equation:

$$\nabla^2 \varphi = 0 \tag{4.1}$$

An example of a parabolic equation is the heat equation, or diffusion equation:

$$-\frac{\partial \varphi}{\partial t} + \beta \nabla^2 \varphi = 0 \tag{4.2}$$

The wave equation is an example of a hyperbolic equation:

$$-\frac{\partial^2 \varphi}{\partial t^2} + \beta \nabla^2 \varphi = 0 \tag{4.3}$$

The parameter $\beta$ has no particular significance in the previous equations. In general, the coefficients of each term can be functions of position and time.

For the equations of hydrodynamics presented in Chapter 3, the method of characteristics is applied to four fields $(p, \vec{v})$, each a function of four independent variables, $x^\mu$, $\mu = 0, 1, 2, 3$. In the author's experience one of the most complete treatments of the method of characteristics is Courant and Hilbert, Volume II [2]. The reader is directed to the following sections for more details related to this presentation: Chapter III, Section 2.3; Chapter V, Section 3; and Chapter VI, Part I, Section 3a. For completeness, some parts of these sections are repeated here. Consider a general first-order PDE system with $n$ fields, $u_n$, and $\nu$ variables, $x^\nu$, written in the following form:

$$A^\nu_{mn} \frac{\partial u_n}{\partial x^\nu} + b_m = 0 \tag{4.4}$$

In (4.4) summation over $n$ and $\nu$ is implied. The matrix of coefficients $A^\nu_{nm}$ and the vector $b_m$ are functions of $x^\nu$ and $u_n$, but not the derivatives of $u_n$. A surface is defined in the space $x^\nu$, $S : \varphi(x^\nu) = 0$. From this surface the characteristic determinant is formed:

$$A = \det\left(A^\nu \frac{\partial \varphi}{\partial x^\nu}\right) \tag{4.5}$$

Setting the determinant equal to zero defines the characteristic equation, a first-order non-linear PDE obeyed by the surface, $\varphi(x^\nu)$. In this case the surface is called a characteristic surface, and a solution to the characteristic equation defines a family of such surfaces: $\varphi = \text{constant}$.

The equations of hydrodynamics are expressed in terms of a generic vector $u_n$, $u_1 = p$, $u_2 = v_x$, $u_3 = v_y$, and $u_4 = v_z$. For example, the equation of continuity becomes

$$\frac{\partial u_1}{\partial t} + \left(u_2 \frac{\partial u_1}{\partial x} + \cdots\right) + \rho c^2 \left(\frac{\partial u_2}{\partial x} + \cdots\right) = 0 \tag{4.6}$$

The ellipsis indicates more terms in the other variables. The first parenthesis is $\vec{v} \cdot \vec{\nabla} p$, while the second parenthesis is $\vec{\nabla} \cdot \vec{v}$. It is left as an exercise for the reader to express the other equation in terms of these variables. The matrix $A^0$ contains the coefficients of all the time derivatives and by an appropriate arrangement of terms becomes the identity:

$$A^0 = \begin{bmatrix} 1 & 0 & 0 & 0 \\ 0 & 1 & 0 & 0 \\ 0 & 0 & 1 & 0 \\ 0 & 0 & 0 & 1 \end{bmatrix} \tag{4.7}$$

The others are as follows:

$$A^1 = \begin{bmatrix} v_x & \rho c^2 & 0 & 0 \\ \rho^{-1} & v_x & 0 & 0 \\ 0 & 0 & v_x & 0 \\ 0 & 0 & 0 & v_x \end{bmatrix} \tag{4.8}$$

$$A^2 = \begin{bmatrix} v_y & 0 & \rho c^2 & 0 \\ 0 & v_y & 0 & 0 \\ \rho^{-1} & 0 & v_y & 0 \\ 0 & 0 & 0 & v_y \end{bmatrix} \tag{4.9}$$

$$A^3 = \begin{bmatrix} v_z & 0 & 0 & \rho c^2 \\ 0 & v_z & 0 & 0 \\ 0 & 0 & v_z & 0 \\ \rho^{-1} & 0 & 0 & v_z \end{bmatrix} \tag{4.10}$$

With the previous equations defined, the characteristic matrix for the system is

$$\begin{bmatrix} D\varphi & \rho c^2 \left( \vec{\nabla} \varphi \right)^T \\ \rho^{-1} \vec{\nabla} \varphi & \mathbf{I} D\varphi \end{bmatrix} \tag{4.11}$$

where $\mathbf{I}$ is the $3 \times 3$ identity. The characteristic conditions are satisfied by setting the determinant of this matrix equal to zero:

$$(D\varphi)^2 \left( (D\varphi)^2 - c^2 \vec{\nabla} \varphi \cdot \vec{\nabla} \varphi \right) = 0 \tag{4.12}$$

The factors of (4.12) define partial Hamiltonians for the system and are treated separately:

$$H_1 = -(D\varphi)^2 \tag{4.13}$$

$$H_2 = -\left((D\varphi)^2 - c^2 \vec{\nabla}\varphi \cdot \vec{\nabla}\varphi\right) \tag{4.14}$$

The sign convention is for convenience in later discussions and for consistency with previous definitions. Each Hamiltonian defines a set of characteristics. The first Hamiltonian, $H_1$, is related to steady flow, while $H_2$ is related to acoustics. Referring to the metric structure introduced in Chapter 3, (4.14) can be expressed as the magnitude of a vector in 4-dimensional (4-dim) space–time coordinates:

$$H_2 = g^{\mu\nu}\frac{\partial\varphi}{\partial x^\mu}\frac{\partial\varphi}{\partial x^\nu} = 0 \tag{4.15}$$

The metric appearing in (4.15) is similar to that introduced in Chapter 3, except for an overall scale factor:

$$g^{\mu\nu} = \frac{1}{c^2}\begin{bmatrix} -1 & -\vec{v}^{\mathrm{T}} \\ -\vec{v} & \mathbf{I}c^2 - \vec{v}\,\vec{v}^{\mathrm{T}} \end{bmatrix} \tag{4.16}$$

Equation (4.16) defines the contravariant metric tensor. The corresponding covariant tensor, the inverse of Equation (4.16), is given in (4.17):

$$g_{\mu\nu} = \begin{bmatrix} -c^2 + v^2 & -\vec{v}^{\mathrm{T}} \\ -\vec{v} & \mathbf{I} \end{bmatrix} \tag{4.17}$$

A factor of $c^{-2}$ has been included in (4.16) to match the top row of table 3.1 because of the characteristic condition, (4.15), the metric can be scaled by any positive scalar function of coordinates and time. In other words, all PDE systems with a characteristic Hamiltonian proportional to $H_2$ have the same set of characteristics and the same geometry. A set of parameterized curves in space–time called bicharacteristics is defined, which serves to push the characteristic surface from an initial configuration to another configuration. The curves are denoted $x^\mu(\lambda)$, and their tangent vectors $dx^\mu/d\lambda$, a 4-dim generalization of velocity. The parameter $\lambda$, not to be confused with wavelength, is arbitrary and not necessarily related to path arc length or time. The 4-dim gradient of the characteristic is identified with a momentum variable conjugate to $dx^\mu/d\lambda$:

$$p_\nu = \partial_\nu\varphi \tag{4.18}$$

The equations governing the characteristic can be derived using the Hamiltonian formulation of mechanics. Hamilton's equations arising from $H_2$ are as follows:

$$\frac{dx^\mu}{d\lambda} = \frac{1}{2}\frac{\partial H_2}{\partial p_\mu} = g^{\mu\nu}p_\nu \tag{4.19}$$

$$\frac{dp_\mu}{d\lambda} = -\frac{1}{2}\frac{\partial H_2}{\partial x^\mu} = -\frac{1}{2}p_\beta p_\alpha \frac{\partial g^{\alpha\beta}}{\partial x^\mu} \tag{4.20}$$

These can be combined to yield a second-order ordinary differential equation for the bicharacteristics. Equation (4.19) can be inverted using the covariant metric:

$$p_\mu = g_{\mu\nu}\frac{dx^\nu}{d\lambda} \tag{4.21}$$

The derivatives of the contravariant metric in (4.20) can be replaced with derivatives of the covariant metric by differentiating the identity $g_{\alpha\mu}g^{\mu\beta} = \delta_\alpha^\beta$:

$$0 = \frac{\partial\delta_\alpha^\beta}{\partial x^\nu} = \frac{\partial g_{\alpha\mu}g^{\mu\beta}}{\partial x^\nu} = g_{\alpha\mu}\frac{\partial g^{\mu\beta}}{\partial x^\nu} + \frac{\partial g_{\alpha\mu}}{\partial x^\nu}g^{\mu\beta} \tag{4.22}$$

From (4.22) the desired result is obtained:

$$\frac{\partial g^{\xi\beta}}{\partial x^\nu} = -g^{\alpha\xi}\frac{\partial g_{\alpha\mu}}{\partial x^\nu}g^{\mu\beta} \tag{4.23}$$

To complete the second-order equation, the derivative of $g_{\mu\nu}$ with respect to $\lambda$ is required:

$$\frac{dg_{\mu\nu}}{d\lambda} = \frac{dx^\alpha}{d\lambda}\frac{\partial g_{\mu\nu}}{\partial x^\alpha} \tag{4.24}$$

Using these results, (4.20) becomes

$$\frac{d^2x^\mu}{d\lambda^2} + \frac{1}{2}g^{\mu\nu}\left(\frac{\partial g_{\nu\alpha}}{\partial x^\beta} + \frac{\partial g_{\nu\beta}}{\partial x^\alpha} - \frac{\partial g_{\alpha\beta}}{\partial x^\nu}\right)\frac{dx^\alpha}{d\lambda}\frac{dx^\beta}{d\lambda} = 0 \tag{4.25}$$

The factor in the second term is the Christoffel symbol of the second kind:

$$\Gamma^\mu_{\alpha\beta} = \frac{1}{2}g^{\mu\nu}\left(\frac{\partial g_{\nu\alpha}}{\partial x^\beta} + \frac{\partial g_{\nu\beta}}{\partial x^\alpha} - \frac{\partial g_{\alpha\beta}}{\partial x^\nu}\right) \tag{4.26}$$

The first of Hamilton's equations can be used to write the characteristic equation in the Lagrangian formalism:

$$L = g_{\alpha\beta}\frac{dx^\alpha}{d\lambda}\frac{dx^\beta}{d\lambda} = 0 \tag{4.27}$$

The combined result of all this information is that the bicharacteristics of the equations of fluid dynamics, and more generally any hyperbolic PDE, are geodesics and that these geodesics have zero length. This last comment deserves discussion. In the study of the theory of surfaces embedded in $\mathbf{R}^3$, the concept of a metric is introduced. It also appears in the treatment of general curvilinear coordinate systems, such as spherical coordinates. The curvature of the surface and the geodesics of the surface are encrypted in the metric [3]. Inner products are defined using the metric tensor. The lengths of curves in a surface are positive definite, and arc length is generally considered a good parameter for curves. The study of such structures is referred to as Riemannian geometry, a generalization of Euclidean geometry. The metric in (4.16) contains a negative eigenvalue. In the limit of a constant medium with no background flow, this metric is approximated by diag($-c^2, 1, 1, 1$). In special relativity, this type of metric is introduced in the description of light photon trajectories [4]. In this type of space, vectors fall into three classes: space-like, time-like, and null. The choice to make the time component negative is referred to as the signature of the metric. Relative to this signature, space-like vectors have a positive magnitude and time-like have negative magnitude. These classes of vectors never mix and are separated by the space containing null vectors, vectors of zero length. The bicharacteristics are thus referred to as null geodesics. The study of curved spaces with this type of metric structure is referred to as Lorentzian or pseudo-Riemannian differential geometry and is a generalization of Riemannian geometry. Much of what is known about differential geometry generalizes nicely, but the null space is an exception. Unlike in Riemannian geometry, in this case "length" has no meaning and cannot be used to parameterize curves. The parameter $\lambda$ is called an affine parameter and is really a member of an equivalence class of parameterizations. Membership in this class is defined by invariance of the geodesic equation, (4.25), under and change in parameter.

Expanding (4.27) and using the explicit form of the metric gives the following first-order equation for the bicharacteristics:

$$g_{\alpha\beta}\frac{dx^\alpha}{d\lambda}\frac{dx^\beta}{d\lambda} = -\left(c^2 - v^2\right)\left(\frac{dt}{d\lambda}\right)^2 - 2\,\vec{v}\cdot\frac{d\vec{x}}{d\lambda}\frac{dt}{d\lambda} + \frac{d\vec{x}}{d\lambda}\cdot\frac{d\vec{x}}{d\lambda} = 0 \qquad (4.28)$$

Dividing through by $(dt/d\lambda)^2$, defining $d\vec{x}/dt = \left(d\vec{x}/d\lambda\right)(d\lambda/dt)$, and applying some factorization, (4.28) can be rewritten as

$$c^2 = \left(\frac{d\vec{x}}{dt} - \vec{v}\right)\cdot\left(\frac{d\vec{x}}{dt} - \vec{v}\right) \qquad (4.29)$$

Defining a unit vector, $\hat{n}$, in the direction of $d\vec{x}/dt - \vec{v}$, (4.29) can be interpreted as

$$c\hat{n} + \vec{v} = \frac{d\vec{x}}{dt} \qquad (4.30)$$

The new degree of freedom $\hat{n}$ is the normal to the wavefront, as a surface in $\mathbf{R}^3$.

Orthogonality is a relative term. The 4-dim tangent vector is "orthogonal" to the four-dimensional gradient of the wavefront relative to the metric in (4.17). In (4.30) $\hat{n} \propto \vec{\nabla}\,\varphi$, and the 3-dim tangent vector is not orthogonal to the three-dimensional wavefront relative to the dot product. The results of this section carry over to the linearized equation for the perturbed fields with the background fields replacing the corresponding terms in the metric. They are also equivalent to rays in the high frequency limit.

For reference the Christoffel symbols derived from (4.17), (4.16), and (4.26) are provided here. Each index runs from $\mu = 0, \ldots, 3$, with $\mu = 0$ being the time coordinate:

$$\Gamma^0{}_{00} = \frac{1}{2c^2}\left(\frac{\partial c^2}{\partial t} + \vec{v} \cdot \vec{\nabla}\left(v^2 - c^2\right)\right) \tag{4.31}$$

$$\Gamma^0{}_{0k} = \frac{1}{2c^2}\left(\frac{\partial}{\partial x^k}\left(c^2 - v^2\right) + v_i\left(\frac{\partial v_i}{\partial x^k} - \frac{\partial v_k}{\partial x^i}\right)\right) \tag{4.32}$$

$$\Gamma^k{}_{00} = -\frac{\partial v_k}{\partial t} - \frac{1}{2}\frac{\partial}{\partial x^k}\left(v^2 - c^2\right) + v_k\Gamma^0{}_{00} \tag{4.33}$$

$$\Gamma^0{}_{ik} = \frac{1}{2c^2}\left(\frac{\partial v_k}{\partial x^i} + \frac{\partial v_i}{\partial x^k}\right) \tag{4.34}$$

$$\Gamma^k{}_{i0} = \frac{1}{2}\left(\frac{\partial v_i}{\partial x^k} - \frac{\partial v_k}{\partial x^i}\right) + v_k\Gamma^0{}_{0i} \tag{4.35}$$

$$\Gamma^k{}_{ij} = v_k\Gamma^0{}_{ij} \tag{4.36}$$

The section ends with a brief comment on covariant differentiation, a follow-up to the last section of Chapter 3. Equation (4.26) is used to define a new type of derivative on manifolds, the covariant derivative. Consider a vector field, $V^\mu(x^\nu)$, with components being functions of coordinates. The covariant derivative of this field is defined as

$$D_\alpha V^\mu = \frac{\partial V^\mu}{\partial x^\alpha} + \Gamma^\mu{}_{\alpha\beta}V^\beta \tag{4.37}$$

The value of this definition is in the fact that $D_\alpha V^\mu$ behaves as a proper tensor under coordinate transformations, whereas the ordinary partial derivative will not. Consequently, the Christoffel symbols are not proper tensors. They transform in such a way as to cancel out the extra terms generated by the partial derivative under coordinate transformation. The geodesic equation, (4.25), can be written in terms of the covariant derivative, where the definition $T^\mu = dx^\mu/d\lambda$ is used:

$$T^\alpha D_\alpha T^\mu = 0 \tag{4.38}$$

In Equation (4.38) the covariant derivative of the vector field $T^\mu$ is projected onto $T^\mu$ and the result is set equal to zero. In general, when this requirement is imposed, the field being differentiated is said to be parallel transported along the direction of $T^\mu$. This process is the way that vectors with different base points are brought together for comparison. The geodesic equation defines curves that parallel transport their own velocity vectors. This is also a generalization of acceleration and the equation states that geodesics are curves with no acceleration. The covariant derivative of a lower index vector is $D_\alpha V_\mu = \partial V_\mu/\partial x^\alpha - \Gamma^\beta{}_{\alpha\mu}V_\beta$. As a final note the metric and its inverse are both "constant" relative to this type of differentiation, $D_\alpha g_{\mu\nu} = D_\alpha g^{\mu\nu} = 0$.

## 4.3  Separation of Variables

Separation of variables is a method designed to reduce the full homogeneous PDE into a set of ordinary differential equations. The goal is to reduce the multidimensional differential equation into several one-dimensional differential equations. The original equation is expressed as

$$F\left(x^{\mu},\psi,\frac{\partial\psi}{\partial x^{\mu}},\ ...,\ \frac{\partial^{n}\psi}{\partial x^{\mu n}},\ ...\text{etc}\right)=0 \tag{4.39}$$

This method requires the equation to be linear but the coefficients may be function of position and time. It is assumed that the field can be factored, such that each factor depends on only one coordinate:

$$\psi=T(t)X(x)Y(y)Z(z) \tag{4.40}$$

The partial derivatives of the original function are proportional to ordinary derivatives:

$$\frac{\partial^{n}\psi}{\partial t^{n}}=\frac{d^{n}T(t)}{dt^{n}}X(x)Y(y)Z(z) \tag{4.41}$$

$$\frac{\partial^{n}\psi}{\partial x^{n}}=T(t)\frac{d^{n}X(x)}{dx^{n}}Y(y)Z(z) \tag{4.42}$$

Expressions similar to (4.42) exist for the other derivatives and the reader can derive expressions for the mixed partials. After applying differential operators to (4.40), a factor of $\psi$ is divided through. For example,

$$\frac{1}{\psi}\frac{\partial^{n}\psi}{\partial t^{n}}=\frac{1}{T(t)}\frac{d^{n}T(t)}{dt^{n}} \tag{4.43}$$

If the resulting equation can be grouped into additive factors, each dependent on a single variable, then each factor can be set equal to a constant:

$$F_{t}\left(t,T,...,T^{(n)}\right)+F_{x}\left(x,X,...,X^{(n)}\right)+F_{y}\left(y,Y,...,Y^{(n)}\right)+F_{z}\left(z,Z,...,Z^{(n)}\right)=0 \tag{4.44}$$

$$F_{t}\left(t,T,...,T^{(n)}\right)=\lambda_{t} \tag{4.45}$$

$$F_{x}\left(x,X,...,X^{(n)}\right)=\lambda_{x} \tag{4.46}$$

$$F_{y}\left(y,Y,...,Y^{(n)}\right)=\lambda_{y} \tag{4.47}$$

$$F_{z}\left(z,Z,...,Z^{(n)}\right)=\lambda_{z} \tag{4.48}$$

The superscript notation in equations (4.44) – (4.48), (n), refers to order of differentiation with respect to the independent variable of each function. Each $\lambda_\mu$ is called a separation constant and these obey a constraint imposed by the original equation:

$$\lambda_t + \lambda_x + \lambda_y + \lambda_z = 0 \tag{4.49}$$

The functions $F_\mu$ will not necessarily have the same form as the original $F$. In general, it is not easy to massage an equation into a form like (4.44) without strict requirements on the coefficients.

The solutions to each equation will depend on the separation constants, $T(t, \lambda_t)$, etc. Each specific choice of parameters obeying (4.49) is a possible solution. A general solution to the original equation can be expressed as a sum, or integral, over these parameters:

$$\psi(x, y, z, t) = \sum A\left(\lambda_t, \lambda_x, \lambda_y, \lambda_z\right) T(t, \lambda_t) X(x, \lambda_x) Y\left(y, \lambda_y\right) Z(z, \lambda_z) \tag{4.50}$$

As an example, consider the wave equation in an inhomogeneous medium described by a sound speed profile, $U(x^\mu) = 1/c^2(x^\mu)$:

$$\nabla^2 \psi - U(x^\mu) \frac{\partial^2 \psi}{\partial t^2} = 0 \tag{4.51}$$

Applying this procedure leads to the following for (4.44):

$$\frac{1}{X} \frac{d^2 X}{dx^2} + \frac{1}{Y} \frac{d^2 Y}{dy^2} + \frac{1}{Z} \frac{d^2 Z}{dz^2} - U(x^\mu) \frac{1}{T} \frac{d^2 T}{dt^2} = 0 \tag{4.52}$$

The last term poses a possible impediment. Assuming a time-independent $U(x^\mu)$ allows the time dependence to be separated:

$$\frac{1}{T} \frac{d^2 T}{dt^2} = -\omega^2 \tag{4.53}$$

$$\frac{1}{X} \frac{d^2 X}{dx^2} + \frac{1}{Y} \frac{d^2 Y}{dy^2} + \frac{1}{Z} \frac{d^2 Z}{dz^2} + U(x^k) \omega^2 = 0 \tag{4.54}$$

The first equation can be solved to yield $T = T_0 \exp(\pm i\omega t)$. It is clear by inspection that separability can be achieved if $U(x^k) = u(x) + v(y) + w(z)$:

$$\left\{ \frac{1}{X} \frac{d^2 X}{dx^2} + u(x)\omega^2 \right\} + \left\{ \frac{1}{Y} \frac{d^2 Y}{dy^2} + v(y)\omega^2 \right\} + \left\{ \frac{1}{Z} \frac{d^2 Z}{dz^2} + w(z)\omega^2 \right\} = 0 \tag{4.55}$$

## 4.4  Homogeneous Solution in Separable Coordinates

Solutions for the Helmholtz equation with a constant sound speed can be found for a class of coordinate systems called separable coordinates. These solutions allow for a description of the free field that can be fit to boundary surfaces that coincide with the coordinate surfaces.

This section provides a review of the three most common separable coordinates: Cartesian, cylindrical, and spherical [5].

## 4.4.1   Cartesian Coordinates

When the speed of sound is constant, the function $U(x^k)$ in (4.54) is a constant, and the factor $\omega^2/c^2 = k^2$ can be grouped with any one of the three derivative terms. In anticipation of defining the wavevector, the separation constants for $X$ and $Y$ are defined as $-k_x^2$ and $-k_y^2$, respectively. Combining this with the equation for $Z$ leads to the following set of separated equations:

$$\frac{d^2X}{dx^2} + k_x^2 X = 0 \tag{4.56}$$

$$\frac{d^2Y}{dy^2} + k_y^2 Y = 0 \tag{4.57}$$

$$\frac{d^2Z}{dz^2} + \left(k^2 - k_x^2 - k_y^2\right)Z = 0 \tag{4.58}$$

In (4.58) the separation constants for $X$ and $Y$ have been used to replace terms with $x$ and $y$ dependence in the original equation. Identifying the term in parenthesis in (4.58) with $k_z^2$, the constraint equation, (4.49), becomes

$$k^2 = k_x^2 + k_y^2 + k_z^2 \tag{4.59}$$

Equations (4.56)–(4.58) are recognized as the equation for a simple harmonic oscillator. The two independent solutions for each can be expressed in terms of sine and cosine or complex exponentials. The latter is presented here as

$$X = \exp(\pm ik_x x) \tag{4.60}$$

$$Y = \exp(\pm ik_y y) \tag{4.61}$$

$$Z = \exp(\pm ik_z z) \tag{4.62}$$

Each solution represents a sinusoidal wave traveling along one of the Cartesian axes. Combing the results into a complete solution leads to the familiar expression for a plane wave solution:

$$\psi = \exp\left(i\left(\vec{k} \cdot \vec{x} - \omega t\right)\right) \tag{4.63}$$

The previous expression also reinforces a standard convention that waves are traveling in the direction of $\vec{k}$. This is a good place to mention conventions. The choice of sign on the time exponential is completely arbitrary. Here, the choice of minus sign is made. With this comes the convention that the spatial phase be positive and waves travel in the direction of $\vec{k}$. Had the other choice of sign for time been made the spatial phase would acquire the opposite sign and the interpretation that wave travel in the $-\vec{k}$ direction. The choice made here is sometimes referred to as the physics convention, while the opposite is the electrical engineering convention (Knott et al. [6]). At the end of the day, the convention does not matter as long as it is used consistently throughout one's work. In (4.63) the physics convention is used.

## 4.4.2 Cylindrical Coordinates

For describing waves traveling in cylindrical waveguides or in an environment with cylindrical symmetry, it is convenient to express the wave equation in cylindrical coordinates. Cylindrical coordinates $\rho, \varphi, z$ are related to Cartesian coordinates by the following transformation:

$$x = \rho \cos \varphi \tag{4.64}$$

$$y = \rho \sin \varphi \tag{4.65}$$

$$\rho = \sqrt{x^2 + y^2} \tag{4.66}$$

$$\varphi = \arctan \frac{y}{x} \tag{4.67}$$

The Laplacian operator in cylindrical coordinates may be expressed as follows:

$$\nabla^2 = \frac{1}{\rho} \frac{\partial}{\partial \rho} \left( \rho \frac{\partial}{\partial \rho} \right) + \frac{1}{\rho^2} \frac{\partial^2}{\partial \varphi^2} + \frac{\partial^2}{\partial z^2} \tag{4.68}$$

Using (4.68), the Helmholtz equation is expressed in cylindrical coordinates:

$$\left\{ \frac{1}{\rho} \frac{\partial}{\partial \rho} \left( \rho \frac{\partial}{\partial \rho} \right) + \frac{1}{\rho^2} \frac{\partial^2}{\partial \varphi^2} + \frac{\partial^2}{\partial z^2} \right\} \psi + \frac{\omega^2}{c^2} \psi = 0 \tag{4.69}$$

Applying separation of variables with $\psi = R(\rho)\Phi(\varphi)Z(z)$ yields

$$\frac{1}{\rho} \frac{1}{R} \frac{d}{d\rho} \left( \rho \frac{dR}{d\rho} \right) + \frac{1}{\rho^2} \frac{1}{\Phi} \frac{d^2\Phi}{d\varphi^2} + \frac{1}{Z} \frac{d^2Z}{dz^2} + \frac{\omega^2}{c^2} = 0 \tag{4.70}$$

The sound speed is constant and the wavenumber is defined as $k = \omega/c$. Separation of variables leads to three equations:

$$\frac{1}{Z} \frac{d^2Z}{dz^2} = -\left(k^2 - \alpha^2\right) \tag{4.71}$$

$$\frac{1}{\Phi} \frac{d^2\Phi}{d\varphi^2} = -\beta^2 \tag{4.72}$$

$$\frac{1}{\rho} \frac{1}{R} \frac{d}{d\rho} \left( \rho \frac{dR}{d\rho} \right) - \frac{\beta^2}{\rho^2} + \alpha^2 = 0 \tag{4.73}$$

For a constant sound speed, the equation for Z can be solved exactly:

$$Z = \exp\left( \pm i\sqrt{k^2 - \alpha^2} z \right) \tag{4.74}$$

For free waves, $\alpha < k$ so this solution is a plane wave traveling in the $z$ direction. The solution to (4.72) is

$$\Phi = \exp(\pm i\beta\varphi) \tag{4.75}$$

These are traveling waves in the $\pm \varphi$ directions. For these waves to be continuous, periodic boundary conditions are applied, $\Phi(\varphi) = \Phi(\varphi + 2\pi)$:

$$1 = \exp(\pm i\beta 2\pi) \tag{4.76}$$

To satisfy (4.76), the separation constant is restricted to be an integer, $\beta = n = 0, 1, \ldots$. The remaining equation for $R$ is

$$\rho^2 R'' + \rho R' + \left( \alpha^2 \rho^2 - n^2 \right) R = 0 \tag{4.77}$$

The second derivative has been expanded and the equation multiplied through by $\rho^2 R$. Defining a new variable, $\xi = \alpha\rho$, this equation becomes Bessel's equation:

$$\xi^2 R'' + \xi R' + \left( \xi^2 - n^2 \right) R = 0 \tag{4.78}$$

This equation has two independent solutions for each $n$, referred to as Bessel's function and Neumann's function. Each is defined by an infinite series:

$$J_n(\xi) = \sum_{m=0}^{\infty} \frac{(-1)^m}{m!(m+n)!} \left( \frac{\xi}{2} \right)^{n+2m} \tag{4.79}$$

$$\pi Y_n(\xi) = 2J_n(\xi)\ln\left( \frac{\xi}{2} \right) - \sum_{m=0}^{n-1} \frac{(n-m-1)!}{m!} \left( \frac{\xi}{2} \right)^{2m-n} - \sum_{m=0}^{\infty} (-1)^m \frac{[\psi(m+1) + \psi(m+n+1)]}{m!(m+n)!} \left( \frac{\xi}{2} \right)^{n+2m} \tag{4.80}$$

In (4.80) $\psi(x)$ is the psi function [7]. Linear combinations of these functions define incoming and outgoing traveling waves propagating along the $\rho$ direction:

$$H_n^{(1)}(\xi) = J_n(\xi) + iY_n(\xi) \tag{4.81}$$

$$H_n^{(2)}(\xi) = J_n(\xi) - iY_n(\xi) \tag{4.82}$$

The function $H_n^{(1)}$ represents outward traveling waves in the physics convention. To illustrate this, consider the limiting behavior of these functions for large values of $\xi$:

$$H_n^{(1)}(\xi) \approx \sqrt{\frac{2}{\pi}} i^{-(n+1/2)} \frac{\exp(i\xi)}{\sqrt{\xi}} \tag{4.83}$$

$$H_n^{(2)}(\xi) \approx \sqrt{\frac{2}{\pi}} i^{(n+1/2)} \frac{\exp(-i\xi)}{\sqrt{\xi}} \tag{4.84}$$

A complete solution can be built up from these solutions. The solution for $n = 0$, and $\alpha = 0$ gives 2-dim Green's function, $\alpha H_0^{(1,2)}(\xi)$, where $\alpha$ is a constant. Following the same convention as for plane waves, the outgoing solution is $H_n^{(1)}$, while the incoming solution is $H_n^{(2)}$. The former can be thought of as circular waves emanating from the origin and traveling out to infinity, while the latter travels toward the origin from infinity.

### 4.4.3   Spherical Coordinates

Spherical coordinates $r, \theta, \varphi$ are given by the following transformation:

$$x = r\,\sin\theta\,\cos\varphi \tag{4.85}$$

$$y = r\,\sin\theta\,\sin\varphi \tag{4.86}$$

$$z = r\cos\theta \tag{4.87}$$

$$r = \sqrt{x^2 + y^2 + z^2} \tag{4.88}$$

$$\theta = \arctan\frac{\sqrt{x^2 + y^2}}{z} \tag{4.89}$$

$$\varphi = \arctan\frac{y}{x} \tag{4.90}$$

The Laplacian operator in spherical coordinates has the following form:

$$\nabla^2 = \frac{1}{r^2}\frac{\partial}{\partial r}\left(r^2\frac{\partial}{\partial r}\right) + \frac{1}{r^2\,\sin\theta}\frac{\partial}{\partial\theta}\left(\sin\theta\frac{\partial}{\partial\theta}\right) + \frac{1}{r^2\,\sin^2\theta}\frac{\partial^2}{\partial\varphi^2} \tag{4.91}$$

A solution of the form $\psi = R(r)\Theta(\theta)\Phi(\varphi)$ is assumed and used to separate the wave equation:

$$\frac{1}{r^2}\frac{1}{R}\frac{d}{dr}\left(r^2\frac{dR}{dr}\right) + \frac{1}{r^2\sin\theta}\frac{1}{\Theta}\frac{d}{d\theta}\left(\sin\theta\frac{d\Theta}{d\theta}\right) + \frac{1}{r^2\sin^2\theta}\frac{1}{\Phi}\frac{d^2\Phi}{d\varphi^2} + k^2 = 0 \tag{4.92}$$

The equation for $\Phi$ is separated first and is identical to the cylindrical case:

$$\frac{1}{\Phi}\frac{d^2\Phi}{d\varphi^2} = -m^2 \tag{4.93}$$

$$\frac{1}{R}\frac{d}{dr}\left(r^2\frac{dR}{dr}\right) + \frac{1}{\sin\theta}\frac{1}{\Theta}\frac{d}{d\theta}\left(\sin\theta\frac{d\Theta}{d\theta}\right) - \frac{1}{\sin^2\theta}m^2 + r^2k^2 = 0 \tag{4.94}$$

Equation (4.94) separates into two equations, defining a second separation constant $\lambda$:

$$\frac{1}{\sin\theta}\frac{1}{\Theta}\frac{d}{d\theta}\left(\sin\theta\frac{d\Theta}{d\theta}\right) - \frac{1}{\sin^2\theta}m^2 = -\lambda \tag{4.95}$$

$$\frac{1}{R}\frac{d}{dr}\left(r^2\frac{dR}{dr}\right) + r^2k^2 = \lambda \tag{4.96}$$

Equation (4.95) is rearranged to give equation (4.97).

$$\sin\theta\frac{d}{d\theta}\left(\sin\theta\frac{d\Theta}{d\theta}\right) + \lambda\Theta\sin^2\theta - \Theta m^2 = 0 \tag{4.97}$$

Defining a new variable $\xi = \cos\theta$, the previous equation transforms to

$$\left(1-\xi^2\right)\frac{d}{d\xi}\left(\left(1-\xi^2\right)\frac{d\Theta}{d\xi}\right) + \left(\lambda\left(1-\xi^2\right)-m^2\right)\Theta = 0 \tag{4.98}$$

This is Legendre's equation, for $\lambda = l(l+1)$ with $l$ a positive integer, and the solutions are associated Legendre polynomials, $P_l^m(\xi)$. The superscript $m$, is constrained, $-l \leq m \leq l$, $l = 0, 1, 2, \ldots$. Given a polynomial for positive $m$, the corresponding solution for $-m$ can be determined:

$$P_l^{-m}(\xi) = (-1)^m\frac{(l-m)!}{(l+m)!}P_l^m(\xi) \tag{4.99}$$

A few examples of associated Legendre polynomials are provided in Table 4.1.

**Table 4.1**   Associated Legendre polynomials

| $(m, l)$ | $P_l^m(\xi)$ |
|---|---|
| $(0, 0)$ | $1$ |
| $(0, 1)$ | $\xi$ |
| $(1, 1)$ | $\sqrt{1-\xi^2}$ |
| $(0, 2)$ | $(3\xi^2-1)/2$ |
| $(1, 2)$ | $3\xi\sqrt{1-\xi^2}$ |
| $(2, 2)$ | $3(1-\xi^2)$ |

The angular functions can be combined into one set of normalized functions on the unit sphere called spherical harmonics:

$$Y_l^m(\theta, \varphi) = (-1)^m \sqrt{\frac{(2l+1)(l-m)!}{4\pi(l+m)!}} P_l^m(\theta) \exp(im\varphi) \tag{4.100}$$

The radial equation is presented:

$$\frac{1}{R}\frac{d}{dr}\left(r^2\frac{dR}{dr}\right) + r^2k^2 = l(l+1) \tag{4.101}$$

For $l=0$ the solution is proportional to the spherical Green's function:

$$R = \alpha\frac{\exp(\pm ikr)}{r} \tag{4.102}$$

The positive sign represents outgoing spherical waves from a point at $r=0$, while the minus sign represents spherical waves coming from infinity and converging on $r=0$. For other values of $l$, the dimensionless variable $\eta = kr$ is defined, and (4.101) becomes the spherical Bessel's equation:

$$\eta^2\frac{2R}{d\eta^2} + 2\eta\frac{dR}{d\eta} + (\eta^2 - l(l+1))R = 0 \tag{4.103}$$

The solutions of (4.103), spherical Bessel functions $j_n(\eta)$, are related to the Bessel functions of half integer order. Let $B_n(\eta)$ represent any of the following functions: Bessel, Neumann, or Henkel. Then

$$j_n(\eta) = \sqrt{\frac{\pi}{2\eta}}B_{n+1/2}(\eta) \tag{4.104}$$

The three coordinate systems presented here are the most common. There are other separable coordinates. The reader is referred to Morse and Feshbach for additional details [8].

## 4.5 Boundary Conditions

The modes described in the last two sections are defined over all of space; their domain is unrestricted. Practical problems involve boundaries that restrict the domain of the solution and require specific behavior at the boundary. There are three types of boundary conditions to consider. Dirichlet boundary conditions restrict the value of the function at the boundary. Neumann boundary conditions place a constraint on the value of the normal derivative of the function evaluated at the boundary. Cauchy conditions impose a constraint on the value of the function and the normal derivatives at the boundary. An example of a Dirichlet boundary condition is a pressure release surface, $p(\vec{x}_S) = 0$, where $\vec{x}_S$ describes points on the boundary. The condition for reflection of pressure waves from an ideally hard surface is an example of Neumann boundary conditions, $\hat{n} \cdot \vec{\nabla} p(\vec{x}_S) = 0$, where $\hat{n}$ is the normal to the surface. In general Cauchy conditions come in a variety of forms. Another type of boundary condition is the Robin condition:

$$c_1 \psi(\vec{x}_S) + c_2 \hat{n} \cdot \vec{\nabla} \psi(\vec{x}_S) = f(\vec{x}_S) \tag{4.105}$$

One can also have mixed boundary conditions where the field and its normal derivate are restricted on disjoint regions of a boundary.

The application of ideal boundary conditions to cases with Cartesian symmetry is presented as an illustrative example. Consider two parallel infinite planes with normal vectors along the $z$ direction, leaving waves free to propagate in the $x$–$y$ plane. Boundary conditions are applied to a linear combination of the two independent solutions in the $z$ coordinate.

$$\psi(\vec{x}) = \exp i(k_x x + k_y y)\{a_0 \exp(ik_z z) + a_1 \exp(-ik_z z)\} \tag{4.106}$$

For this example, the boundaries are at $z_{BC} = 0, L_z$. There are three distinct choices of ideal boundary conditions: (1) two ideal pressure release surfaces, (2) two ideal hard boundaries, and (3) one soft boundary and the other hard. To simplify equations the following variable is defined:

$$p = \exp(ik_z L_z) \tag{4.107}$$

---

**Case 4.1:   Two Soft Boundaries**

Pressure release boundary conditions on both planes, $\psi|_{z=z_{BC}} = 0$, lead to

$$a_0 + a_1 = 0 \tag{4.108}$$

$$a_0 p + a_1 \bar{p} = 0 \tag{4.109}$$

The first constraint implies $a_1 = -a_0$ and the functions satisfying this are $\sin(k_z z)$. The second constraint, after applying the results of the first constraint, leads to the following condition:

$$\sin(k_z L_z) = 0 \tag{4.110}$$

This is satisfied only for $k_z L_z = n\pi$, where $n$ is an integer. The application of the first boundary condition fixes a node at $z = 0$, while the second places a restriction on the allowed wavenumber producing a discrete spectrum.

---

**Case 4.2:   Two Hard Boundaries**

Ideal hard boundary conditions are imposed on both planes, $\hat{n} \cdot \nabla \psi |_{z=z_{BC}} = 0$, leading to

$$-ik_z\{a_0 - a_1\} = 0 \qquad (4.111)$$

$$ik_z\{a_0 p - a_1 \bar{p}\} = 0 \qquad (4.112)$$

The first constraint implies $a_1 = a_0$ and the appropriate function satisfying this is cos $(k_z L_z)$. The second constraint, evaluated at $a_1 = a_0$, leads to the same result for the discrete spectrum.

---

**Case 4.3:   One Pressure Release and One Hard Surface**

Fixing the pressure release surface at $L_z$ and the hard surface at 0, leads to the following for the boundary conditions:

$$-ik_z\{a_0 - a_1\} = 0 \qquad (4.113)$$

$$a_0 p + a_1 \bar{p} = 0 \qquad (4.114)$$

Once again, the first constraint leads to a cosine function, while the second implies

$$\cos(k_z L_z) = 0 \qquad (4.115)$$

The constraint is satisfied only for $k_z L_z = (2n+1)\pi/2$.

The entire exercise of finding appropriate eigenfunctions for ideal boundary conditions in Cartesian coordinates boils down to fitting the trigonometric functions sine and cosine into the region between the boundaries such that nodes occur at a soft boundary and antinodes at a hard boundary. Results for all three cases are summarized in Table 4.2.

This completes the development of eigensolutions for these three cases of boundary conditions in one of three dimensions. The complete solution for any of the three cases may be expressed as follows:

$$\psi(\vec{x}) = A_0 \exp i(k_x x + k_y y) Z_n(z) \qquad (4.116)$$

Following the eigenvalue indexing discussed in the previous section, eigenfunctions for the cases described thus far are as follows:

$$\psi(\vec{x}; \vec{k}) = A_0 \exp\left(i\vec{k} \cdot \vec{x}\right), \quad \text{No boundary} \qquad (4.117)$$

$$\psi_n\left(\vec{x};\vec{k}_T\right) = A_0 \exp i\left(\vec{k}_T \cdot \vec{x}\right) Z_n(z), \quad BC \text{ in the } z \text{ direction} \tag{4.118}$$

The shorthand $\vec{k}_T = \left[k_x, k_y\right]^T$ for the 2-dim propagation vector in the $x$–$y$ plane has been introduced.

The analysis presented earlier applies to pairs of surfaces in the other two Cartesian directions. There is no need to repeat the details. Following the same convention, boundaries are placed at $x_{BC} = 0, L_x$ and $y_{BC} = 0, L_y$; all the results of Table 4.2 carry over with the appropriate variable replacing $z$. As an example, consider ideally hard boundaries applied to pairs of boundary plains in the $y$ and $z$ directions. Indexing the discrete wavenumbers by $m$ and $n$, respectively, leads to the following for the discrete wavenumber:

$$\left(k_{ym}, k_{zn}\right) = \left(m\frac{\pi}{L_y}, n\frac{\pi}{L_z}\right) \tag{4.119}$$

In this case the wavenumber in the $x$ direction is a continuous parameter, taking any real value, $k_x \in (-\infty, \infty)$. The eigenfunctions for this case are denoted:

$$\psi_{m,n}\left(\vec{x};k_x\right) = A_0 \exp i(k_x x) Y_m(y) Z_n(z) \tag{4.120}$$

The functions $Y_m(y)$ and $Z_n(z)$ are taken from case 2 of Table 4.2. Continuing in this manner another pair of hard boundaries is added in the $x$ direction. Introducing a third index and following all the same conventions leads to

$$\left(k_{xl}, k_{ym}, k_{zn}\right) = \left(l\frac{\pi}{L_x}, m\frac{\pi}{L_y}, n\frac{\pi}{L_z}\right) \tag{4.121}$$

$$\psi_{l,m,n}\left(\vec{x}\right) = A_0 X_l(x) Y_m(y) Z_n(z) \tag{4.122}$$

These modes might be ideal for describing sound in an enclosed rectangular volume of space, such as a room, with ideal hard walls, floor, and ceiling. For simplicity, an eigenfunction and its set of eigenvalues are denoted by the triple index $(l, m, n)$. The reader can develop other examples. Samples of the 1-dim modes for $n = 1$ are presented in Figure 4.1.

**Table 4.2** List of eigenfunctions and eigenvalues for three ideal cases

| Case | Eigenvalues, $k_{zn}$ | Eigenfunctions, $Z_n$ |
|------|----------------------|----------------------|
| 1 | $n\frac{\pi}{L_z}$ | $\sin(k_{zn}z)$ |
| 2 | $n\frac{\pi}{L_z}$ | $\cos(k_{zn}z)$ |
| 3 | $\left(n+\frac{1}{2}\right)\frac{\pi}{L_z}$ | $\cos(k_{zn}z)$ |

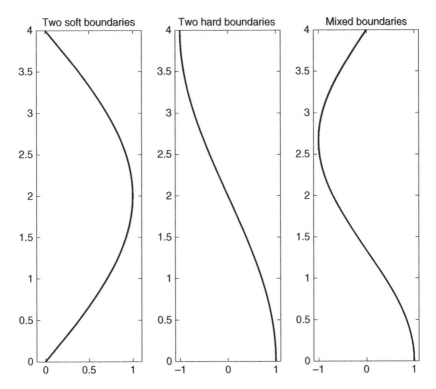

**Figure 4.1** Sample mode, $n=1$, for two soft (left), two hard (center), and mixed (right) boundary conditions

## 4.6 Representing Functions with the Homogeneous Solutions

Solving the homogeneous wave equation provides a set of functions indexed by either a continuous parameter or a discrete index set. While these functions may appear to be limited at first glance, they form a complete set for a basis of a function space. As a consequence, any function in the same space can be expressed as a linear superposition of the solutions to the homogeneous equation. This makes them an invaluable tool in theory and practice. This section introduces the basic ingredients required for applying the basis functions to practical situations. A compact vector notation commonly used in quantum mechanics is also introduced [9]. The notation makes lengthy expressions easier to write and manipulate and is also common in mathematical descriptions of finite element and boundary element methods.

The individual solutions to the homogeneous equation can be compared with the linearly independent directions in a vector space. The fundamental difference is that the number of independent directions in a function space can be either countably or uncountably infinite. Countably infinite refers to the size of the natural numbers, or integers, while uncountably infinite refers to the size of the real number set. Examples of countably infinite sets of functions include the results from the last section, for example, $\sin(n\pi z/L)$, $0 \le z \le L$, and $n$ an integer. Examples of an uncountable set of functions are the solutions to the wave equation without boundary conditions, $\exp(ikz)$, $0 \le |z|, |k| < \infty$. In the latter case the "index" $k$ is continuous. In general, the solution set to the homogeneous equation will be linearly independent and span the space where the equation is defined. However, they are not guaranteed to the orthogonal or normalized relative to an inner product. The inner product defined on the function space is a generalization of the dot product between vectors in a vector space and provides a measure of the magnitude of a function and a means to project one function onto another. For solutions to the Helmholtz equation, a suitable inner product is

$$\int \bar{\varphi}(\vec{x}) \psi(\vec{x}) d^3 \vec{x} \tag{4.123}$$

The integration is over the entire domain of the problem. For free problems, it is all of space, while for problems with boundaries the domain will be restricted in one or possibly all dimensions. Solutions to the wave equation are frequently expressed in complex number representation for convenience, even though physically meaningful quantities result in taking either the real part of a solution or its magnitude. The bar over the function $\varphi$ indicates complex conjugation. Equation (4.123) can be thought of as a continuously infinite sum of terms of the form $\bar{z}_1 z_2$, where $z_{1,2}$ are complex numbers. When $\psi = \varphi$, Equation (4.123) gives the square magnitude of the function:

$$|\varphi|^2 = \int \bar{\varphi}(\vec{x}) \varphi(\vec{x}) d^3 \vec{x} \tag{4.124}$$

At times, it will be convenient to define an inner product with a weighting factor:

$$\int w(\vec{x}) \bar{\varphi}(\vec{x}) \psi(\vec{x}) d^3 \vec{x}$$

For this presentation weighting factors will not be included, but their use is pointed out when necessary in later chapters. A convenient notation for (4.123) is the bra–ket notation. The inner product is written in compact form as

$$\langle \varphi | \psi \rangle = \int \bar{\varphi}(\vec{x}) \psi(\vec{x}) d^3 \vec{x} \tag{4.125}$$

Notice the lack of complex conjugation on the first function; the "bra" applies the conjugation. Expressed in this manner one thinks of the $|\psi\rangle$ as an element of a vector space, or an abstraction of a vector space, and $\langle\varphi|$ as a vector in the dual space. A canonical example is taking the ordinary dot product of two vectors with complex numbers as elements. The vectors are equivalent to a column, while the dual vector is a row. In linear algebra terms, the dual would be a complex conjugate transpose:

$$[\bar{u} \ \bar{v}] \begin{bmatrix} w \\ z \end{bmatrix} = \bar{u}w + \bar{v}z$$

When attached to each other, $\langle \varphi | \psi \rangle$, the sum over individual components is represented by the integration over space. The "bra" refers to the dual vector and the "ket" to the vector, and together they make a bracket, or bra–ket. The reader should note that whatever is in the "bra" gets conjugated. The notation is very convenient since the operations are all linear. In the end integrations over space are needed to complete a calculation, but this succinct notional makes getting to that point a little easier. The following properties hold for the bra–ket, given a complex number, $c$:

$$\langle \varphi | c\psi \rangle = c \langle \varphi | \psi \rangle \tag{4.126}$$

$$\langle c\varphi | \psi \rangle = \bar{c} \langle \varphi | \psi \rangle \tag{4.127}$$

$$|\varphi + \psi \rangle = |\varphi \rangle + |\psi \rangle \tag{4.128}$$

$$\langle \varphi + \psi | = \langle \varphi | + \langle \psi | \tag{4.129}$$

$$\overline{\langle \varphi | \psi \rangle} = \langle \psi | \varphi \rangle \tag{4.130}$$

Two functions are orthogonal if $\langle \varphi | \psi \rangle = 0$, and a function is unit length if $\langle \varphi | \varphi \rangle = 1$. Equation (4.124) can be used to normalize functions. As an example, this is applied to the 1-dim free modes and modes with ideal boundary conditions. The case of 1-dim modes with boundary conditions is considered first. For two soft or two hard boundaries, the following integrals need to be performed:

$$\int_0^L \sin(k_n z)\sin(k_m z)dz \tag{4.131}$$

$$\int_0^L \cos(k_n z)\cos(k_m z)dz \tag{4.132}$$

The following trig identity is used to evaluate these integrals:

$$\cos(a \pm b) = \cos a \cos b \mp \sin a \sin b \tag{4.133}$$

With these each of the previous integrals can be written as

$$\frac{1}{2}\int_0^L \{\cos((k_n - k_m)z) - \cos((k_n + k_m)z)\}dz \tag{4.134}$$

$$\frac{1}{2}\int_0^L \{\cos((k_n - k_m)z) + \cos((k_n + k_m)z)\}dz \tag{4.135}$$

These can be integrated to yield

$$\frac{L}{2}\left\{\frac{\sin((k_n - k_m)L)}{(k_n - k_m)L} \mp \frac{\sin((k_n + k_m)L)}{(k_n + k_m)L}\right\} \tag{4.136}$$

The argument of each term is an integer times $\pi$; hence for any pair of modes, $n \neq m$, the result is zero. From this it can be concluded that the modes for two soft or two hard boundaries are orthogonal. To evaluate the case $n = m$, one can go back to the original integrals and reevaluate or use the fact that the first term in (4.136) is the sinc function and is equal to 1 when $x = 0$:

$$\mathrm{sinc}(x) = \frac{\sin x}{x} \tag{4.137}$$

The first term in (4.136) is 1 when $n = m$, while the second term vanishes. These results are summed up as follows:

$$\int_0^L \sin(k_n z)\sin(k_m z)dz = \int_0^L \cos(k_n z)\cos(k_m z)dz = \frac{L}{2}\delta_{nm} \tag{4.138}$$

The results in (4.138) can be used to normalize the modes by scaling them by a factor of $\sqrt{2/L}$.

To normalize free modes one requires that the inner product of two independent states yield a Dirac delta function. The Dirac delta function has the following representation in 1-dim:

$$\delta(x - x') = \frac{1}{2\pi}\int_{-\infty}^{\infty} \exp(ik(x - x'))dk \tag{4.139}$$

With this definition, the inner product of two free states is worked out:

$$\left\langle \varphi\left(\vec{k}\right) \middle| \varphi\left(\vec{k'}\right) \right\rangle = \iiint \exp\left(i\vec{x} \cdot \left(\vec{k'} - \vec{k}\right)\right)dxdydz = (2\pi)^3 \delta\left(\vec{k'} - \vec{k}\right)$$

The result states that states with different wavevectors are orthogonal and that the individual plane wave states are normalized by dividing by $(2\pi)^{3/2}$ (Liboff [9]).

## 4.7 Green's Function

### 4.7.1 Green's Function in Free Space

So far attention has been given to the homogeneous equation. This section deals with the inhomogeneous equation and describes Green's function. A time harmonic source, $S(\vec{x}, t) = s(\vec{x})\exp(-i\omega t)$, added to the wave equation leads to the inhomogeneous Helmholtz equation

$$\nabla^2 \psi + k^2 \psi = -s(\vec{x}) \tag{4.140}$$

In (4.140) the value of $k$ is determined by the source and is not an eigenvalue. This is the inhomogeneous wave equation in the frequency domain. A method for solving this involves finding Green's function of the Helmholtz operator. Green's function is a solution of the inhomogeneous equation for a point source at a specific location $\vec{x}'$:

$$\nabla^2 G(\vec{x} - \vec{x}') + k^2 G(\vec{x} - \vec{x}') = -\delta(\vec{x} - \vec{x}') \tag{4.141}$$

Green's function depends on the difference between an arbitrary position and the source position and on the harmonic frequency of the source. Some authors write $G_k(\vec{x} - \vec{x}')$ or $G(\vec{x} - \vec{x}'; k)$. Given a solution to (4.141) for Green's function, a solution to (4.140) is known for an arbitrary source:

$$\psi = \int G(\vec{x} - \vec{x}') s(\vec{x}') d^3 \vec{x}' \tag{4.142}$$

To prove this, apply the operator $\nabla^2 + k^2$ to (4.142):

$$\{\nabla^2 + k^2\}\psi = \{\nabla^2 + k^2\} \int G(\vec{x} - \vec{x}') s(\vec{x}') d^3 \vec{x}'$$

$$= \int \{\nabla^2 G(\vec{x} - \vec{x}') + k^2 G(\vec{x} - \vec{x}')\} s(\vec{x}') d^3 \vec{x}'$$

$$= \int \{-\delta(\vec{x} - \vec{x}')\} s(\vec{x}') d^3 \vec{x}'$$

$$= -s(\vec{x})$$

The second equal sign is possible because the differential operator acts on $\vec{x}$ while the integral is over $\vec{x}'$ and because Green's function is the only factor of the integrand that depends on $\vec{x}$. The next step is merely an application of (4.141), and the final step follows from the definition of the Dirac delta function.

In the section on separable coordinates, Green's functions were presented in spherical and cylindrical coordinate for 3-dim and 2-dim propagation, respectively. They are repeated here for convenience with appropriate scale factors included. For convenience $R = |\vec{x} - \vec{x}'|$ for both 2-dim and 3-dim relative position vectors in the following [8]:

$$G = \frac{\exp(\pm ikR)}{4\pi R} \tag{4.143}$$

$$G = \frac{i}{4} H_0^{(1,2)}(kR) \tag{4.144}$$

Green's function is symmetric with respect to the exchange, $\vec{x} \leftrightarrow \vec{x}'$, a fact that is obvious from the explicit form of (4.143) and (4.144) but can be proven without appealing to an explicit form of the function.

Green's functions presented describe the field of a time harmonic source at a single location in space. Consider the time domain equivalent of (4.141) in 3-dim:

$$\nabla^2 G - \frac{1}{c^2}\frac{\partial^2 G}{\partial t^2} = -\delta\left(\vec{x}-\vec{x}'\right)\delta(t-t') \tag{4.145}$$

Green's function is now a function of position and time, $G\left(\vec{x}-\vec{x}',t-t'\right)$. The solution to (4.145) is

$$G = \frac{\delta(R/c-(t-t'))}{4\pi R} \tag{4.146}$$

This Green's function represents a pulse traveling through the medium caused by a sudden impulse applied to the medium at location $\vec{x}'$ and time $t'$.

The outgoing Green's functions in the frequency domain obey the Sommerfeld radiation condition (Schot [10]). The condition states that as one takes the limit of $r \to \infty$, the function should die sufficiently fast. Given a solution to the Helmholtz equation, $\psi$, in $n=2,3$ dimensions, the condition states

$$\lim_{r\to\infty}\left(r^{(n-1)/2}\left(\frac{\partial\psi}{\partial r}-ik\psi\right)\right)=0 \tag{4.147}$$

## 4.7.2   Mode Expansion of Green's Functions

Green's functions listed earlier are valid solutions when the sound speed is constant and there are no boundaries present. The ability to find Green's function and apply (4.142) to more general situations is highly desirable. It turns out that it is possible to construct a representation of Green's function from the normal modes of a specific problem. This allows for the construction of Green's function for describing wave propagation in the presence of boundaries and more generally for cases where the refractive effects of the environment are nontrivial. The example of standing waves in an enclosure will serve as motivation for the description but the results hold in general. After developing an explicit form of the mode representation on Green's function for discrete modes, the case of continuous modes will be discussed. The reader can then combine these results for cases that contain a mixture of discrete and continuous modes. Assuming that a complete set of solutions for the wave equation that satisfies the boundary conditions of the problem has been found, Green's function is expressed as a linear superposition of these solutions:

$$G\left(\vec{x}-\vec{x}'\right) = \sum_{(l,m,n)} a_{l,m,n}\psi_{l,m,n}\left(\vec{x}\right) \tag{4.148}$$

Equation (4.148) is inserted into (4.141) and expanded:

$$\{\nabla^2 + k^2\} \sum_{(l,m,n)} a_{l,m,n}\Psi_{l,m,n}(\vec{x}) = \sum_{(l,m,n)} a_{l,m,n}\{\nabla^2\Psi_{l,m,n}(\vec{x}) + k^2\Psi_{l,m,n}(\vec{x})\}$$

$$= \sum_{(l,m,n)} a_{l,m,n}\{-k_e^2\Psi_{l,m,n}(\vec{x}) + k^2\Psi_{l,m,n}(\vec{x})\}$$

$$= \sum_{(l,m,n)} a_{l,m,n}\{k^2 - k_e^2\}\Psi_{l,m,n}(\vec{x})$$

$$= -\delta(\vec{x} - \vec{x}')$$

In the second equality, the shorthand $k_e^2 = k_{xl}^2 + k_{ym}^2 + k_{zn}^2$ is introduced, where the subscript $e$ indicates eigenvalue corresponding to the index $(l, m, n)$. The next step is to take the inner product of the last equality with an arbitrary basis function, $\Psi_{l',m',n'}(\vec{x})$, or $\Psi_{e'}(\vec{x})$:

$$\int_V \bar{\Psi}_{e'}(\vec{x})\left\{\sum_e a_e\{k^2 - k_e^2\}\Psi_e(\vec{x})\right\}d^3\vec{x} = -\int_V \delta(\vec{x} - \vec{x}')\bar{\Psi}_{e'}(\vec{x})d^3\vec{x}$$

The r.h.s. is evaluated using the definition of the Dirac delta function:

$$\int_V \delta(\vec{x} - \vec{x}')\bar{\Psi}_{e'}(\vec{x})d^3\vec{x} = \bar{\Psi}_{e'}(\vec{x}')$$

The l.h.s. is reduced using the orthonormality of the basis functions:

$$\int_V \bar{\Psi}_{e'}(\vec{x})\left\{\sum_e a_e\{k^2 - k_e^2\}\hat{\Psi}_e(\vec{x})\right\}d^3\vec{x} = \sum_e a_e\{k^2 - k_e^2\}\int_V \bar{\Psi}_{e'}(\vec{x})\Psi_e(\vec{x})d^3\vec{x}$$

$$= \sum_e a_e\{k^2 - k_e^2\}\delta_{ll'}\delta_{mm'}\delta_{nn'} = a_{e'}\{k^2 - k_{e'}^2\}$$

Combining these results gives an expression for the expansion coefficients:

$$a_{e'} = -\frac{\bar{\Psi}_{e'}(\vec{x}')}{k^2 - k_{e'}^2} \tag{4.149}$$

Since the index set is arbitrary, the primes can be eliminated, giving the final expression for Green's function:

$$G(\vec{x} - \vec{x}') = -\sum_e \frac{\bar{\Psi}_e(\vec{x}')\Psi_e(\vec{x})}{k^2 - k_e^2} \tag{4.150}$$

The derivation did not make explicit use of a specific set of boundary conditions and the result is generic. By picking the eigenfunctions appropriate for a specific problem, Green's function may be constructed using (4.150). The process holds for the continuum as well but the form is slightly different. It is worth going through the derivation. Writing Green's function as an expansion over the mode basis,

$$G(\vec{x} - \vec{x}') = \int a\left(\vec{k}_e\right)\psi_e\left(\vec{x}\right)d^3\vec{k}_e \tag{4.151}$$

Inserting this into (4.141) and performing the same steps as in the previous case leads to the following:

$$\int a\left(\vec{k}_e\right)\{k^2 - k_e^2\}\psi_e\left(\vec{x}\right)d^3\vec{k}_e = -\delta(\vec{x} - \vec{x}') \tag{4.152}$$

Projecting both sides onto another arbitrary eigenfunction,

$$\int a\left(\vec{k}_e\right)\{k^2 - k_e^2\}\delta\left(\vec{k}_e - \vec{k}_{e'}\right)d^3\vec{k}_e = a\left(\vec{k}_{e'}\right)\{k^2 - k_{e'}^2\} = -\bar\psi_{e'}\left(\vec{x}'\right) \tag{4.153}$$

Solving for $a\left(\vec{k}_{e'}\right)$ provides the last step in deriving Green's function:

$$a\left(\vec{k}_{e'}\right) = -\frac{\bar\psi_{e'}\left(\vec{x}'\right)}{k^2 - k_{e'}^2} \tag{4.154}$$

$$G(\vec{x} - \vec{x}') = -\int \frac{\bar\psi_e\left(\vec{x}'\right)\psi_e\left(\vec{x}\right)}{k^2 - k_e^2} d^3\vec{k}_e \tag{4.155}$$

There is a physical interpretation of Green's function that nicely expresses the physical significance of the function and its use as well as the unique value linear systems have in physics. Linearity means that:

1. Any linear combination of solutions to the homogeneous equation is also a solution to the homogeneous equation.
2. Given solutions to the inhomogeneous equation for several sources, a solution for a combination of the sources can be built from a linear combination of the solutions for the individual sources treated separately.

The first point has been used in the development of a mode representation of Green's function. To illustrate the second point, consider a family of equations indexed by $n$, where $n$ indicates a specific inhomogeneous wave equation:

$$\nabla^2\psi_n + k^2\psi_n = -s_n\left(\vec{x}\right) \tag{4.156}$$

Consider the arbitrary sum

$$\sum_n c_n \psi_n \tag{4.157}$$

Applying the Helmholtz operator to this function gives the following:

$$\{\nabla^2 + k^2\} \sum_n c_n \psi_n = \sum_n c_n \{\nabla^2 + k^2\} \psi_n = -\sum_n c_n s_n(\vec{x}) \tag{4.158}$$

This demonstrates that (4.157) is a solution to the inhomogeneous wave equation for the source:

$$\sum_n c_n s_n(\vec{x}) \tag{4.159}$$

Going the other way, one can always dissect a distributed source into small pieces or a complicated function into a sum over simpler terms, each a function of position. The latter case may correspond to expressing a given source in a Taylor series or expanded in terms of a set of special functions. Then, in principle, the more complex system has an exact solution expressed as a sum, possibly infinite, over the various components. This is exactly how Green's function is being used. Here the source is a point source at a fixed location, and the general solution can be thought of as being an integral, or sum, over a distribution of point sources smeared throughout a finite volume of space in the propagation region. This is the physical meaning of Green's function, a solution for unit strength point source.

For a wide class of problems, a tractable path toward modeling solutions for any source function is expressed by the following procedure:

1. Solve the homogeneous wave equation.
2. Ensure that the eigenfunction set obeys the boundary conditions of the problem.
3. Estimate Green's function for the problem.
4. Evaluate the integral of Green's function over the source distribution.

For simple Cartesian boundary conditions, item 1 is complete, and there is a general procedure for completing step 2 for simple boundaries. All that remains is evaluating the field by item 3 and integrating over a suitable representation of Green's function over the source. The source may also be expanded in terms of the eigenfunctions of the system:

$$s_e = \int_{V_s'} \bar{\psi}_e(\vec{x}') s(\vec{x}') dV' \tag{4.160}$$

In (4.160), $V_s'$ represents the region of space where the source is defined. With (4.160) the solution in the free field case, that is, continuum of eigenvalues, is

$$\psi(\vec{x}) = -\int \frac{\psi_e(\vec{x}) s(\vec{k_e})}{k^2 - k_e^2} d^3 \vec{k_e} \tag{4.161}$$

A similar result holds for the discrete and mixed spectra. The factor of $(k^2 - k_e^2)^{-1}$ is referred to as a propagator and is the equivalent of Green's function in a momentum space representation. In fact, if (4.161) is projected onto the basis set, the result is

$$\langle \psi_e | \psi \rangle = -\frac{s(\vec{k_e})}{k^2 - k_e^2} \tag{4.162}$$

Readers familiar with signal processing will recognize the propagator as being proportional to the Fourier transform of Green's function. The propagator has a singularity at $k^2 = k_e^2$ requiring some care.

The closed-form expression for Green's function in 3-dim expanded in plane waves is

$$G(\vec{x} - \vec{x}') = -\frac{1}{(2\pi)^3} \int \frac{\exp\left(-i(\vec{x} - \vec{x}') \cdot \vec{k_e}\right)}{k^2 - k_e^2} d^3 \vec{k_e} \tag{4.163}$$

The integral is over 3-dim space in the variables $\vec{k_e}$. It is worth noting that the integral on the r.h.s of (4.163) evaluates to (4.143) and is left to the reader to investigate [7].

## 4.8　Method of Images

When propagation occurs in a trivial medium in the presence of ideal boundaries, the method of images allows one to determine Green's function satisfying boundary conditions. Consider a boundary surface that separates space into two simply connected domains, for example, an infinite plane or sphere. The region where sources exist is called the physical space, and the other domain, inaccessible to measurements, is called unphysical, imaginary, or image space. Finding a suitable Green's function means finding a solution to the Helmholtz equation for a point source valid in the physical space that satisfies the boundary conditions. Since the medium is trivial, an exact solution exists in physical space. Near each source the field should look like Green's function for a point source in free space. Since the imaginary space is not accessible, anything can be placed in that space in order to satisfy boundary conditions. Application of the method involves placing point sources, or extended source distributions, of various strengths and locations in the imaginary region and imposing boundary conditions on the total field due to all sources. This will lead to an equation of constraint relating the parameters of the imaginary source(s) to the physical source(s). If a solution can be found for the strength and position of each imaginary source satisfying the constraint, then Green's function satisfying the boundary conditions has been found. To illustrate the process, consider a point source in the

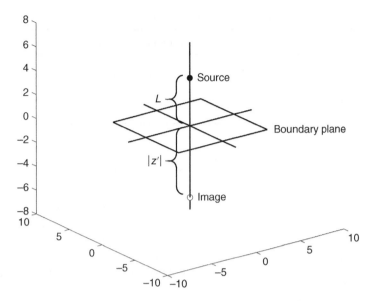

**Figure 4.2** Image method defined

presence of an infinite plane with either ideal pressure release or hard boundary condition. The configuration is illustrated in Figure 4.2.

The boundary plane is the $x - y$ plane at $z = 0$, physical space the upper half space $z > 0$, and imaginary space the lower half space. The monochromatic source, filled sphere, is located at $z = L$. An imaginary source, hollow sphere, is placed in the lower half space. Due to the symmetry of the configuration, it makes sense to place the imaginary source on the same vertical axis as the true source, but its vertical position and strength, denoted $s$, will be assumed unknown. Without loss of generality the sources are placed on the $z$ axis. If the boundary were absent and these two sources present in space, the solution to the Helmholtz equation is

$$\varphi = G_s + s G_i = \frac{\exp(-ikR)}{R} + s\frac{\exp(-ikR')}{R'} \tag{4.164}$$

with

$$R = \sqrt{x^2 + y^2 + (z-L)^2} \tag{4.165}$$

$$R' = \sqrt{x^2 + y^2 + (z + |z'|)^2} \tag{4.166}$$

Pressure release boundary conditions $\varphi(z=0) = 0$ are satisfied by the choice $|z'| = L$ and $s = -1$. The relative strength can be expressed as a 180 degree phase shift, that is, the source and its image are out of phase. For reflection from a hard surface, the boundary condition is $\partial_z \varphi(z=0) = 0$. Taking the derivative of (4.164), evaluating it at $z = 0$, and setting the result to zero give

**Table 4.3**  Modified Green's functions for a single boundary

| Green's functions for half space with soft/hard boundary conditions | |
| --- | --- |
| $\dfrac{\exp(-ikR)}{R} - \dfrac{\exp(-ikR')}{R'}$ | Pressure release |
| $\dfrac{\exp(-ikR)}{R} + \dfrac{\exp(-ikR')}{R'}$ | Hard surface |

$$\frac{-L}{R}\frac{\partial G_s}{\partial R}\bigg|_{z=0} + \frac{|z'|}{R'}s\frac{\partial G_s}{\partial R'}\bigg|_{z=0} = 0 \tag{4.167}$$

The derivative of Green's function with respect to $R$ is easily worked out, but the form is irrelevant to this example, suffice to say that it is a function of $R$, or $R'$ only, and identical in form for both real and imaginary sources. By inspection this boundary condition can be satisfied by again placing the image at $|z'| = L$, and by choosing $s = 1$. In this case the real and imaginary sources are in phase. The result, referred to sometimes as a modified Green's function, is presented in Table 4.3 for each case.

The method is easily extended to cases involving multiple boundaries. This is first extended to the problem of a point source in a space between two parallel planes, followed by the case of a point source in a rectangular enclosure, for a variety of boundary conditions. Figure 4.3 illustrates the situation for a point source between two vertical planes, one at $z = 0$, and the other at $z = L$. The point source is located between the planes at $z = z_s$. Also shown are the images created by the boundaries. To extend the method to two parallel boundaries, one would expect to need an imaginary source directly opposite to each boundary to satisfy boundary conditions with the source. This poses a situation where each imaginary source created needs to be balanced with another image, leading to the requirement of an infinite number of images. The location of images is discussed first, followed by the relative sign that depends on the type of boundary conditions. Referring to Figure 4.3, to satisfy boundary conditions with the source and the bottom plane, an image is placed at $z = -z_s$. To satisfy boundary conditions at the upper plane requires an image placed at $z = 2L - z_s$,; this is the position of the source relative to $z = 0$. Now the image at $-z_s$ must be balanced by another image behind the upper plane, and likewise the image at $2L - z_s$ must be balanced by an image behind the lower plane. Figure 4.3 shows several of these reflections. With a little effort, a set of formula for the location of these images can be worked out; the result is provided in Table 4.4.

The amplitudes are now discussed. There are three independent cases to consider: (1) two pressure release surfaces, (2) two hard boundaries, and (3) one hard boundary and one pressure release. The case of two hard boundaries is the easiest to consider. In this case, there is no relative phase between the source and its image. This will also hold for the other images created by multiple reflections. Therefore, in this case all images have amplitude $s = 1$. For two pressure release surfaces, the images will alternate in sign. Consider the first images at $-z_s$ and $2L - z_s$. The amplitude for each of these will be $s = -1$. The image of each of these negative amplitude images behind the other boundary will pick up another negative sign relative to the first image, for example, they will now be in phase with the source. For the third case, consider the bottom boundary to be hard and the upper to be pressure release. The image at $-z_s$ will have $s = 1$, while that at $2L - z_s$ will have $s = -1$. The image at $-z_s$ is now reflected in the pressure release surface,

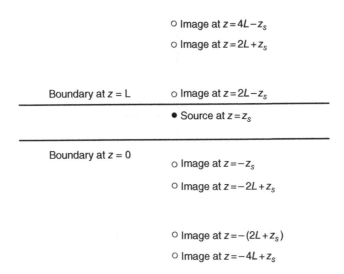

○ Image at $z = 4L - z_s$

○ Image at $z = 2L + z_s$

| Boundary at $z = L$ | ○ Image at $z = 2L - z_s$ |
| | • Source at $z = z_s$ |

| Boundary at $z = 0$ | |

○ Image at $z = -z_s$

○ Image at $z = -2L + z_s$

○ Image at $z = -(2L + z_s)$

○ Image at $z = -4L + z_s$

**Figure 4.3**   Multiple images for satisfying two boundary conditions

**Table 4.4**   Locations and relative phase of images for three cases

| Images | Placement | Relative sign | | |
| --- | --- | --- | --- | --- |
| | | SS | HH | SH |
| First lower | $-z_s$ | $-1$ | $+1$ | $+1$ |
| | $2nL - z_s$ | $-1$ | | |
| | $-2nL + z_s$ | $+1$ | | |
| $n$th group of four | $2nL + z_s$ | $+1$ | $+1$ | $(-1)^n$ |
| | $-2nL - z_s$ | $-1$ | | |

so its image at $2L + z_s$ will have $s = -1$. Similarly, the image at $2L - z_s$, which is negative, is reflected by a hard boundary, so its image at $-2L + z_s$ will also have $s = -1$. And the process continues. There is a discernable pattern in these cases, which is provided in Table 4.4.

Given the complete set of images, the field at any point inside the free space between the boundaries can be determined by

$$\psi = \frac{\exp(-ikR_s)}{R_s} + \sum_{j=1}^{\infty} A_j \frac{\exp(-ikR_j)}{R_j} \tag{4.168}$$

The notation in (4.168) is as follows:

$$R_s = \left| \vec{x} - \vec{x}_s \right| \tag{4.169}$$

$$R_j = \left| \vec{x} - \vec{x}_j \right| \tag{4.170}$$

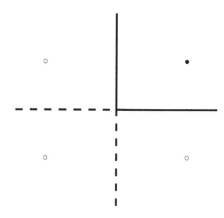

**Figure 4.4**  Images for a source in a right-angle corner

The subscript $s$ labels the source, while $j$ labels the $j$th image.

A simpler case is that of two intersecting infinite planes. Consider two semi-infinite planes attached at a common edge, or think of this as a bent infinite plane. The physical space is defined in the first quadrant and the imaginary space is the complement of that region. A source is placed somewhere in physical space near the corner (see Figure 4.4).

Assume that the source coordinates are $(x, y, z)$ and that the $z$ direction is out of the paper. Images are required at $(-x, y, z)$ and $(x, -y, z)$ to account for the reflections of the source by the physical boundary. However, these two alone will not satisfy the boundary conditions. To complete the analysis, the physical boundaries are extended into the unphysical region and images found for these boundaries (represented by dashed lines). In this case, it is clear that one more image at $(-x, -y, z)$ is needed to complete the picture. The amplitudes are determined by the type of boundary conditions applied. If both boundaries are hard, then all images will have $s = 1$. If both are soft, then the images will alternate in sign going counterclockwise starting with the source $s = (1, -1, 1, -1)$. For a mixed boundary condition, consider the vertical boundary to be hard and the horizontal boundary to be soft. Then the image in the upper left is positive and both lower images are negative.

The last case considered is that of a point source in a rectangular enclosure. For the purpose of illustration, the problem will be presented in two dimensions or as a waveguide in three dimensions. The reader can extrapolate to accommodate a full three-dimensional rectangular enclosure. The physical region is the interior of a rectangle. For a point source contained in this region, the results of the last two examples may be combined. The results for the infinite parallel boundaries can be used to generate the pattern of images along the two axes through pairs of opposite boundaries. Then each of these sets of images is reflected multiple times in the other set of boundaries. The result is shown in Figure 4.5. This process can be described by a periodic lattice of images, each a repeat of the corner reflections considered in the last case.

Simulating the field by images using these procedures works well for estimates of the sound in rooms and waveguides with simple shapes. Simulations based on images are fairly easy to write. The only drawback is that the true solution requires an infinite series. Calculations based on images require a cutoff and will have truncation error. Allen and Berkley apply this method to small room acoustics [11].

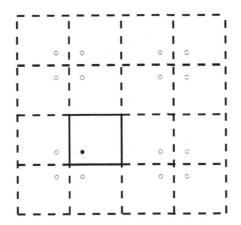

**Figure 4.5**   Images for a point source in a rectangular enclosure

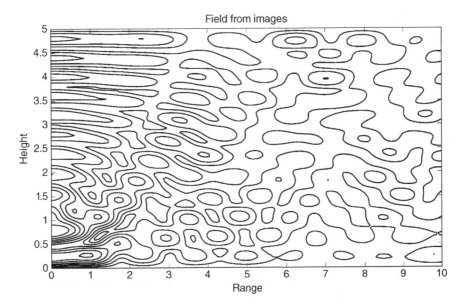

**Figure 4.6**   Contour plot of |$p$| built from images

## 4.9   Comparison of Modes to Images

As an illustrative example, the sound field for a point source in the presence of boundaries is presented. The field is built from a mode expansion and separately from images. A point source is located at $z = 1$ m and at the origin in the $x$–$y$ plane, in between two vertical pressure release boundaries at $z = 0$ and $z = 5$ m. The frequency of the source is 440 Hz and the sound speed 345.6 m/s. The mode expansion uses the expression for modes in cylindrical coordinates with the 2-dim Green's function serving as the radial part. Both results are calculated at a transverse slice at $y = 0.5$ m. Figure 4.6 shows a contour plot of the field built from the method of images,

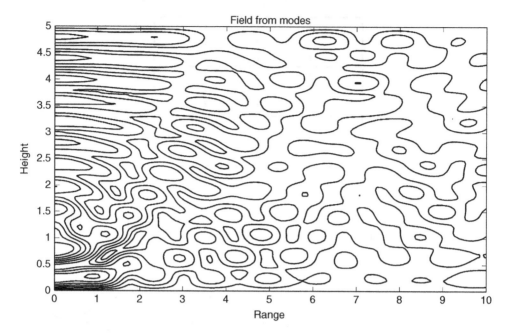

**Figure 4.7**   Contour plot of |p| built from modes

and Figure 4.7 shows the same for the field built from a mode expansion. The relative difference between peak values, that is, maximum field strength, is $\sim 5 \times 10^{-5}$. Larger values of relative difference are encountered near nodes of the field but never rise above $\sim 3.3\%$ in this case.

## 4.10   Exercises

1.  Work out an image solution for a point source in a 3-dim rectangular enclosure for two sets of boundary conditions (more if desired).
    (a)  Five hard boundaries + 1 pressure release
    (b)  Hard boundaries on all 6 sides
2.  Work out the images required for a point source between two planes intersecting at an arbitrary angle for the three independent combinations of boundary conditions.
3.  Write a procedure, in any software language, for evaluating the pressure field using the image solution developed for Exercise 1.
4.  Using the mode solutions in Table 4.2, develop a 3-dim mode solution for the same rectangular enclosure described in Exercise 1.

## References

[1]  John, F., Partial Differential Equations, Courant Institute of Mathematical Sciences Lecture Notes, New York University, New York, 1952
[2]  Courant, R. and Hilbert, D., Methods of Mathematical Physics, Volumes I and II, John Wiley & Sons, Inc., New York, 1962.

[3] O'Neill, B., Elementary Differential Geometry, Academic Press, San Diego, 1966.

[4] Misner, C. W., Thorne, K. S., and Wheeler, J. A., Gravitation, W. H. Freeman, New York, 1973.

[5] Arfkin, G. B. and Weber, H. J., Mathematical Methods for Physicists, Fifth Edition, Harcourt Academic Press, San Diego, 2001.

[6] Knott, E. F., Shaeffer, J. F., and Tully, M. T., Radar Cross Section, Second Edition, SciTech Publishing Inc., Raleigh, NC, 2004.

[7] Gradshteyn, I. S. and Ryzhik, I. M., Table of Integrals, Series, and Products, Fifth Edition, A. Jeffery, Editor, Academic Press, San Diego, 1994.

[8] Morse, P. M. and Feshbach, H., Methods of Theoretical Physics, Part I & Part II, McGraw-Hill, New York, 1953.

[9] Liboff, R. L., Introductory Quantum Mechanics, Addison Wesley, Reading, 1980.

[10] Schot, S. H., Eighty Years of Sommerfeld's Radiation Condition, Hist. Math., Vol. 19, pp. 385–401, 1992.

[11] Allen, J. B. and Berkley, D. A., Image method for efficiently simulation small-room acoustics, J. Acoust. Soc. Am., Vol. 65, No. 4, pp. 943–950, 1979.

# 5

# Wave Propagation

## 5.1 Introduction

In the previous chapter, several methods for developing solutions and descriptions of the behavior of the wave equation were presented. Included in that presentation was mode analysis in separable coordinates, Green's functions, method of characteristics, and the effect of ideal boundary conditions on the spectrum of modes. This chapter covers topics related to the description of waves in general and behavior at the boundary separating different media. The Fourier transform is introduced with the purpose of decomposing time-dependent sources into separate frequencies clustered about a central frequency to facilitate the use of solutions in the frequency domain for building time-dependent solutions through synthesis. A similar process in coordinate space is discussed. The next section introduces dispersion and the group and phase velocities associated with waves. Following this is a treatment of the reflection and transmission of plane waves through an interface separating two fluid media and a fluid–solid interface. This treatment introduces more general conditions for handling waves at boundaries between media and the concept of transmission and reflection coefficients. The last section discusses attenuation, sources of attenuation, and the modeling effect of bulk attenuation loss on a wave.

## 5.2 Fourier Decomposition and Synthesis

The techniques presented in the last chapter were mostly in the frequency domain, that is, Helmholtz equation rather than the full wave equation for a time harmonic source. Thus it would seem that one only has solution methods for the behavior of a continuous wave (c.w.). More realistic situations require modeling of pulses or wave packets, chopped in time.

*Computational Acoustics: Theory and Implementation*, First Edition. David R. Bergman.
© 2018 John Wiley & Sons Ltd. Published 2018 by John Wiley & Sons Ltd.

Sources emitting sound will have a definite starting and ending time and produce a packet with a finite duration and hence a finite length in space. The finite duration of the pulse produces a spectrum of frequencies clustered about a center frequency called the carrier. The width of the frequencies is referred to as the bandwidth. The longer the time duration, the narrower the spectrum, and conversely the shorter the pulse time, the wider the spectrum. These are referred to as narrow and wideband pulses, respectively. Since the wave equation is linear, independent solutions can be added together to produce new solutions. If one has a source with multiple frequencies, either continuous or due to the source producing a packet, the source can be decomposed into individual frequency components. Each of these components will act as a time harmonic source. The solution to the full wave equation can then be built up by combining the results of each frequency component weighted by the spectrum amplitude. This process is called synthesis. A detailed account of the theory behind Fourier analysis can be found in Körner [1].

Given a function of time, $f(t)$, the Fourier transform is a function of frequency:

$$F(\omega) = \int_{-\infty}^{\infty} f(t)\exp(i\omega t)dt \tag{5.1}$$

The inverse of (5.1) is

$$f(t) = \frac{1}{2\pi}\int_{-\infty}^{\infty} F(\omega)\exp(-i\omega t)d\omega \tag{5.2}$$

Equation (5.2) states that the behavior of the function $f(t)$ can be expressed as a superposition of monochromatic waves. The function $F(\omega)$ gives the amount of each frequency present in $f(t)$. A few simple examples are presented. The first case, almost trivial, is $f(t) = \exp(-i\omega_0 t)$. From the definition of a Dirac delta function, the transform results in $F(\omega) = 2\pi\delta(\omega - \omega_0)$. The spectrum peaks at $\omega = \omega_0$ and is zero otherwise, reflecting the fact that the original function is monochromatic. Inserting $F(\omega)$ into (5.2) recovers the original function. Equally trivial is the case of an impulse at $t_0$, expressed as $f(t) = \delta(t - t_0)$. Inserting into (5.1) gives $F(\omega) = \exp(i\omega t_0)$. In the first case an input that existed for all time had a single frequency component, whereas in the second case an input that existed for just one instant has all frequency components present in its spectrum. As the final example consider the case of a chopped pulse. This is a more realistic example in which a monochromatic source is turned on and then off after some duration:

$$f(t) = \exp(-i\omega_0 t), \quad t_0 - \frac{T}{2} \le t \le t_0 + \frac{T}{2} \tag{5.3}$$

Inserting (5.3) into (5.1) yields the following integral:

$$\int_{t_0-T/2}^{t_0+T/2} \exp(i(\omega-\omega_0)t)dt = T\exp(i(\omega-\omega_0)t_0)\frac{\sin((\omega-\omega_0)T/2)}{(\omega-\omega_0)T/2} \tag{5.4}$$

The result here is a more interesting spectrum and illustrates a few features of the Fourier transform. The frequency behavior is contained in the sinc function. The spectrum power is defined as $|F(\omega)|^2$. The spectrum has a maximum at $\omega = \omega_0$, the carrier frequency, and periodic side lobes. While the spectrum does have values as $\omega \to \infty$, the strength of these modes is weak compared to the main lobe. The pulse width $T$ produces missing frequencies at $\omega - \omega_0 = 2n\pi T^{-1}$.

The decomposition of time dependence through Fourier transform can be applied to functions of space and time. Consider a source depending on spatial coordinates and time $s(\vec{x}, t)$. This can be transformed into the frequency domain:

$$S(\vec{x}, \omega) = \int_{-\infty}^{\infty} s(\vec{x}, t) \exp(i\omega t) dt \qquad (5.5)$$

Likewise, one may assume a relationship between the solution to the wave equation and Helmholtz equation:

$$\varphi(\vec{x}, \omega) = \int_{-\infty}^{\infty} \psi(\vec{x}, t) \exp(i\omega t) dt \qquad (5.6)$$

Inserting these into the wave equation leads to the Helmholtz equation for each frequency component:

$$\nabla^2 \varphi(\vec{x}, \omega) + k^2 \varphi(\vec{x}, \omega) = -S(\vec{x}, \omega) \qquad (5.7)$$

Given a solution to (5.7), the field in time domain can be developed via application of (5.2):

$$\psi(\vec{x}, t) = \frac{1}{2\pi} \int_{-\infty}^{\infty} \varphi(\vec{x}, \omega) \exp(-i\omega t) d\omega \qquad (5.8)$$

While the approach yields a solution in theory, in practice there are practical issues. In some rare cases the transform of the source term may yield a simple function for which Helmholtz can be solved. In general, the solution to Helmholtz will be an infinite sum or integral over spatial modes. Once this is known, the inverse Fourier transform will have to be performed. In the most general case, this will not be possible analytically, and numerical integration will be required. There is a technique for performing discrete Fourier transforms rapidly called the fast Fourier transform (FFT). The real issue is not whether or not one can practically do this transform but the effect that finite sampling of the power spectrum will have on the result. Recall the example that produced (5.4). A pulse of finite duration in time has side lobes in the frequency spectrum. The same result will occur when transforming from frequency to time domain. Transforming a finite window of the frequency spectrum will produce side lobes in time. These side lobes are symmetrically placed about the peak value. Just as the transform to the frequency domain produced side lobes in $\omega$, the transform of a chopped frequency function to the time domain will produce side lobes in time, both before and after the peak time value. These are an unphysical artifact of the process of estimating these infinite integrals with finite blocks of data. Techniques exist to suppress side lobes by windowing the chopped data with a smooth function [2].

The same technique is applied to the spatial part of the field. Many problems are easily solved for a plane wave input. The solutions can be used to model more complex inputs by integrating the plane wave solution over a series of different input directions. The spatial Fourier transform is

$$F\left(\vec{k}\right) = \int f\left(\vec{x}\right)\exp\left(i\,\vec{x}\cdot\vec{k}\right)d^3\,\vec{x} \tag{5.9}$$

The inverse of (5.9) is

$$f\left(\vec{x}\right) = \frac{1}{(2\pi)^{3/2}}\int F\left(\vec{k}\right)\exp\left(-i\,\vec{x}\cdot\vec{k}\right)d^3\,\vec{k} \tag{5.10}$$

The statements regarding the time–frequency transform and the examples presented here carry over to the spatial transform.

## 5.3   Dispersion

The relationship between frequency and wavenumber, $\omega(k)$, defines the dispersion relation of a wave. Associated with the wave are a group velocity and a phase velocity. For 3-dimensional (3-dim) propagation, the group velocity is

$$\vec{v}_g = \vec{\nabla}_{\vec{k}}\omega \tag{5.11}$$

The gradient operator in (5.11) is with respect to coordinates in $\vec{k}$ space

$$\vec{\nabla}_{\vec{k}} = \hat{k}_x\frac{\partial}{\partial k_x} + \hat{k}_y\frac{\partial}{\partial k_y} + \hat{k}_z\frac{\partial}{\partial k_z} \tag{5.12}$$

Another type of wave velocity is the phase velocity:

$$\vec{v}_p = \hat{k}\frac{\omega}{k} \tag{5.13}$$

For free waves the dispersion relation is

$$\omega = ck \tag{5.14}$$

In this case $\vec{v}_g = \vec{v}_p = c\hat{k}$. As a second example consider waves propagating in a waveguide created by ideal boundaries. For waves free to travel along the z-axis but confined by hard boundaries in the x and y directions, the dispersion relation is

$$\omega = c\sqrt{k_z^2 + k_{n,m}^2} \tag{5.15}$$

The shorthand $k_{n,m}^2$ is introduced:

$$k_{n,m}^2 = \left(\frac{\pi}{L}\right)^2 (n^2 + m^2)$$

(5.16)

With boundaries constraining the wave in $x$ and $y$ directions, producing a standing wave for these modes, the remaining portion of the solution is free to propagate in the $z$ direction, resulting in an effective 1-dim problem. Equation (5.15) can be thought of as a dispersion relation for the remaining propagating mode, $\omega(k_z)$, which yields the following phase and group velocity, in which $k_0 = \omega/c$:

$$v_p = \frac{c}{\sqrt{1 - (k_{n,m}/k_0)^2}}$$

(5.17)

$$v_g = c\sqrt{1 - \left(\frac{k_{n,m}}{k_0}\right)^2}$$

(5.18)

These become imaginary for $k_0 < k_{n,m}$, or $\omega < ck_{n,m}$, which imposes a frequency cutoff for propagating modes to occur. For values below this cutoff, the wavenumber $k_z$ becomes imaginary, and the solution transitions from oscillating to a decaying exponential. These modes are referred to as evanescent modes. In the high frequency limit, the ratio $k_{n,m}/k_0 \rightarrow 0$, and both group and phase velocity approach the wave speed on the medium, $c$. Figure 5.1 shows the group and phase velocity curves for the example worked out in this section. Quantities are scaled for the asymptote at 1 (dashed line). The group (phase) velocity curves are below (above) the asymptote.

A description of the propagation of pulses in dispersive media can be found in Tolstoy and Clay [3].

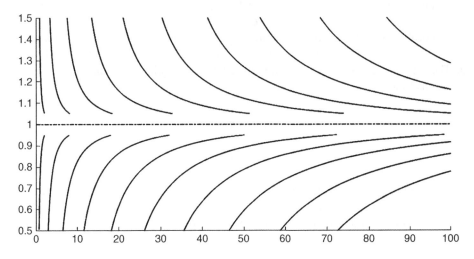

**Figure 5.1**  Group velocity and phase velocity curves for a normal mode problem

## 5.4 Transmission and Reflection

Boundary conditions and their effect on solutions of the wave equation were discussed in the last chapter. In that case the boundary conditions placed a restriction on the field, its normal derivative, or both. Also of interest is the interaction of a wave with a boundary separating two media, like air and water, or fluid and solid. This is a classic problem in wave theory and leads to the introduction of transmission and reflection coefficients. A detailed account of problems like this in general can be found in Brekhovskikh [4]. In this presentation, the fields will be in the frequency domain and the boundary separating two media will be flat. The properties of each medium are constant and stationary. A plane wave in one of the media will be incident on the interface. The medium containing the incident wave is referred to as medium 1. The waves in each region are solutions to the Helmholtz equation with a constant sound speed. Upon hitting the boundary, the wave will initiate movement of the other medium, referred to as medium 2. The wave in medium 2 is referred to as the transmitted wave. Some of the incident wave will be reflected by the boundary back into the initial medium. This wave is called the reflected wave. Each will be a plane wave with different propagation directions. If medium 2 is a solid, then more than one type of transmitted wave will be produced. Each wave is a plane wave traveling in a specific direction, $\vec{k}$. Boundary conditions at the interface are discussed first followed by a solution to one type of problem to illustrate the technique.

For a fluid–fluid interface, the acoustic pressure and the component of particle velocity perpendicular to the interface must be continuous:

$$p_1 = p_2 \tag{5.19}$$

$$\vec{v}_1 \cdot \hat{n} = \vec{v}_2 \cdot \hat{n} \tag{5.20}$$

At the interface between a fluid and a solid, similar conditions hold but are more general to account for S- and P-waves. The normal component of the particle displacement must be continuous at the interface, likewise for the normal components of the particle velocity. Also, the normal and tangential components of the stress tensor evaluated on the boundary must be continuous. Local coordinates are defined at the boundary such that one axis is along $\hat{n}$ and the other two define a local tangent plane at the point where the boundary condition is to be evaluated. These two tangential directions are labeled $\hat{t}_1$, $\hat{t}_2$. Continuity of the stresses may be expressed as follows:

$$\hat{n}^T (\sigma_1 \cdot \hat{n}) = \hat{n}^T (\sigma_2 \cdot \hat{n}) \tag{5.21}$$

$$\hat{t}_j^T (\sigma_1 \cdot \hat{n}) = \hat{t}_j^T (\sigma_2 \cdot \hat{n}), \ j = 1, 2 \tag{5.22}$$

It is understood that $\sigma_{1,2}$ are the $3 \times 3$ stress matrices in media 1 and 2. For the case when one medium, say, medium 1, is a fluid, the normal component of the stress is continuous and equal to the pressure in the fluid medium. The tangential components of the stress vanish in the fluid:

$$p_1 = \hat{n}^T (\sigma_2 \cdot \hat{n}) \tag{5.23}$$

$$0 = \hat{t}_j^T (\sigma_2 \cdot \hat{n}), \; j = 1,2 \tag{5.24}$$

$$\vec{v}_1 \cdot \hat{n} = \vec{v}_2 \cdot \hat{n} \tag{5.25}$$

At the interface of two solid media that are welded, that is, connected by some means that can be considered a holonomic constraint, we have continuity of all components of particle displacement and normal stress:

$$\hat{n}^T (\sigma_1 \cdot \hat{n}) = \hat{n}^T (\sigma_2 \cdot \hat{n}) \tag{5.26}$$

$$\hat{t}_j^T (\sigma_1 \cdot \hat{n}) = \hat{t}_j^T (\sigma_2 \cdot \hat{n}), \; j = 1,2 \tag{5.27}$$

$$\vec{v}_1 = \vec{v}_2 \tag{5.28}$$

Two examples are presented to illustrate the application of these boundary conditions and introduce the concept of transmission and reflection coefficients: (1) fluid–fluid interface and (2) fluid–solid interface. These are commonly found in introductory texts on acoustics and with some modification for optics and quantum mechanics. The reader is referred to Ref. [5] for more elaborate examples. Following the notation of Ref. [5] the scalar field will be the velocity potential associated with the acoustic field. In these examples, each medium will fill a half space, $z > 0$ being medium 1 and $z < 0$ medium 2, with a plane interface at $z = 0$. The properties of each medium are constant, fluids being described by a constant sound speed, and the solid medium is assumed to be homogeneous and isotropic. A plane wave in medium 1 is incident on the interface at an arbitrary angle. This incident wave will be partly reflected back into medium 1, and some of it will be refracted into medium 2. The refraction is due to the process of particle movement of medium 1 initiating movement in medium 2, transferring energy to that medium in the process. Each wave is described by a plane wave solution. The wave in medium 1 is a superposition of incident and reflected waves. The waves in each medium are labeled by the medium index, for example, $\varphi_1$ is the wave in medium 1. The amplitude and wavenumber of each component are labeled $I$ for incident, $R$ for reflected, and $T$ for transmitted. These are provided below for a fluid–fluid interface:

$$\varphi_1 = A_I \exp \left( i \vec{k}_I \cdot \vec{x} \right) + A_R \exp \left( i \vec{k}_R \cdot \vec{x} \right) \tag{5.29}$$

$$\varphi_2 = A_T \exp \left( i \vec{k}_T \cdot \vec{x} \right) \tag{5.30}$$

The following statements follow from symmetry. The three vectors—$\vec{k}_I$, $\vec{k}_R$, and $\vec{k}_T$—will be coplanar, and the horizontal component of $\vec{k}_I$ will equal that of $\vec{k}_R$. Without loss of generality the problem can be reduced to a 2-dim problem in the $x$–$z$ plane. The incident and refracted directions are given by their angle with the $z$-axis (see Figure 5.2) $k_x = k \sin\theta$, $k_z = k \cos\theta$ for any wavevector. Angles are labeled the same as the corresponding amplitudes and wavevectors. The law of reflection, stated earlier, can be expressed as $\theta_I = \theta_R$. Applying boundary

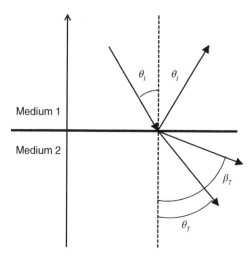

**Figure 5.2** Incident, reflected, and refracted waves at an interface between two media

conditions appropriate for a fluid–fluid interface leads to the two equations relating the wave functions:

$$A_I + A_R = \frac{\rho_2}{\rho_1} A_T \exp(i(k_{Tx} - k_{Ix})x) \tag{5.31}$$

$$k_{Iz} A_I + k_{Rz} A_R = k_{Tz} A_T \exp(i(k_{Tx} - k_{Ix})x) \tag{5.32}$$

In both equations, the phase on the l.h.s. has been divided through after evaluating at $z = 0$. The occurrence of factors of $k_z$ in (5.32) comes from taking the normal derivative of the waves. The same symmetry argument that led to $k_{Ix} = k_{Rx}$ implies that $k_{Iz} = -k_{Rz}$. The r.h.s. of each equation contains a function of $x$, while the l.h.s. of each is independent of $x$. In order for these equations to hold for all $x$, $k_{Tx} = k_{Ix}$ is required. This is Snell's law of the law of refraction:

$$\frac{\sin\theta_I}{c_1} = \frac{\sin\theta_T}{c_2} \tag{5.33}$$

Equation (5.32) is divided through by a factor of $k_{Iz}$, and both equations divided through by the initial amplitude, $A_I$. The reflection and transmission coefficients are defined as $R = A_R/A_I$ and $T = A_T/A_I$. Rearranging the terms leads to the following $2 \times 2$ system for the coefficients $T$ and $R$:

$$\begin{bmatrix} 1 \\ 1 \end{bmatrix} = \begin{bmatrix} \rho_2/\rho_1 & -1 \\ k_{Tz}/k_{Iz} & 1 \end{bmatrix} \begin{bmatrix} T \\ R \end{bmatrix} \tag{5.34}$$

Solving (5.34) leads to the following for the coefficients:

$$T = \frac{\rho_1}{\rho_2} \frac{2Z_2}{Z_1 + Z_2} \tag{5.35}$$

$$R = \frac{Z_1 - Z_2}{Z_1 + Z_2} \tag{5.36}$$

The final results are expressed in terms of the impedance:

$$Z_1 = \frac{\rho_1 c_1}{\cos \theta_I} \tag{5.37}$$

$$Z_2 = \frac{\rho_2 c_2}{\cos \theta_T} \tag{5.38}$$

When the second medium is a solid, there are two types of waves that are excited: compression waves (P-waves) and shear waves (S-waves). A solid with maximum symmetry is described by two Lamé parameters, $\lambda$, and $\mu$. If the density, $\rho$, of the solid is constant, the speed of each type of wave is related to the material constants:

$$c_P = \sqrt{\frac{\lambda + 2\mu}{\rho}} \tag{5.39}$$

$$c_S = \sqrt{\frac{\mu}{\rho}} \tag{5.40}$$

The incident wave is assumed to be in medium 1, and the wave takes the same form as in (5.29). In the second medium, the displacement field is expressed in terms of two potentials:

$$\vec{v}_2 = \vec{\nabla} \varphi_2 + \vec{\nabla} \times \vec{\psi}_2 \tag{5.41}$$

The same symmetry arguments hold in this case, and the problem can be treated as a 2-dim problem in the $x - z$ plane, with $\vec{\psi}_2 = \hat{j}\psi_2$. The nonzero components of the velocity field in medium 2 are

$$v_{2x} = \frac{\partial \varphi_2}{\partial x} - \frac{\partial \psi_2}{\partial z} \tag{5.42}$$

$$v_{2z} = \frac{\partial \varphi_2}{\partial z} + \frac{\partial \psi_2}{\partial x} \tag{5.43}$$

The normal components of the stress tensor are defined in terms of the material displacement, $\vec{u}$, that is, the strain tensor, through the generalized Hooke's law:

$$\sigma_{zz} = \lambda \left( \frac{\partial u_x}{\partial x} + \frac{\partial u_z}{\partial z} \right) + 2\mu \frac{\partial u_z}{\partial z} \tag{5.44}$$

$$\sigma_{xz} = \mu\left(\frac{\partial u_x}{\partial z} + \frac{\partial u_z}{\partial x}\right) \tag{5.45}$$

For a monochromatic plane wave, the velocity and displacement field are related by a factor of $i\omega$; hence the displacement fields can be replaced by the velocity fields and an overall factor of $i\omega$ divided out. A fluid is described by the same stress strain relations with $\psi = 0$ and $\mu = 0$. Direct substitution of these expressions and a little algebra lead to the following set of boundary conditions:

$$\rho_1 c_1^2 \nabla^2 \varphi_1 = \rho_2 c_P^2 \nabla^2 \varphi_2 + 2\rho_2 c_S^2\left(\frac{\partial^2 \psi_2}{\partial z \partial x} - \frac{\partial^2 \varphi_2}{\partial x^2}\right) \tag{5.46}$$

$$0 = 2\frac{\partial^2 \varphi_2}{\partial z \partial x} + \frac{\partial^2 \psi_2}{\partial x^2} - \frac{\partial^2 \psi_2}{\partial z^2} \tag{5.47}$$

$$\frac{\partial \varphi_1}{\partial z} = \frac{\partial \varphi_2}{\partial z} + \frac{\partial \psi_2}{\partial x} \tag{5.48}$$

Equations (5.46) and (5.47) are continuity of the normal components of the stress tensor, while (5.48) is continuity of the normal component of velocity. As before the incident, reflected, and transmitted fields are assumed to have a plane wave form:

$$\varphi_1 = A_I \exp\left(i\vec{k_I} \cdot \vec{x}\right) + A_R \exp\left(i\vec{k_R} \cdot \vec{x}\right) \tag{5.49}$$

$$\varphi_2 = A_T \exp\left(i\vec{k_T} \cdot \vec{x}\right) \tag{5.50}$$

$$\psi_2 = B_T \exp\left(i\vec{\sigma_T} \cdot \vec{x}\right) \tag{5.51}$$

A second transmitted wave has been added to account for the shear waves produced in the solid. The same symmetry arguments hold for the incident and reflected waves as in the fluid–fluid case. A new angle, $\beta_T$, relative to the z-axis is defined for the propagation of the transmitted field $\psi_2$. The wavevectors are expressed in terms of the angles:

$$\vec{k_I} = k_I[\sin\theta_I, 0, -\cos\theta_I]^T \tag{5.52}$$

$$\vec{k_R} = k_I[\sin\theta_I, 0, \cos\theta_I]^T \tag{5.53}$$

$$\vec{k_T} = k_T[\sin\theta_T, 0, -\cos\theta_T]^T \tag{5.54}$$

$$\vec{\sigma_T} = \sigma_T[\sin\beta_T, 0, -\cos\beta_T]^T \tag{5.55}$$

Inserting these expressions into (5.48) leads to the following:

$$k_{Iz}A_I + k_{Rz}A_R = k_{Tz}A_T \exp(i(k_{Tx} - k_{Ix})x) + \sigma_{Tx}B_T \exp(i(\sigma_{Tx} - k_{Ix})x) \tag{5.56}$$

For this condition to hold for all $x$, the wavenumbers must satisfy the conditions $k_{Tx} = k_{Ix}$ and $\sigma_{Tx} = k_{Ix}$, which states that Snell's law of refraction holds independently among the two transmitted waves and the incident wave:

$$\frac{\sin\theta_I}{c_1} = \frac{\sin\theta_T}{c_P} = \frac{\sin\beta_T}{c_S} \tag{5.57}$$

Evaluating (5.56) at this constraint and dividing through by $A_I$ leads to the following equation for the coefficients $R = A_R/A_I$, $P = A_T/A_I$, $S = B_T/A_I$:

$$1 - R = \frac{k_{Tz}}{k_{Iz}} P + \frac{\sigma_{Tx}}{k_{Iz}} S \tag{5.58}$$

Next, the expressions for the assumed solution are inserted in (5.47):

$$0 = k_T^2 \sin2\theta_T P + \sigma_T^2 \cos2\beta_T S \tag{5.59}$$

And finally, (5.46) leads to the following equation:

$$\frac{\rho_1}{\rho_2}(1 + R) = \left(1 - 2\frac{c_S^2}{c_P^2}\sin^2\theta_T\right)P - \sin2\beta_T S \tag{5.60}$$

Equations (5.58) through (5.60) can be solved for $R$, $P$, and $S$. Impedances are defined for each type of wave:

$$Z_1 = \frac{\rho_1 c_1}{\cos\theta_I} \tag{5.61}$$

$$Z_P = \frac{\rho_2 c_P}{\cos\theta_T} \tag{5.62}$$

$$Z_S = \frac{\rho_2 c_S}{\cos\beta_T} \tag{5.63}$$

The final form of the reflection and transmission coefficients is presented in terms of the impedances and the angle $\beta_T$, as in Ref. [5]:

$$P = \frac{\rho_1}{\rho_2} \frac{2Z_P \cos2\beta_T}{Z_P \cos^2 2\beta_T + Z_S \sin^2 2\beta_T + Z_1} \tag{5.64}$$

$$S = -\frac{\rho_1}{\rho_2} \frac{2Z_S \sin2\beta_T}{Z_P \cos^2 2\beta_T + Z_S \sin^2 2\beta_T + Z_1} \tag{5.65}$$

$$R = \frac{Z_P \cos^2 2\beta_T + Z_S \sin^2 2\beta_T - Z_1}{Z_P \cos^2 2\beta_T + Z_S \sin^2 2\beta_T + Z_1} \tag{5.66}$$

Two cases of interest are (1) normal incidence, $\theta_I = \theta_T = \beta_T = 0$, and (2) when $\beta_T = \pi/4$. In the case of normal incidence, the coefficients are

$$P = \frac{\rho_1}{\rho_2} \frac{2Z_P}{Z_P + Z_1} \tag{5.67}$$

$$S = 0 \tag{5.68}$$

$$R = \frac{Z_P - Z_1}{Z_P + Z_1} \tag{5.69}$$

For case (2) the coefficients become

$$P = 0 \tag{5.70}$$

$$S = -\frac{\rho_1}{\rho_2} \frac{2Z_S}{Z_S + Z_1} \tag{5.71}$$

$$R = \frac{Z_S - Z_1}{Z_S + Z_1} \tag{5.72}$$

In the case of normal incidence, only compression waves are excited, whereas in the second case only shear waves are excited.

## 5.5 Attenuation

Attenuation refers to a weakening of the wave amplitude. For cylindrical or spherical waves, the amplitude will weaken with distance from the source. This weakening is referred to as geometric spread. Interaction with the medium can lead to other types of attenuation by the processes of absorption or scattering. Mathematically this is accounted for by a complex wavenumber, $\kappa = k + i\alpha$. Fluid media can exhibit a variety of attenuation strengths. At the micro level, the cause of attenuation is due to the interaction of waves and the particles in the medium. At a macro level, the effects are rolled up into the attenuation coefficient, $\alpha$. This coefficient can depend on the frequency of the sound as well as on environmental factors such as temperature, relative humidity, and pressure. Attenuation acts opposite the direction of propagation, like air resistance for a moving object. In 3-dim a plane wave with attenuation is expressed as

$$\exp\left(i\vec{k} \cdot \vec{x} - \alpha s\right) \tag{5.73}$$

where $s$ is the distance traveled along the propagation path or direction. The same result applies to a spherical wave, $\exp(-\alpha R)$. The unit of the attenuation coefficient is m$^{-1}$. Models of the atmospheric attenuation of air and attenuation in water can be found in Refs. [6] and [4, 7], respectively.

## 5.6   Exercises

1.  Sound in a fluid medium is incident on a thin layer of a solid medium of thickness $L$, with two Lamé parameters.
    (a) Determine the boundary conditions at each interface of the solid medium.
    (b) Solve for the transmission and reflection coefficients and the coefficients of the wave in the solid medium.
2.  Calculate the 2-dim Fourier transform of the following spatial window functions.
    (a) Step function, 1 for $|x| < L_x/2$ and $|y| < L_y/2$, zero otherwise
    (b) Circular step function, 1 for $\rho < a$, zero otherwise
    (c) Radial quadratic window, $1 - (\rho/a)^2$ for $\rho < a$, zero otherwise
    (d) What effect does the change in window have on the spectrum?
3.  Given the transmission and reflection coefficient for a plane wave incident at the boundary of two media,
    (a) Develop a procedure for modeling the transmitted and reflected field from a point source a distance, $D$, above the plane separating the medium.
    (b) If possible work out the results analytically; otherwise write a procedure in the language of your choice.

## References

[1] Körner, T. W., Fourier Analysis, Cambridge University Press, Cambridge, 1990.

[2] Press, W. H., Teukolsky, S. A., Vetterling, W. T., and Flannery, B. P., Numerical Recipes in C++, The Art of Scientific Computing, Second Edition, Cambridge University Press, Cambridge, 2005.

[3] Tolstoy, I. and Clay, C. S., Ocean Acoustics, Theory and Experiment in Underwater Sound, McGraw-Hill, New York, 1966.

[4] Ainslie, M. A. and McColm, J. G., A simplified formula for viscous and chemical absorption in sea water, J. Acoust. Soc. Am., Vol. 103, No. 3, pp. 1671–1672, 1998.

[5] Brekhovskikh, L. M., Waves in Layered Media, Academic Press, New York, 1960.

[6] Rickley, E. J., Fleming, G. G., and Roof, C. J., Simplified procedure for computing the absorption of sound by the atmosphere, Noise Control Eng. J., Vol. 55, No. 6, pp. 482–494, 2007.

[7] Fisher, F. H. and Simmons, V. P., Sound absorption in sea water, J. Acoust. Soc. Am., Vol. 62, No. 3, pp. 558–564, 1977.

# 6

# Normal Modes

## 6.1 Introduction

In many practical situations, the refractive properties of the medium depend on one coordinate. In such cases the wave equation can be separated in a suitable coordinate system, and the solution described by a factor that represents a propagating wave in two dimensions and time and a factor that is a solution to a one-dimensional ordinary differential equation containing the index of refraction. This equation is similar in form to Schrödinger's time-independent eigenvalue equation for the wave function of a particle in a potential well. The set of solutions to this equation are referred to as normal modes and can be used to build up a complete solution to the acoustic wave equation for a collection of point sources embedded in the medium. Due to the nature of the potential well and possible boundary conditions, the spectrum of these modes may be continuous or discrete. There exist a handful of example refractive index functions for which exact solutions to this equation may be found in terms of special functions. More general cases require approximation or numerical methods. This chapter introduces normal mode theory applied to the second-order wave equation for the pressure field with a depth-dependent refractive index. Some cases leading to exact solutions are presented to illustrate the general features of the modes. To solve more complex problems, this chapter presents perturbation theory and applies it to several problems with boundary conditions. Afterward the relaxation technique is presented and discussed as a numerical method for solving the eigenvalue problem with boundary conditions and nontrivial refractive index. The final section contains a brief description of the theory behind coupled modes and their use in extending mode theory to more general environments.

*Computational Acoustics: Theory and Implementation*, First Edition. David R. Bergman.
© 2018 John Wiley & Sons Ltd. Published 2018 by John Wiley & Sons Ltd.

## 6.2 Mode Theory

Consider an environment described by $c = c(z)$, $\vec{v}_0 = 0$, and $\rho_0 = $ constant. The pressure field obeys the wave equation (6.1):

$$\nabla^2 p - \frac{1}{c^2}\frac{\partial^2 p}{\partial t^2} = 0. \qquad (6.1)$$

Assuming time harmonic behavior, $p = \varphi(\vec{x})\exp(-i\omega t)$, for the pressure field, (6.1) is converted into the Helmholtz equation for $\varphi(\vec{x})$:

$$\nabla^2 \varphi + \frac{\omega^2}{c^2}\varphi = 0. \qquad (6.2)$$

Applying separation of variables in $\vec{x}$, $\varphi(\vec{x}) = X(x)Y(y)Z(z)$, leads to a set of ordinary differential equations (ODE) for each factor:

$$X'' = -k_x^2 X, \qquad (6.3)$$

$$Y'' = -k_y^2 Y, \qquad (6.4)$$

$$Z'' + \left\{\frac{\omega^2}{c^2} - k_T^2\right\}Z = 0. \qquad (6.5)$$

The separation constants $k_x^2$ and $k_y^2$ have been introduced, having the physical interpretation of wavevector components in the $x$ and $y$ directions, respectively. The transverse wavenumber, $\vec{k}_T = [k_x \ k_y \ 0]^T$, $k_T^2 \equiv k_x^2 + k_y^2$, appears in (6.5). The solution for the transverse propagation is given by a plane wave:

$$XY = \exp\left(i\vec{k}_T \cdot \vec{x}\right) \qquad (6.6)$$

When $c = $ constant, the term in brackets in (6.5) is identified as the wavenumber $k_z^2$, and the separation constants obey the dispersion relation

$$k_x^2 + k_y^2 + k_z^2 = \frac{\omega^2}{c^2} \qquad (6.7)$$

A complete solution is a plane wave traveling in the $\vec{k}$ direction. In the general case a position-dependent wavenumber is defined:

$$k_z^2 \equiv \left\{\frac{\omega^2}{c^2} - k_T^2\right\} \qquad (6.8)$$

With this definition (6.5) is written in compact form:

$$\frac{d^2Z}{dz^2} + k_z^2(z)Z = 0 \tag{6.9}$$

Solutions to (6.9) may have bound (decaying) or unbound (oscillatory) behavior depending on the sign of $k_z^2$. Since this is no longer constant, there may be locations where the behavior changes from one type to the other. The zeros of $k_z^2$ will be determined by the specific nature of the sound speed function $c(z)$. Consider a single transition point at some arbitrary value, $z = z'$, for which $k_z^2(z') = 0$. From (6.9) it follows that this is a point of inflection of the function Z. For values of $z$ such that $k_z^2 > 0$, the solution will have oscillatory behavior, and for $k_z^2 < 0$ it will have exponential behavior.

## 6.3 Profile Models

Physical models of refraction in underwater acoustics frequently involve mathematical models of speed profiles that tend to infinity at $|z| \to \infty$. Simple models with this behavior include the following:

$$c(z) = c_0 + \varepsilon|z| \tag{6.10}$$

$$c(z) = c_0 + \varepsilon z^{2n}, \quad n = 1, 2, 3, \ldots \tag{6.11}$$

$$c(z) = c_0 \left( 1 + \varepsilon \cosh\left(\frac{z}{L}\right) \right) \tag{6.12}$$

A specific example from underwater acoustics is the SOFAR channel described by the Garrett–Munk profile:

$$c(z) = c_0[1 + \varepsilon(\eta - 1 + \exp(-\eta))] \tag{6.13}$$

The variable $\eta = 2(z - z_0)/z_0$ is defined in (6.13). This model has free parameters, namely, $c_0$, $\varepsilon$, and $z_0$. Typical values found in the literature are $c_0 = 1500$ m/s, $\varepsilon = 0.00737$, and $z_0 = 1300$ m. For the profile presented in (6.13), the positive z-axis is directed downward, a common definition used in underwater acoustics.

For very large values of $|z|$, all of these profiles approach infinity and the modes vanish. At these extreme limits, sound is trapped in a waveguide-like structure due solely to the refractive nature of the sound speed. At the other extreme one often encounters the purely mathematical situation where $c(z') = 0$ at some value, $z'$. In this case $k_z^2 \to \infty$ and the field must vanish as $z \to z'$. This is a mathematical artifact of certain models and does not represent anything physical. An example with this behavior is the linear sound speed profile:

$$c(z) = c_0 + \varepsilon z \tag{6.14}$$

The sound speed equals zero at $z' = -c_0/\varepsilon$ and physically relevant solutions do not exist for $z' < -c_0/\varepsilon$.

It is possible to find examples of $c(z)$ for which (6.9) becomes one of the well-known equations defining a special function. Consider a physical model where the sound speed profile can be written as a constant, $c_0$, plus a function of position that is small relative to $c_0$, at least in some interval $z \in [z_1, z_2]$:

$$c(z) = c_0 + \tilde{c}(z), \tag{6.15}$$

Under this assumption (6.8) is approximated by a first-order Taylor expansion in $\tilde{c}(z)/c_0$:

$$k_z^2 = \left\{ \frac{\omega^2}{c^2} - k_T^2 \right\} \approx \frac{\omega^2}{c_0^2} - 2\frac{\omega^2}{c_0^3}\tilde{c}(z) - k_T^2 \equiv k_{z0}^2 - 2\frac{\omega^2}{c_0^3}\tilde{c}(z) \tag{6.16}$$

The term $k_{z0}^2$, the wavenumber when $\tilde{c}(z) = 0$, is defined as

$$k_{z0}^2 = \frac{\omega^2}{c_0^2} - k_T^2 \tag{6.17}$$

Using (6.16) and (6.17) in (6.9) and a little rearranging gives the following form for the vertical mode equation:

$$-\frac{d^2Z}{dz^2} + U(z)Z = E_z Z, \tag{6.18}$$

The "potential energy" function, $U(z)$, and energy, $E_z$, appearing in (6.18) are defined as

$$U(z) = 2\frac{\omega^2}{c_0^3}\tilde{c}(z) \tag{6.19}$$

$$E_z = k_{z0}^2 \tag{6.20}$$

The special form of (6.18) is reminiscent of Schrödinger's time-independent equation. It should be pointed out that the expansion in (6.16) is not necessary to massage (6.9) into this form. The identification of $U$ and $E$ with potential and energy level is for illustrative purposes, the quantities do not necessarily have units of energy.

A reasonable question is whether or not the modes predicted using (6.16) will describe solutions to the original form, (6.8). It is clear that in general the two sets will not match since at some point the condition $\tilde{c}(z) \ll c_0$ may be violated. This is not a concern if ideal boundary conditions are applied in the interval $[z_1, z_2]$ where the approximation holds. To better understand the difference, (6.9) is cast in the same form as (6.9) by adding $-k_{z0}^2 Z$ to both sides:

$$-\frac{d^2Z}{dz^2} + \left(k_{z0}^2 - k_z^2(z)\right)Z = k_{z0}^2 Z \tag{6.21}$$

Comparing (6.21) with (6.18) with (6.16) serving as the potential, it is clear that the differences (and similarities) can be understood by comparing the following two effective potentials:

$$U_2(z)/k_0^2 = 2\tilde{c}(z)/c_0 \tag{6.22}$$

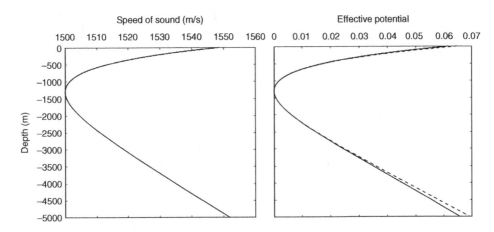

**Figure 6.1**  The Garrettt–Munk sound speed profile (left) and its effective potentials (right)

$$U_1(z)/k_0^2 = \frac{(1+\tilde{c}(z)/c_0)^2 - 1}{(1+\tilde{c}(z)/c_0)^2} \tag{6.23}$$

Figure 6.1 illustrates the difference for the Garrett–Munk profile. The left pane is the sound speed profile plotted using (6.13). The right pane shows both effective potentials, (6.23) solid and (6.22) dashed. The potentials are scaled by $k_0^2 = \omega^2/c_0^2$ to avoid frequency dependence and provide a clear view of the functional form of these potentials. The relative difference between the two potential functions is less than 1% for $z \in [585\,\text{m}, 2417\,\text{m}]$ and less than 5% down to 4500 m. The figure clearly demonstrates the similarities between the two models for a range of values that correspond to expected ocean depths. Two more examples are shown to further illustrate some of these features. In all examples, solid lines represent the unperturbed potential, while dashed lines represent the first-order expansion given in (6.22). The first is for the linear profile in (6.14). The vertical axis is height in meters and the horizontal potential (no units). The pair of potentials is plotted for three values of $\varepsilon$, that is, 0.001, 0.005, and 0.01, which are plotted in Figure 6.2, shown from left to right. Attention should be given to the horizontal axis. It is clear that for increasing values of the sound speed deviation, the mismatch between the two becomes more severe. For $\varepsilon = 0.001$, the relative difference is less than 1% for heights less than 6.5 m and less than 1.6% for those up to 10 m.

The last example is that of a quadratic refractive index. The sound speed profile for this case will be presented in the next section, (6.24). Plots of the two effective potentials are shown in Figure 6.3. In this case the shape parameter is fixed at $\varepsilon = 0.001$. The potentials are plotted for three distances away from the waveguide axis—100, 500, and 1000 m—from left to right. For the first case a relative difference is less than 0.1% over the entire range of position. It is clear that the two potentials fall far out of sync at larger distances, as to be expected since $U_2$ is based on a first-order expansion. This example also illustrates how the approximate potential mimics the behavior of the sound speed profile. What these examples illustrate is that when ideal boundary conditions are included in the problem, (6.22) may be a good model for (6.23) and, as illustrated in the next section, may be easier to solve.

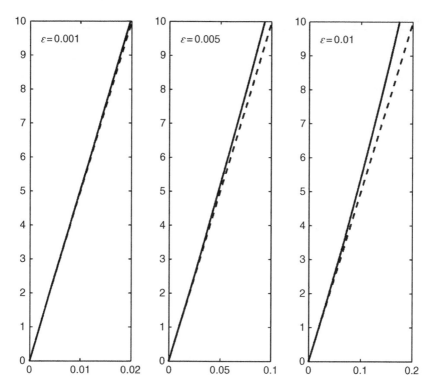

**Figure 6.2** Effect potentials for the linear sound speed profile at various strengths

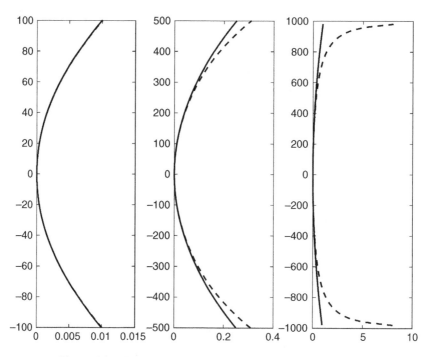

**Figure 6.3** Effective potentials for an ideal refractive waveguide

Why go through the effort to cast the equation in a new form? A matter of convenience. One may encounter a scenario where the modes based on (6.23) cannot be exactly solved for, while those for the form based on (6.22) can. This may make the job of modeling a specific problem a little easier. Of course, one still has to be aware of the limitations and not use the approximate solution outside the range of validity. The rest of this chapter is devoted to solving (6.18) and using those solutions to model sources of sound.

## 6.4   Analytic Examples

This subsection briefly discusses cases where an exact solution for the modes can be found. There exist only a handful of potential functions for which the wave equation will provide an exact solution. These are worth the effort to work through and understand as they provide toy models for describing the various behaviors of the acoustic field as a function of the environmental parameters. Working through examples with exact solutions helps with familiarity of the process and exposes some new physics associated with propagation in refractive waveguides. Two such examples are presented here for illustration. The references provide more detailed information on special functions and their differential equations in general [1–3].

### 6.4.1   Example 1: Harmonic Oscillator

The first example presented, (6.24), produces a quadratic potential function for (6.23):

$$c = \frac{c_0}{\sqrt{\left(1 - (\xi z)^2\right)}} \tag{6.24}$$

Substitution in (6.21) yields the equation for the function $Z(z)$:

$$\frac{d^2Z}{dz^2} + \left(k_0^2 - k_T^2\right)Z - (k_0 \xi z)^2 Z = 0 \tag{6.25}$$

With the change of variables, $\eta = \sqrt{k_0 \xi} z$, and a redefinition of the function $Z = \tilde{Z}/(k_0\xi)$ (6.25) is converted into the familiar form in (6.26):

$$\frac{d^2\tilde{Z}}{d\eta^2} + \lambda\tilde{Z} - \eta^2\tilde{Z} = 0 \tag{6.26}$$

$$\lambda = \frac{\left(k_0^2 - k_T^2\right)}{k_0 \xi} \tag{6.27}$$

Equation (6.26) is identical to the equation describing the wave function of a simple harmonic oscillator in nonrelativistic quantum mechanics. The solution is developed by assuming a solution of the form

$$\tilde{Z}(\eta) = \exp\left(-\frac{\eta^2}{2}\right)H(\eta) \tag{6.28}$$

In (6.28) the factor $H(\eta)$ is an undetermined function. Inserting the form of this solution in (6.26), an equation for $H(\eta)$ is determined:

$$\frac{d^2H}{d\eta^2} - 2\eta\frac{dH}{d\eta} + (\lambda-1)H = 0 \tag{6.29}$$

With a redefinition of the last term, $2\bar{\lambda} = (\lambda-1)$, (6.29) can be identified as Hermite's equation, one of a family of equations whose solutions are special functions. A treatment familiar to many readers involves integer eigenvalues:

$$\lambda_n - 1 = 2n, \quad n = 0, 1, 2, \dots \tag{6.30}$$

In this case the corresponding solutions, $H_n$, are Hermite polynomials. Each Hermite polynomial is of finite order, $n$. For each $n$, the Hermite polynomial $H_n$ and the associated value of $\lambda_n$ form an eigenfunction–eigenvalue pair. A second solution to (6.29) may be found in Ref. [3]. A complete series solution analysis will not be presented here, but the interested reader can find it in Arfkin and Weber [1]. For the integer case the Hermite polynomials can be generated by a recursion relation and the first two polynomials, $H_0$ and $H_1$:

$$H_{n+1}(\eta) = 2\eta H_n(\eta) - 2nH_{n-1}(\eta) \tag{6.31}$$

$$H_0(\eta) = 1 \tag{6.32}$$

$$H_1(\eta) = 2\eta \tag{6.33}$$

Table 6.1 lists these polynomials up to order 8, starting at $n = 2$, as a reference. The Hermite polynomials are normalized by the following coefficient:

$$\sqrt{\frac{1}{2^n n!\sqrt{\pi}}} \tag{6.34}$$

Figure 6.4 shows the first three of these modes for the parameters $c_0 = 340\,\mathrm{m/s}$, $\xi = 0.01, f = 500\,\mathrm{Hz}$.

Table 6.1 Hermite polynomials

| $n$ | $H_n$ |
| --- | --- |
| 2 | $4\eta^2 - 2$ |
| 3 | $8\eta^3 - 12\eta$ |
| 4 | $16\eta^4 - 48\eta^2 + 12$ |
| 5 | $32\eta^5 - 160\eta^3 + 120\eta$ |
| 6 | $64\eta^6 - 480\eta^4 + 720\eta^2 - 120$ |
| 7 | $128\eta^7 - 1344\eta^5 + 3360\eta^3 - 1680\eta$ |
| 8 | $256\eta^8 - 3584\eta^8 + 13440\eta^4 - 13440\eta^2 - 1680$ |

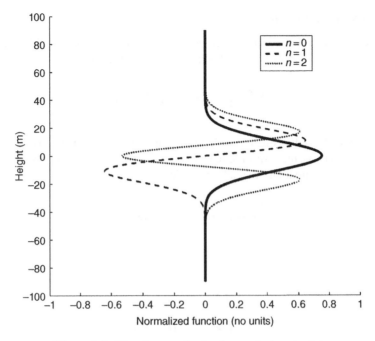

**Figure 6.4**   First three modes for the quadratic potential

From (6.27) the dispersion relation can be determined. Recall that the separation constant, $k_T^2$, represents the wavenumber in the transverse direction. For separation in Cartesian coordinates, this would be $k_T^2 = k_x^2 + k_y^2$, whereas in cylindrical coordinates this would be the radial wavenumber:

$$k_T = k_0 \sqrt{1 - \xi k_0^{-1}(2n+1)} \tag{6.35}$$

It is clear from (6.35) that if $n$ becomes too large, $k$ will become imaginary. Setting (6.35) to zero and solving for $n$ provides a cutoff value for the index, denoted $n_c$. Perhaps a more appropriate term would be a transition value. For values of $n$ below the cutoff, the transverse part of the solution will propagate. For values of $n$ above (or at) the cutoff, the transverse part of the solution will decay. The decaying modes are referred to as evanescent modes. They are present but typically decay rapidly and become negligible in the far field. The cutoff parameter depends on the environmental parameters, $c_0$ and $\xi$, as well as the source frequency, $\omega$. From the definition of $k_0$, the cutoff may be expressed in terms of a minimum frequency, $\omega_m = c_0 \xi$:

$$n_c = \text{ceil}\left(\frac{\omega/\omega_m - 1}{2}\right) \tag{6.36}$$

Using the Garrett–Munk parameters as an example, the minimum sound speed is $c_0 = 1500\,\text{m/s}$. Taylor expanding near the SOFAR channel axis gives $\xi \approx 1.321 \times 10^{-4}$. For these parameters $\omega_m \approx 0.2\,\text{Hz}$.

### 6.4.2 Example 2: Linear

This example involves a linear sound speed profile, $c = c_0(1 + \xi z)$. Inserting this into (6.9) and performing a change of variables to $\eta = 1 + \xi z$ yields the following:

$$\frac{d^2 Z}{d\eta^2} + \left( \frac{\xi^{-2} k_0^2}{\eta^2} - \xi^{-2} k_T^2 \right) Z = 0 \tag{6.37}$$

This is a form of Bessel's equation with an imaginary index. While this is well defined in terms of special functions, these specific functions are difficult to pin down. A simpler equation results when the approximation of small $\xi$ is made:

$$\frac{d^2 Z}{dz^2} - 2k_0^2 \xi z Z + \left( k_0^2 - k_T^2 \right) Z \approx 0 \tag{6.38}$$

Equation (6.38) can be simplified by making the following substitutions:

$$\eta = \left( 2k_0^2 \xi \right)^{1/3} z \tag{6.39}$$

$$\beta = \left( 2k_0^2 \xi \right)^{-2/3} \left( k_0^2 - k_T^2 \right) \tag{6.40}$$

$$\frac{d^2 Z}{d\eta^2} - (\eta - \beta) Z \approx 0 \tag{6.41}$$

With one final redefinition of the coordinate, $\sigma = (\eta - \beta)$, $d\sigma = d\eta$. Equation (6.41) can now be identified with Airy's equation (Gradshteyn and Ryzhik [3]):

$$\frac{d^2 Z}{d\sigma^2} - \sigma Z \approx 0 \tag{6.42}$$

The solutions to (6.42), Airy's functions, are expressed in terms of Bessel functions:

$$\mathrm{Ai}(\sigma) = \frac{\sqrt{\sigma}}{3} \left( I_{-1/3} \left( \frac{2}{3} \sigma^{3/2} \right) - I_{1/3} \left( \frac{2}{3} \sigma^{3/2} \right) \right) \tag{6.43}$$

$$\mathrm{Ai}(-\sigma) = \frac{\sqrt{\sigma}}{3} \left( J_{1/3} \left( \frac{2}{3} \sigma^{3/2} \right) + J_{-1/3} \left( \frac{2}{3} \sigma^{3/2} \right) \right) \tag{6.44}$$

$$\mathrm{Bi}(\sigma) = \left( \frac{\sigma}{3} \right)^{1/2} \left( I_{1/3} \left( \frac{2}{3} \sigma^{3/2} \right) + I_{-1/3} \left( \frac{2}{3} \sigma^{3/2} \right) \right) \tag{6.45}$$

$$\mathrm{Bi}(-\sigma) = \left( \frac{\sigma}{3} \right)^{1/2} \left( J_{-1/3} \left( \frac{2}{3} \sigma^{3/2} \right) - J_{1/3} \left( \frac{2}{3} \sigma^{3/2} \right) \right) \tag{6.46}$$

What happened to the eigenvalue? Nothing, the equation is invariant under a shift in the variable. Thus, the only function that the constant serves is to align the zeros of the solution with a particular point on the coordinate axis. The continuum of values $k_r^2$ is allowed, up to the point where the wavenumber becomes imaginary and the solution decays. These special functions, Ai($\sigma$) and Bi($\sigma$), are the two linearly independent solutions of (6.42) and are referred to as the Airy function of the first and second kind, or the Airy and Bairy functions, respectively. They are somewhat of a generalization of the sine and cosine solutions to the free wave equation. For values of $\sigma < 0$, the solution oscillates, and for values of $\sigma > 0$, the solution exhibits exponential behavior. The Airy function goes to zero for $\sigma \to \infty$, while Bairy diverges in the same limit. Because the wavenumber is unrestricted, these can be fitted to various boundary conditions like sine and cosine functions.

The derivatives of the Airy functions can be expressed in terms of Bessel functions as well:

$$\frac{d\text{Ai}(\sigma)}{d\sigma} = -\frac{\sigma}{3}\left(I_{-2/3}\left(\frac{2}{3}\sigma^{3/2}\right) - I_{2/3}\left(\frac{2}{3}\sigma^{3/2}\right)\right) \tag{6.47}$$

$$\frac{d\text{Ai}(-\sigma)}{d\sigma} = -\frac{\sigma}{3}\left(J_{-2/3}\left(\frac{2}{3}\sigma^{3/2}\right) - J_{2/3}\left(\frac{2}{3}\sigma^{3/2}\right)\right) \tag{6.48}$$

$$\frac{d\text{Bi}(\sigma)}{d\sigma} = \frac{\sigma}{\sqrt{3}}\left(I_{-2/3}\left(\frac{2}{3}\sigma^{3/2}\right) + I_{2/3}\left(\frac{2}{3}\sigma^{3/2}\right)\right) \tag{6.49}$$

$$\frac{d\text{Bi}(-\sigma)}{d\sigma} = \frac{\sigma}{\sqrt{3}}\left(J_{-2/3}\left(\frac{2}{3}\sigma^{3/2}\right) + J_{2/3}\left(\frac{2}{3}\sigma^{3/2}\right)\right) \tag{6.50}$$

Burying the parameter $\beta$ in the variable produces a shift in the function. Shifted Airy functions can be shown to satisfy an equivalent shifted Airy equation. Applying either type of boundary condition at one value of $z$, say, $z = 0$, merely requires shifting the functions such that an appropriate node or antinode appears at the boundary. For problems with a single boundary, the radiation condition may force one to choose only one of the two functions. However, for problems with two boundaries, application of the second boundary will affect the spectrum of the allowed modes. In this case both sets of functions may be necessary, and neither would be ruled out since the limit $z \to \infty$ will never be reached. The general solution to (6.42) may be written in terms of both functions:

$$Z = a\text{Ai}(\sigma) + b\text{Bi}(\sigma) \tag{6.51}$$

For a soft or hard boundary at $z = 0$, or $\sigma = -\beta$, the constants $a$ and $b$ are related. Redefining the overall constant leads to the following expressions for the modes for soft and hard boundary, respectively:

$$Z = C\left(\frac{\text{Ai}(\sigma)}{\text{Ai}(-\beta)} - \frac{\text{Bi}(\sigma)}{\text{Bi}(-\beta)}\right) \tag{6.52}$$

$$Z = C\left(\frac{\text{Ai}(\sigma)}{\text{Ai}'(-\beta)} - \frac{\text{Bi}(\sigma)}{\text{Bi}'(-\beta)}\right) \tag{6.53}$$

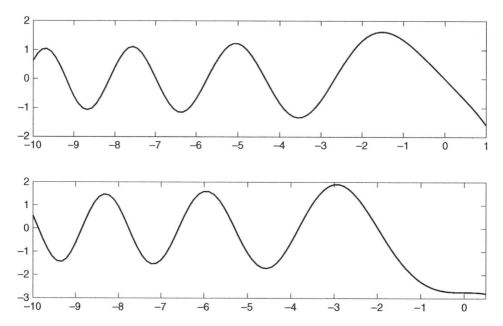

**Figure 6.5**   Modes of the linear potential for pressure release (top) and hard (bottom) boundary conditions

The boundary condition at $z=L$ will impose a restriction on the free parameters of the problem. The value of the argument at $z=L$ is denoted $\sigma_L$:

$$\frac{\mathrm{Ai}(\sigma_L)}{\mathrm{Ai}(-\beta)} = \frac{\mathrm{Bi}(\sigma_L)}{\mathrm{Bi}(-\beta)} \tag{6.54}$$

$$\frac{\mathrm{Ai}(\sigma_L)}{\mathrm{Ai}'(-\beta)} = \frac{\mathrm{Bi}(\sigma_L)}{\mathrm{Bi}'(-\beta)} \tag{6.55}$$

An example of fitting Airy functions to a single boundary is provided in Figure 6.5. The mode depicted on top satisfies the pressure release condition at $x=0$, while that on the bottom satisfies the hard boundary condition.

## 6.5   Perturbation Theory

The previous section provided some examples of normal modes for special cases that involved known special functions. There are very few cases of refractive index that lead to solvable problems and those that do often involve special functions that require infinite series solutions, for example, Bessel functions, Mathieu functions, and so on. It is also sometimes difficult to fit these functions to specific boundaries. It was mentioned that refractive profiles can be expressed as a constant plus a small perturbation. Thus, it would make sense to develop solutions to refractive problems involving boundaries using a perturbation series approach.

A detailed treatment of perturbation theory can be found in Morse and Feshbach [2]. The technique is commonly presented in quantum mechanics texts. This presentation follows closely that found in Liboff [4]. Equation (6.18) is the starting point for developing a perturbative solution to the normal mode problem. The approach presented here will start with a complete set of solutions to the free wave equation with boundary conditions and then introduce the refractive changes as a perturbation in the effective potential function.

The solutions to the unperturbed system are denoted $\varphi_n$ and $E_n$. These are known exactly in terms of sine and cosine functions. With the refractive changes included, the new wave equation is written as follows:

$$-\frac{d^2\varphi}{dz^2} - E\varphi + \varepsilon U(z)\varphi = 0 \tag{6.56}$$

In (6.56) the factor $\varepsilon$ is a small parameter, which will be used to develop the perturbative expansion. The solutions of (6.56) are denoted $\widetilde{\varphi}_n$ and $\widetilde{E}_n$. A solution is sought that approaches $\varphi_n$ and $E_n$ in the limit as $\varepsilon \to 0$. The solutions of the original equation form a complete set for the given boundary conditions. Any function that satisfies the same boundary conditions can be written as a linear superposition of the original set of solutions. It is assumed that the eigenfunctions and eigenvalues of the new system may be written in the following form:

$$\widetilde{\varphi}_n = \sum_{j=0}^{\infty} \varepsilon^j \varphi_n^{(j)} \tag{6.57}$$

$$\widetilde{E}_n = \sum_{j=0}^{\infty} \varepsilon^j E_n^{(j)} \tag{6.58}$$

Each term, $\varphi_n^{(j)}$, $E_n^{(j)}$, represents a correction to the solution or order $\varepsilon^j$. Equations (6.57) and (6.58) are inserted into (6.56) to give the following:

$$\sum_{j=0}^{\infty} \left[ \hat{H}_0 \varepsilon^j \varphi_n^{(j)} + \hat{H}' \varepsilon^{j+1} \varphi_n^{(j)} \right] - \sum_{i,j=0}^{\infty} \varepsilon^{i+j} E_n^{(i)} \varphi_n^{(j)} = 0 \tag{6.59}$$

We define operators $(H_0)$ and $H'$. In equation (6.59) the following operators have been introduced: $\hat{H}_0 = -d^2/dz^2$, $\hat{H}' = U$.

For this equation to hold for arbitrary $\varepsilon$, the factor of each power of $\varepsilon$ much vanish independently. This requirement leads to a hierarchy of equations. Following are the equations for each power of $\varepsilon$, from 0 to 3 explicitly, followed by a generic formula for order $\varepsilon^j$:

$$\left[ \hat{H}_0 - E_n^{(0)} \right] \varphi_n^{(0)} = 0 \tag{6.60}$$

$$\left[ \hat{H}_0 - E_n^{(0)} \right] \varphi_n^{(1)} + \left[ \hat{H}' - E_n^{(1)} \right] \varphi_n^{(0)} = 0 \tag{6.61}$$

$$\left[ \hat{H}_0 - E_n^{(0)} \right] \varphi_n^{(j)} + \hat{H}' \varphi_n^{(j-1)} - \sum_{m=1}^{j-1} E_n^{(m)} \varphi_n^{(j-m)} - E_n^{(j)} \varphi_n^{(0)} = 0 \tag{6.62}$$

The lowest-order equation is the original unperturbed equation, that is, the lowest-order estimate to the perturbed system is simply the solution to the unperturbed system. The higher-order corrections are constrained to be orthogonal to the unperturbed solutions:

$$\left\langle \varphi_n^{(m)} | \varphi_n^{(0)} \right\rangle = 0, \ \forall n, m > 0 \tag{6.63}$$

It is now further assumed that each correction may be written as a superposition of the unperturbed functions:

$$\varphi_n^{(m)} = \sum_{j=0}^{\infty} C_{nj}^{(m)} \varphi_j^{(0)} \tag{6.64}$$

First consider the constraint. Inserting the expansion into (6.63) gives the following:

$$\left\langle \varphi_n^{(m)} | \varphi_n^{(0)} \right\rangle = \sum_{j=0}^{\infty} C_{nj}^{(m)} \left\langle \varphi_j^{(0)} | \varphi_n^{(0)} \right\rangle = C_{nn}^{(m)} \tag{6.65}$$

The second equality holds due to the fact that the basis function forms an orthonormal basis. The constraint requires $C_{nn}^{(m)} = 0$ for all corrections. Next, consider the first-order equation. Using this expansion for the first-order equation leads to

$$\sum_{j=0}^{\infty} C_{nj}^{(1)} \left[ \hat{H}_0 - E_n^{(0)} \right] \varphi_j^{(0)} + \left[ \hat{H}' - E_n^{(1)} \right] \varphi_n^{(0)} = 0 \tag{6.66}$$

To solve for the coefficients, this equation is projected into an arbitrary basis function $\varphi_m^{(0)}$:

$$-C_{nm}^{(1)} \Delta E_{nm}^{(0)} + H_{mn}' - E_n^{(1)} \delta_{nm} = 0 \tag{6.67}$$

In deriving (6.67) use was made of the fact that $H_0 \varphi_j^{(0)} = E_j^{(0)} \varphi_j^{(0)}$, and the fact that $\varphi_j^{(0)}$ are orthonormal set of functions. Shorthand notations are introduced in the first and second terms of (6.67):

$$\Delta E_{nm}^{(0)} = E_n^{(0)} - E_m^{(0)} \tag{6.68}$$

$$H_{mn}' = \int \bar{\varphi}_m^{(0)} \hat{H}' \varphi_n^{(0)} dz \tag{6.69}$$

For $n = m$ the first-order correction to the eigenvalue is obtained:

$$E_n^{(1)} = H_{nn}' \tag{6.70}$$

For $n \neq m$ the coefficients are obtained:

$$C_{nm}^{(1)} = \frac{H'_{mn}}{\Delta E_{nm}^{(0)}} \tag{6.71}$$

At this point all higher-order equations have the same form. Consider a generic order, $k > 1$. Following the same steps as with the first-order equation leads to the following:

$$-C_{nm}^{(k)} \Delta E_{nm}^{(0)} + \left\langle \varphi_m^{(0)} | \hat{H}' \varphi_n^{(k-1)} \right\rangle - \sum_{j=1}^{k-1} E_n^{(j)} \left\langle \varphi_m^{(0)} | \varphi_n^{(k-j)} \right\rangle - E_n^{(k)} \delta_{nm} = 0 \tag{6.72}$$

The first term has been expanded in the zeroth-order basis to arrive at (6.72), while the second and third terms are left as is for the time being. The correction to the eigenvalue spectrum is determined by setting $n = m$. This immediately leads to the following result:

$$E_n^{(k)} = \left\langle \varphi_m^{(0)} | \hat{H}' \varphi_n^{(k-1)} \right\rangle = \sum_{i \neq n} C_{ni}^{(k-1)} H'_{ni} \tag{6.73}$$

The coefficients for this order expansion are determined by $n \neq m$:

$$C_{nm}^{(k)} = \frac{1}{\Delta E_{nm}^{(0)}} \left[ \left\langle \varphi_m^{(0)} | \hat{H}' \varphi_n^{(k-1)} \right\rangle - \sum_{j=1}^{k-1} E_n^{(j)} \left\langle \varphi_m^{(0)} | \varphi_n^{(k-j)} \right\rangle \right] \tag{6.74}$$

Using the expansion of the lower-order corrections for the eigenfunctions leads to

$$C_{nm}^{(k)} = \frac{1}{\Delta E_{nm}^{(0)}} \left[ \sum_{i \neq n} C_{ni}^{(k-1)} H'_{mi} - \sum_{j=1}^{k-1} E_n^{(j)} C_{nm}^{(k-j)} \right] \tag{6.75}$$

This procedure is carried out for each order. The second-order corrections are presented as follows:

$$E_n^{(2)} = \sum_{m \neq n} \frac{H'_{mn} H'_{nm}}{\Delta E_{nm}^{(0)}} \tag{6.76}$$

$$C_{nm}^{(2)} = \frac{1}{\Delta E_{nm}^{(0)}} \left[ \sum_{j \neq n} \frac{H'_{mj} H'_{jn}}{\Delta E_{nj}^{(0)}} - \frac{H'_{nn} H'_{mn}}{\Delta E_{nm}^{(0)}} \right] \tag{6.77}$$

Equations (6.73) and (6.75) can be used recursively to develop the corrections to the spectrum and the coefficients for each correction to the modes.

Taking a perturbative approach to finding normal modes for acoustics problems with boundaries makes practical sense. The approach is now applied to refractive problems with vertical boundary conditions. The zeroth-order functions are provided in Chapter 4. As a specific example the techniques will be applied to finding modes for several well-known profiles in the presence of these conditions. Referring back to the derivation of the eigenvalue corrections and the expansion coefficients for the eigenfunction corrections, the pertinent factor is $H'_{mn}$ in (6.69):

$$H'_{mn} = \frac{2}{L} \int_0^L \sin\left(\frac{m\pi}{L}z\right) \sin\left(\frac{n\pi}{L}z\right) \tilde{U}(z)dz \tag{6.78}$$

$$H'_{mn} = \frac{2}{L} \int_0^L \cos\left(\frac{m\pi}{L}z\right) \cos\left(\frac{n\pi}{L}z\right) \tilde{U}(z)dz \tag{6.79}$$

$$H'_{mn} = \frac{2}{L} \int_0^L \cos\left(\frac{(2m+1)\pi}{2L}z\right) \cos\left(\frac{(2n+1)\pi}{2L}z\right) \tilde{U}(z)dz \tag{6.80}$$

where Equation (6.78) ((6.79) and (6.80)) applies to SS (HH and SH) boundary conditions. Individual steps will not be presented for the following; the reader can consult a table of integrals or symbolic computation software. Suffice to say that by use of trigonometric identities, the previously mentioned integrals can be converted into a sum of integrals over sine and cosine of the sum and difference of the arguments. Then the integrals needed to compute the terms in the perturbation series are as follows:

$$\int_0^\pi \sin(j\xi)\tilde{U}(\xi)d\xi \tag{6.81}$$

$$\int_0^\pi \cos(j\xi)\tilde{U}(\xi)d\xi \tag{6.82}$$

The following change of variables has been introduced, $\xi = \pi z/L$, and the integer $j$ is $n \pm m$. For the purposes of constructing illustrative examples, the following choices for $\tilde{U}(\xi)$ are considered:

$$\tilde{U}(\xi) = \lambda(\xi - \xi_0)^N, \quad N = 1, 2 \tag{6.83}$$

$$\tilde{U}(\xi) = \lambda \exp(\alpha z) \tag{6.84}$$

From these two functions, many practical examples can be constructed, including the fish eye and the Garrett–Munk profile. The offset $\xi_0$ is present to allow one to place the refractive waveguide axis between the boundaries. By another change of variables and the use of trig identities, integrals involving such an offset can be reduced to standard forms in Equations (6.81) and (6.82) with appropriate coefficients. Specific examples will be presented here. These are (1) the linear, (2) quadratic, and (3) exponential.

### Example 6.1 Linear

In this case $\tilde{U}(z) = z$. The other scale factors can be included afterward. For the soft boundary conditions, the matrix elements are as follows:

$$H'_{nm} = \frac{L}{\pi^2}((-1)^{n+m} - 1)\frac{4nm}{(n^2 - m^2)^2} \tag{6.85}$$

$$H'_{nn} = \frac{L}{2} \tag{6.86}$$

The hard boundary conditions lead to the following matrix elements:

$$H'_{nm} = \frac{2L}{\pi^2}((-1)^{n+m} - 1)\frac{n^2 + m^2}{(n^2 - m^2)^2} \tag{6.87}$$

$$H'_{nn} = \frac{L}{2} \tag{6.88}$$

Lastly the matrix elements for the mixed conditions are as follows:

$$H'_{nm} = \frac{L}{\pi^2}\left[\frac{(-1)^{n+m} - 1}{(n-m)^2} - \frac{(-1)^{n+m} + 1}{(n+m+1)^2}\right] \tag{6.89}$$

$$H'_{nn} = \frac{2L}{\pi^2}\left[\frac{\pi^2}{4} - \frac{1}{(2n+1)^2}\right] \tag{6.90}$$

### Example 6.2 Quadratic

In this case $\tilde{U}(z) = (z - z_0)^2$. For the soft boundary conditions, the matrix elements are as follows:

$$H'_{nm} = \frac{2L}{\pi^2}((L - z_0)(-1)^{n+m} + z_0)\frac{4nm}{(n^2 - m^2)^2} \tag{6.91}$$

$$H'_{nn} = \frac{1}{3}\left((L - z_0)^2 - z_0(L - 2z_0)\right) - \frac{1}{2\pi^2 n^2}\frac{L^2}{} \tag{6.92}$$

The hard boundary conditions lead to the following matrix elements:

$$H'_{nm} = \frac{2L}{\pi^2}\left((L-z_0)(-1)^{n+m}+z_0\right)\frac{2(n^2+m^2)}{(n^2-m^2)^2} \tag{6.93}$$

$$H'_{nn} = \frac{1}{3}\left((L-z_0)^2-z_0(L-2z_0)\right)+\frac{1}{2\pi^2 n^2}\frac{L^2}{} \tag{6.94}$$

Lastly the matrix elements for the mixed conditions are as follows:

$$H'_{nm} = \frac{2L}{\pi^2}\left[\frac{(L-z_0)(-1)^{n+m}+z_0}{(n-m)^2} - \frac{(L-z_0)(-1)^{n+m}-z_0}{(n+m+1)^2}\right] \tag{6.95}$$

$$H'_{nn} = \frac{1}{3}\left((L-z_0)^2-z_0(L-2z_0)\right)+\left[\frac{2L^2}{\pi^2(2n+1)^2}\right] \tag{6.96}$$

## Example 6.3   Exponential

The last case considered here is $\tilde{U}(z) = \exp(\alpha z)$, with the following constant defined, $\sigma = \alpha L/\pi$. For the soft boundary conditions the matrix elements are as follows:

$$H'_{nm} = \frac{\sigma}{\pi}(\exp(\alpha L)(-1)^{n+m}-1)\left[\frac{1}{(n-m)^2+\sigma^2} - \frac{1}{(n+m)^2+\sigma^2}\right] \tag{6.97}$$

$$H'_{nn} = \frac{1}{\pi\sigma}(\exp(\alpha L)-1)\frac{4n^2}{4n^2+\sigma^2} \tag{6.98}$$

The hard boundary conditions lead to the following matrix elements:

$$H'_{nm} = \frac{\sigma}{\pi}(\exp(\alpha L)(-1)^{n+m}-1)\left[\frac{1}{(n-m)^2+\sigma^2} + \frac{1}{(n+m)^2+\sigma^2}\right] \tag{6.99}$$

$$H'_{nn} = \frac{1}{\pi\sigma}(\exp(\alpha L)-1)\frac{4n^2+2\sigma^2}{4n^2+\sigma^2} \tag{6.100}$$

Lastly the matrix elements for the mixed conditions are as follows:

$$H'_{nm} = \frac{\sigma}{\pi}\left[\frac{(\exp(\alpha L)(-1)^{n+m}-1)}{(n-m)^2+\sigma^2} - \frac{(\exp(\alpha L)(-1)^{n+m}+1)}{(n+m+1)^2+\sigma^2}\right] \tag{6.101}$$

$$H'_{nn} = \frac{1}{\pi\sigma}(\exp(\alpha L)-1)\frac{(2n+1)^2}{(2n+1)^2+\sigma^2} - \frac{1}{\pi}\frac{2\sigma}{(2n+1)^2+\sigma^2} \tag{6.102}$$

With these matrix elements and the recursion relation for the corrections, one can build up the modes and spectra for these problems with any of the three boundary conditions. To illustrate this technique, Figure 6.6 shows the results of the perturbation expansion applied to a linear perturbation with soft boundary conditions. The original mode is the dashed curve, and the solid curves show the perturbed modes for order 1 through 4. The parameters used for this example are $L=4$ m, $c_0=345$ m/s, $\varepsilon=0.01$, and $f=220$ Hz.

The ability to achieve good results in a small number of steps resides in a measure of how small the perturbation is for a given problem. There are several things to consider in these expansions. First, even if one can develop decent fidelity for a small order, say, 2 or 3, any order greater than 1 requires infinite series over all zeroth-order modes. This is an impediment as one will need to truncate these series and have a reliable criterion for doing so. Hence, right off the bat one has two orders of approximation: the truncation of the perturbation series and the truncation of the infinite series present in each perturbation order greater than 1. In some rare cases, as with those leading to exact solutions, one may get lucky and be able to identify a closed form for the series. The next item to consider is that of how large a perturbation is allowed to ensure convergence. Since the corrections are required to be small compared with the original values, the first-order term provides a criterion for this, $H'_{nm} \ll |E_n^{(0)} - E_m^{(0)}|$ and $H'_{nn} \ll E_n^{(0)}$. A final point to consider is that the spectrum of a bound quantum mechanical system will go to infinity. Here the eigenvalues represent only the discrete $z$-component of the wavevector, and for values above a cutoff, the modes will not propagate. That being said, the application of a perturbation to these modes may push some previously propagating modes above the cutoff, making them evanescent. Conversely it is possible for a perturbation to bring modes that were previously evanescent into the propagating portion of the spectrum.

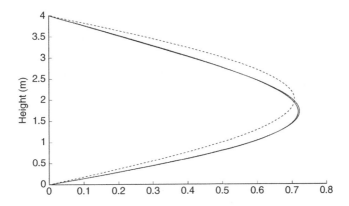

**Figure 6.6** Perturbation applied too linear sound speed and soft boundary conditions

## 6.6 Multidimensional Problems and Degeneracy

The previous presentation focused on the 1-dimensional (1-dim) problem with ideal boundaries. However, the method of perturbation theory applies to the wave equation in general and can be applied to 2-dim and 3-dim problems where the local refractive properties vary in all coordinates. This will be illustrated for modes in rectangular enclosures with ideal boundary conditions. Everything from the last section carries over to this analysis with the extension of the integrals to two and three variables. One issue that is encountered is degeneracy. In the 1-dim case the initial functions had unique eigenvalues. In multidimensional cases, if there is symmetry present in the system, multiple unique solutions of the Helmholtz equation can have the same eigenvalue. Naively applying the prescription of perturbation theory to these degenerate states will result in infinite values for some of the first-order coefficients in (6.71). The key to fixing this problem is to find a new basis for those degenerate states that diagonalizes the portion of the matrix, (6.69), which corresponds to the degenerate states. A prescription for achieving this diagonalization in the first-order solution will be presented. New basis functions are defined for each degenerate subset and merged with the original basis functions, replacing their degenerate counterparts. Once the full set of basis functions is complete, the techniques of the last section can be applied without any issues. As physical motivation for the procedure, consider modes in a cubic enclosure in two or three dimensions with Dirichlet boundary conditions applied to all boundaries. The eigenvalues of each are scaled by $\pi^2/L^2$ to focus on the dependence on mode numbers. Several degenerate states and their scaled eigenvalues are listed in Table 6.2 for the 2-dim rectangle (left) and 3-dim cube (right). Each is sorted by increasing eigenvalue level. It is clear that permutations of the index set will have the same eigenvalue. The 2-dim case will have at most a twofold degeneracy for states with $n \neq m$. The 3-dim case will have threefold degeneracy for states with two of three indices equal and a sixfold degeneracy when all three indices are unique. These are the only degeneracies present for this example.

**Table 6.2**  Mode degeneracy for 2-dim and 3-dim examples

| State degeneracy for 2- and 3-dim rectangular enclosures | | | |
|---|---|---|---|
| 2-dim | | 3-dim | |
| $(m, n)$ | $m^2 + n^2$ | $(l, m, n)$ | $l^2 + m^2 + n^2$ |
| $(1, 1)$ | 2 | $(1, 1, 1)$ | 3 |
| $(1, 2)$ | 5 | $(1, 1, 2)$ | 6 |
| $(2, 1)$ | | $(1, 2, 1)$ | |
| $(2, 2)$ | 8 | $(2, 1, 1)$ | |
| $(1, 3)$ | 10 | $(1, 2, 2)$ | 9 |
| $(3, 1)$ | | $(2, 1, 2)$ | |
| $(2, 3)$ | 13 | $(2, 2, 1)$ | |
| $(3, 2)$ | | $(1, 1, 3)$ | 11 |
| $(1, 4)$ | 17 | $(1, 3, 1)$ | |
| $(4, 1)$ | | $(3, 1, 1)$ | |
| $(3, 3)$ | 18 | $(2, 2, 2)$ | 12 |

To diagonalize the submatrix for each degenerate case, a new basis function is formed from a linear superposition of the degenerate subspace:

$$\bar{\varphi}_n = \sum_{i=1}^{d} a_{nd}\varphi_d^{(0)} \tag{6.103}$$

The summation runs over the indices labeling the degenerate states, symbolically $i = 1, \ldots, d$, where $d$ is the order of the degeneracy. These new basis functions are chosen to diagonalize the submatrix of $H'_{nm}$ corresponding to the original degenerate states:

$$\langle \bar{\varphi}_m | \hat{H}' \bar{\varphi}_n \rangle = \tilde{H}'_{mn} \delta_{mn} \tag{6.104}$$

The value of the matrix element in this new basis is denoted $\tilde{H}'_{mn}$. It turns out that these values will be the first-order corrections predicted by perturbation theory. A condition for this approach to be successful is that all values of $\tilde{H}'_{mn}$ among the degenerate set are distinct. If this does not occur, then a higher-order technique is required to remove the degeneracy. The new states are made to obey the eigenvalue equation of the degenerate submatrix, which leads to an equation for the coefficients, $a_{nd}$:

$$\hat{H}' \sum_{i=1}^{d} a_{nd}\varphi_d^{(0)} = \tilde{E}'_n \sum_{i=1}^{d} a_{nd}\varphi_d^{(0)} \tag{6.105}$$

Taking the inner product of (6.105) with the original degenerate basis functions $\varphi_p^{(0)}$ gives

$$\sum_{i=1}^{d} H'_{pd} a_{nd} = \tilde{E}'_n a_{np} \tag{6.106}$$

It is a straightforward matter to show that for each fixed value of $n$ the vector $a_{np}$ is an eigenvector of the matrix $H'_{pd}$ with eigenvalue $\tilde{E}'_n$. With the eigenvectors of (6.106) solved, the functions given in (6.103) will replace the original degenerate states $\varphi_d^{(0)}$, modulo a renormalization.

The 2-dim case is presented as an example. Referring to Table 6.2, in this case the worst degeneracy will be second order. The states may be arranged, as in Table 6.2, such that the degenerate pairs have consecutive indices. The eigenvalue problem can be solved exactly:

$$\begin{vmatrix} H'_{11} - E & H'_{12} \\ H'_{21} & H'_{22} - E \end{vmatrix} = 0 \tag{6.107}$$

The determinant of (6.107) gives the polynomial whose roots are the eigenvalues:

$$E^2 - (H'_{11} + H'_{22})E - H'_{21}H'_{12} = 0 \tag{6.108}$$

The roots of (6.108) are

$$E_{(1,2)} = \frac{\text{Tr}(H')}{2} \pm \sqrt{\left(\frac{\text{Tr}(H')}{2}\right)^2 + \det(H')} \qquad (6.109)$$

In (6.109) Tr is the trace of the matrix and det the determinant. The coefficients of the expansion are found by inserting each of the two roots in (6.109) into (6.106) and solving for the two $a_{np}$. For the 3-dim case, the eigenvalue problem will need to be solved for $3 \times 3$ and $6 \times 6$ matrices.

## 6.7   Numerical Approach to Modes

The previous sections discussed the normal mode concept in general, illustrating how to build up a solution for the acoustic field of a point source given a set of modes. Some toy models for $c$ $(z)$ were presented that allowed for exact solutions in terms of special functions. This section presents a description of a numerical method for solving (6.18) called the relaxation method. Given a second-order linear ODE with an undetermined free parameter, a solution is sought for both the unknown function, $Z$, and the free parameter, $E$, which satisfies the prescribed boundary conditions. The first task will be discretizing the differential equation. This involves developing finite difference approximations to the derivatives and evaluating (6.18) on a discrete lattice of points. Since the value of $E$ is also a variable to be solved, the differential equation will need to be augmented to include an equation for determining $E$. Finite difference techniques will be described in more detail in a later chapter; here the simplest type of finite difference is applied to solving the one-dimensional wave equation. The end result is that the differential equation in one scalar field is converted to a matrix equation for the sampled values of that scalar field at a set of lattice points. This is not possible in the case of the eigenvalue equation since there is a free parameter. The equations of the relaxation method are derived by forming the Taylor expansion of the finite difference equation about an assumed initial solution, resulting in a matrix equation for the correction to the solution. The method requires an initial guess at a solution. A new guess is made from the initial guess and the correction, and this is used to start the process again. The method refines the solution in a feedback loop until a convergence criterion is met. To help facilitate the application of the method, it will help to apply reduction of order to the second-order equation, defining a first-order system of equations. The derivation presented here follows that in *Numerical Recipes* [5], to which the reader is directed for additional information.

### 6.7.1   Derivation of the Relaxation Equation

The first step is to convert (6.18) into a first-order system by reduction of order. A new variable is introduced, $p \equiv dZ/dz$, and the second-order ODE in one scalar variable becomes a first-order ODE in two scalar variables:

$$\frac{dZ}{dz} = p \qquad (6.110)$$

$$\frac{dp}{dz} = -k_z^2(z)Z \tag{6.111}$$

This pair of equations is turned into one vector equation by defining the variable $\vec{Y} \equiv [Z\, p]^T$:

$$\frac{d\vec{Y}}{dz} = \vec{F}\left(\vec{Y},z\right) \tag{6.112}$$

The right-hand side of (6.112) is defined as follows and sometimes called a force vector borrowing jargon from classical mechanics:

$$\vec{F}\left(\vec{Y},z\right) = \begin{bmatrix} p \\ -k_z^2(z)Z \end{bmatrix} \tag{6.113}$$

The right-hand side depends only on the state. Given the state at any point, the instantaneous rate of change of the state at the same point can be evaluated. The reduction of order technique can be applied to any number of variables and to equations of order higher than 2. For example, the motion of a particle in 3-dim space subjected to external forces that may depend on position, velocity, and time, given by Newton's second law, is converted from a second-order ODE in three variables, the position vector, to a first-order ODE in six variables, position and momentum. The original system is given in (6.114):

$$\frac{d^2\vec{x}}{dt^2} = \vec{f}\left(\vec{x}, \frac{d\vec{x}}{dt}, t\right) \tag{6.114}$$

The replacement $d\vec{x}/dt = \vec{p}$ leads to the enlarged set of equations, $d\vec{x}/dt = \vec{p}$ and $d\vec{p}/dt = \vec{f}\left(\vec{x}, \vec{p}, t\right)$. The introduction of a new variable, $\vec{Y} \equiv \left[\vec{x}^T\ \vec{p}^T\right]^T$, produces (6.112) again:

$$\frac{d\vec{Y}}{dt} = \vec{F}\left(\vec{Y}, t\right) \tag{6.115}$$

Using mechanics of point particles as an example reveals a similarity between this reduction of order and the Hamiltonian description of mechanics in which one describes the evolution of the system in state space, position and momentum, rather than configuration space, position alone. A take-away from this is that any system can be reduced to a larger-dimensional first-order system by this method. Thus, (6.115) represents the most general form of an ODE system. The next issue is that the problem at hand is an eigen value equation and the unknown eigenvalue must be solved for in the process. To proceed, an equation is required for the eigenvalue. This is not difficult. The eigenvalue is added to the unknown variable:

$$\vec{Y} = \begin{bmatrix} Z \\ p \\ E \end{bmatrix} \tag{6.116}$$

To account for this variable in the ODE system, the following equation is added:

$$\frac{dE}{dz} = 0 \tag{6.117}$$

Equation (6.115) is now appropriate for describing this larger set of variables. In addition to the differential equations, boundary conditions are applied to the problem. This introduces another issue in that the boundary conditions will provide two constraints on three variables. What is needed is a third boundary condition. This will be discussed in more detail later.

The system is discretized by defining a grid, in this case uniform, on the $z$-axis. The distance between consecutive values of $z$ is denoted $\Delta$,

$$z_k - z_0 = k\Delta, \quad k = 1, ..., K \tag{6.118}$$

Figure 6.7 illustrates that lattice of points defined in (6.118). Also shown are the sample function points as well as the deviation for later reference. The field defined by (6.116) is evaluated at the lattice points, (6.118), and their values denoted by the same index:

$$\vec{Y}_k = \vec{Y}(z_k) \tag{6.119}$$

With these definitions in place, the first derivative is discretized by a backward difference, for example, a secant line approximation, using the pair of points, $(z_{k-1}, z_k)$:

$$\frac{d\vec{Y}}{dz} \rightarrow \frac{\vec{Y}_k - \vec{Y}_{k-1}}{z_k - z_{k-1}} \tag{6.120}$$

This is where the approach loses uniqueness. One could use a forward difference, defined by the points $(z_k, z_{k+1})$. The simplest differencing scheme is employed here so that more attention can be given to the derivation of the relaxation equation. The derivatives require at least two points to be defined, whereas the function itself is evaluated at a single grid point. The rub is that the derivative expressed in (6.120) will not be valid in general at either end point of the interval. From the mean value theorem, it follows that this value will be achieved for some value of $z$,

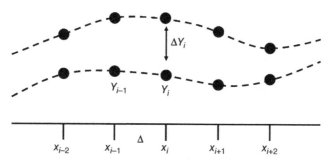

**Figure 6.7**   Definition of relaxation variables and lattice points

$z'_k \in [z_{k-1}, z_k]$. If the grid points are close enough together, the midpoint of the interval is a good enough choice. To complete the discretization of (6.112), the right-hand side is evaluated at the midpoint of each interval:

$$\vec{F}\left(\vec{Y},z\right) \to \vec{F}\left(\frac{\vec{Y}_k+\vec{Y}_{k-1}}{2}, \frac{z_k+z_{k-1}}{2}\right) \tag{6.121}$$

The relaxation method takes an initial guess at a solution and refines it by searching for corrections to the guess in a sequence of steps, which is an iterative process. With a good initial guess and a fine enough mesh, the corrections should become smaller with each iteration, and a solution is reached when the correction is smaller than some threshold. The next step in developing this algorithm is to develop an equation for the correction factors. Denote the initial configuration by $\vec{Y}_k^{(0)}$, and a change in configuration, $\Delta\vec{Y}_k$. Then a new state is developed:

$$\vec{Y}_k = \vec{Y}_k^{(0)} + \Delta\vec{Y}_k \tag{6.122}$$

where $\Delta\vec{Y}_k$ is small. Moving the force term to the left-hand side of (6.112) and defining the result to be $\vec{H}$, the discretized equation is written in the following shorthand notation:

$$\vec{H}_k\left(\vec{Y}_k, \vec{Y}_{k-1}\right) = 0 \tag{6.123}$$

Equation (6.123) holds for $k = 2, \ldots, K$. Inserting (6.122) into (6.123) and expanding to first order gives the following equation for the corrections:

$$\vec{H}_k\left(\vec{Y}_k, \vec{Y}_{k-1}\right) \approx \vec{H}_k\left(\vec{Y}_k^{(0)}, \vec{Y}_{k-1}^{(0)}\right) + \Delta\vec{Y}_k \cdot \vec{\nabla}_{\vec{Y}_k}\vec{H}_k\Big|_{\vec{Y}_k^{(0)}} + \Delta\vec{Y}_{k-1} \cdot \vec{\nabla}_{\vec{Y}_{k-1}}\vec{H}_k\Big|_{\vec{Y}_{k-1}^{(0)}} \tag{6.124}$$

The seemingly odd use of the gradient operator is a shorthand for summing over all partials with respect to all degrees of freedom. This expansion deserves some comments. The reader might be questioning why derivatives with respect to $z$ do not appear. The grid is fixed and a solution to this ODE system is an array of field values at each grid point. The field values are represented by the vector $\vec{Y}(z)$, and the discrete version of this is $\vec{Y}_k = \vec{Y}(z_k)$. In fact, $\vec{Y}_k$ is an array of vectors, one $n$-dim vector for each $k$. The finite difference equation, (6.124), relates the vector of be dependent variables at two and only two consecutive grid points, $k$ and $k + 1$. The variation being performed is of the assumed field values at each fixed grid point, with grid points remaining fixed in the process. Hence, anything with an index $k$ is the value of an object evaluated at the $k$th grid point, and (6.124) really represents $K - 2$ equations, each in $N$ variables. The specific form of (6.113) makes evaluating the partials in (6.124) straightforward. A specific component of (6.124) is denoted by the index $n$:

$$H_{n,k} = Y_{n,k} - Y_{n,k-1} - \Delta_k F_n(Y_{m,k}, Y_{m,k-1}) \tag{6.125}$$

The form of (6.125) reveals that we have $N$ linear equations for each grid point $k$. Furthermore, the set of equations couples $2N$ degrees of freedom represented by the vector of deviations:

$$\delta \vec{Y} = \left[ \Delta \vec{Y}_{k-1}^T \Delta \vec{Y}_k^T \right]^T \tag{6.126}$$

A matrix of partial derivatives is defined as

$$S_{nl,k} = \left[ \frac{\partial H_{n,k}}{\partial Y_{l,k-1}} \quad \frac{\partial H_{n,k}}{\partial Y_{l,k}} \right] \tag{6.127}$$

Each of these matrices is an $N \times 2N$ matrix. With (6.126) and (6.127) the right-hand side of (6.125) is written as

$$H_{n,k} + \left[ \frac{\partial H_{n,k}}{\partial Y_{l,k-1}} \quad \frac{\partial H_{n,k}}{\partial Y_{l,k}} \right] \left[ \begin{array}{c} \Delta Y_{l,k-1} \\ \Delta Y_{l,k} \end{array} \right] \tag{6.128}$$

with summation over the index $l$ implied. It is understood that $H_{n,k}$ and the partials are all evaluated at the initial, or current, state. The initial guess is not going to be a true solution: hence, $H_{n,k} \neq 0$. To get an estimate for the change in state, the first-order expansion, (6.125), is set to zero. While (6.125) holds for interior points, it will not in general hold at the boundary. For $k = 1$ and $k = K$, appropriate boundary conditions are imposed. The number of boundary conditions at $k = 1$ is denoted by $n_1$, and at $k = K$ by $n_K$, with $n_1 + n_K = N$. The upper left and lower right blocks of the operator will be $n_1 \times N$ and $n_K \times N$, respectively. The initial guess may not satisfy the boundary conditions and will need to be refined in the relaxation process. To develop equations for the changes in the solution at the boundary, the boundary conditions are expanded. In the end, the system of equations for the deviations in state at each point on the grid will be an $NK \times NK$ linear matrix equation:

$$S \cdot \delta Y = -H \tag{6.129}$$

The partials take the following form:

$$\frac{\partial H_{n,k}}{\partial Y_{l,k}} = \delta_{n,l} - \Delta_k \frac{\partial F_n}{\partial Y_{l,k}} \tag{6.130}$$

$$\frac{\partial H_{n,k}}{\partial Y_{l,k-1}} = -\delta_{n,l} - \Delta_k \frac{\partial F_n}{\partial Y_{l,k-1}} \tag{6.131}$$

Until now the treatment has been as generic as possible to provide the reader with a path to generalizing the technique to arbitrary systems. At this point things are specialized to particular eigenvalue problem being solved. There are exactly 3 degrees of freedom: $Y_1 = Z$, $Y_2 = p$, and $Y_3 = E$. The vector $\vec{F}$ has components $F_1 = Y_2$, $F_2 = (U - Y_3)Y_1$, and $F_3 = 0$. The following are defined for convenience:

$$A_k = U_k - \frac{1}{2}(E_{k+1} + E_k) \tag{6.132}$$

$$B_k = \frac{1}{2}(Z_{k+1} + Z_k)$$ (6.133)

$$U_k = U\left(\frac{1}{2}(z_{k+1} + z_k)\right)$$ (6.134)

The matrix of partials has the following explicit form:

$$\begin{bmatrix} -1 & -\Delta/2 & 0 & 1 & -\Delta/2 & 0 \\ -A_k\Delta/2 & -1 & B_k\Delta/2 & -A_k\Delta/2 & 1 & B_k\Delta/2 \\ 0 & 0 & -1 & 0 & 0 & 1 \end{bmatrix}$$ (6.135)

The 3-by-6 submatrix in (6.135) couples the fields at neighboring points on the grid. It is customary to stack the variables of the system as follows:

$$\mathbf{Y} = \begin{bmatrix} Y_{1,1} & Y_{2,1} & Y_{3,1} & Y_{1,2} & Y_{1,2} & Y_{1,2} \cdots Y_{1,K-1} & Y_{2,K-1} & Y_{3,K-1} & Y_{1,K} & Y_{1,K} & Y_{1,K} \end{bmatrix}^T$$ (6.136)

For interior points the explicit form of (6.129) can be written as

$$\begin{bmatrix} -1 & -\Delta/2 & 0 & 1 & -\Delta/2 & 0 \\ -A_k\Delta/2 & -1 & B_k\Delta/2 & -A_k\Delta/2 & 1 & B_k\Delta/2 \\ 0 & 0 & -1 & 0 & 0 & 1 \end{bmatrix} \begin{bmatrix} \Delta Y_{1,k-1} \\ \Delta Y_{2,k-1} \\ \Delta Y_{3,k-1} \\ \Delta Y_{1,k} \\ \Delta Y_{2,k} \\ \Delta Y_{3,k} \end{bmatrix} = - \begin{bmatrix} H_{1,k} \\ H_{2,k} \\ H_{3,k} \end{bmatrix}$$ (6.137)

Equation (6.137) is meant to represent one block of the full system.

## 6.7.2 Boundary Conditions in the Relaxation Method

Now that a procedure is in place for discretizing the differential equation on interior points of a grid, a procedure for discretizing the boundary conditions is presented. Three types of ideal boundary conditions are considered here, defined on two infinite planes, at $z = z_0$, and $z = z_0 + L$. The first is two pressure release surfaces, a special case of Dirichlet boundary conditions:

$$Z(z_0) = Z(z_0 + L) = 0$$ (6.138)

The second corresponds to two hard surfaces and is a special case of Neumann boundary conditions:

$$Z'(z_0) = Z'(z_0 + L) = 0$$ (6.139)

**Table 6.3**   Boundary conditions for relaxation

| $(n, k)$ | Dirichlet | Neumann | Cauchy |
|---|---|---|---|
| $(1, 1)$ | $Y_{1,1}=0$ | — | $Y_{1,1}=0$ |
| $(2, 1)$ | — | $Y_{2,1}=0$ | $Y_{2,K}=0$ |
| $(1, K)$ | $Y_{1,K}=0$ | $Y_{1,K}-\alpha=0$ | $Y_{1,K}-\alpha=0$ |
| $(2, K)$ | $Y_{2,K}-\alpha=0$ | $Y_{2,K}=0$ | — |

The third case corresponds to the model of an ocean waveguide mentioned previously and is an example of Cauchy boundary conditions:

$$Z(z_0)=Z'(z_0+L)=0 \tag{6.140}$$

Each of these cases provides two conditions, but there are now three degrees of freedom in the system. To obtain a third boundary condition, note that since the original equation and the boundary conditions are linear and homogeneous, any scalar multiple of a solution will also be a solution. There is freedom to specify the value of the function or its derivative at either boundary. Any of the boundary conditions, (6.138) through (6.140), can be augmented with the following:

$$Z'(z_0+D)-\alpha=0, \quad \text{for Dirichlet} \tag{6.141}$$

$$Z(z_0+D)-\alpha=0, \quad \text{for Neumann or Cauchy} \tag{6.142}$$

On the discrete set of points, using the notation of the previous section, the collection of boundary conditions is listed in Table 6.3.

The factor $\alpha$ is an arbitrary constant. These are included in the existing set of equations generated by discretizing the differential equation. The constraint equations are written in the same notation as the ODE at interior points:

$$H_1=0 \tag{6.143}$$

$$H_{K+1}=0 \tag{6.144}$$

The index $K+1$ is merely symbolic notation to distinguish it from the equation for $H_K$, which couples the variables at the points $z_K$ and $z_{K-1}$. Since there is no point at $k=K+1$ or $k=0$, these extra terms are evaluated only at the end points. The variation, (6.126), is applied to these equations. The $z_1$ boundary conditions are included in the upper left corner of the matrix $S$, and the $z_K$ boundary conditions in the lower right corner. The convention adopted for boundary conditions presented here differ from that of other authors. This convention was chosen such that it (1) preserves the order of the degrees of freedom in all rows and (2) always places one boundary condition at the top of $S$ and two at the bottom, that is, the top row of $S$ is boundary one and the bottom two rows are boundary two. Applying a first-order expansion to each set of boundary conditions, we arrive at the following for the first and last two rows of the system:

$$[1 \quad 0 \quad 0]\begin{bmatrix} \Delta Y_{1,1} \\ \Delta Y_{2,1} \\ \Delta Y_{3,1} \end{bmatrix} = -\begin{bmatrix} H_{1,1} \\ H_{2,1} \\ H_{3,1} \end{bmatrix}, \quad \text{for soft boundary} \tag{6.145}$$

$$[0\ 1\ 0]\begin{bmatrix} \Delta Y_{1,1} \\ \Delta Y_{2,1} \\ \Delta Y_{3,1} \end{bmatrix} = -\begin{bmatrix} H_{1,1} \\ H_{2,1} \\ H_{3,1} \end{bmatrix}, \quad \text{for hard boundary} \tag{6.146}$$

$$\begin{bmatrix} 1 & 0 & 0 \\ 0 & 1 & 0 \end{bmatrix}\begin{bmatrix} \Delta Y_{1,K} \\ \Delta Y_{2,K} \\ \Delta Y_{3,K} \end{bmatrix} = -\begin{bmatrix} H_{1,K+1} \\ H_{2,K+1} \\ H_{3,K+1} \end{bmatrix}, \quad \text{for soft boundary} \tag{6.147}$$

$$\begin{bmatrix} 0 & 1 & 0 \\ 1 & 0 & 0 \end{bmatrix}\begin{bmatrix} \Delta Y_{1,K} \\ \Delta Y_{2,K} \\ \Delta Y_{3,K} \end{bmatrix} = -\begin{bmatrix} H_{1,K+1} \\ H_{2,K+1} \\ H_{3,K+1} \end{bmatrix}, \quad \text{for hard boundary} \tag{6.148}$$

For Dirichlet conditions use (6.145) and (6.147), for Neumann (6.146) and (6.148), and for Cauchy (6.145) and (6.148). This last set of equations along with the block form connecting each pair, (6.137), provides the ingredients needed to build up the matrix $S$.

### 6.7.3   Initializing the Relaxation

Relaxation methods can be very useful, and converge quickly, if a good initial guess is made. Here is where the comments of previous sections and the investment in exact solutions come in handy. The complete set of modes for a propagating wave in a uniform medium, one with a constant refractive index, suitable for several sets of boundary conditions involving ideal surfaces has already been presented in Chapter 4. The presentation of the problem in this chapter would indicate that for refractive media with boundaries, these modes are a good initial guess. This is a particularly good choice when modeling situations where nontrivial effects are small. The free space modes will not have the correct energy levels or the correct behavior. In fact, they are most certainly guaranteed to not satisfy (6.123), but they will have the correct behavior at the boundaries. A subtle point emerges when choosing the initialization. To account for the third variable, an additional boundary condition was included, fixing the value of the function or its derivative at one boundary with a free parameter. An initial value is needed for this quantity. In lieu of any other choice, the value determined by the initial function suffices to get this input.

In some circumstances using the free modes with boundary conditions may not be the best choice for initializing this technique. It was mentioned because the starting point was an enclosure with ideal boundary conditions and a "small" refractive perturbation present. In the discussion on the effective potential, comparisons were made with an overall scale of $k_0^2$ factor removed. This factor is frequency dependent. As the frequency is increased, the effective potential will get stronger. This was overlooked in the examples using perturbation theory, but clearly there will be a threshold at which that technique will not work. The consequence for the relaxation method is that a better first choice may be needed to start the process. For the normal modes, $\sin(k_n z)$ and $\cos(k_n z)$, the lowest mode is always the same, independent of frequency, and its shape determined solely by the size of the enclosure. For weak refraction, a small change can be expected and these modes will likely make a good initializer. For strong refraction, the true shape of the solution may deviate so much from these modes that initializing with them could cause increased run time or, worse, divergence. As an illustrative example, consider a refractive waveguide like (6.12) or (6.13). For these shapes the lowest-order mode will be strongest near the waveguide axis, and at higher frequencies the shape of this mode will become more concentrated on the axis. For this type of expected behavior, the modes of a harmonic oscillator wave function, that is, (6.28), might make a better initial guess. These functions are closer to exhibiting the correct behavior in the waveguide but do not obey the boundary conditions. The relaxation equation contains boundary conditions and should, in theory, produce a final estimate that does as well.

## 6.7.4   Stopping the Relaxation

The relaxation technique is an iterative process in which the output from iteration $n$ is used to initialize the same calculation, producing a refined output. A stop criterion is needed for the process. Assuming that the refinements are converging, the results for $\Delta \vec{Y}_k$ should get smaller with each iteration. An estimate of the error in the solution is used to judge how good the results are. The choice of error estimate is not unique, but a common choice is the mean of the absolute relative errors among the set $\left\{ \vec{Y}_k \right\}$. The error in each variable is $\Delta \vec{Y}_k$, but no actual value of the variable exists for comparison. It is up to the user to decide on a set of "typical values" for the $\vec{Y}_k$ based on some knowledge of the system. Calling these quantities $\left\{ \vec{Y}_k^{\,t} \right\}$, an error estimate can be made:

$$\text{error} = \sum_{i=1}^{\left| \{ \vec{Y}_k \} \right|} \frac{\left| \Delta Y_{m,k} \right|}{\left| Y_{m,k}^t \right|} \tag{6.149}$$

Again, this choice of error estimate is not unique and requires a tunable input vector $Y_{m,k}^t$. The notation in (6.149) is defined as follows: $\left| \vec{Y}_k \right|$ represents the number of elements in $\vec{Y}_k$, while the other terms are absolute values of individual vector elements. After each iteration (6.149) is evaluated and compared to a threshold. If (error < threshold), then the solution has converged to within a prescribed tolerance, or else proceed with another iteration. Either way the new state is constructed as $\vec{Y}_k + \Delta \vec{Y}_k$. When the error is too large to stop, adding the result may cause the result to drift farther from the true solution at the next step. An adjustable parameter, $\beta$, is included to weight the correction for steps that fail to meet the convergence criterion. Denoting the error by $\epsilon$, if $\epsilon > \beta$, then the correction is scaled by $\beta/\epsilon$, or else no weight factor is included. Finally, it is possible for an iterative procedure to spin off in an infinite loop, never satisfying the stopping criterion. For this reason, a maximum number of iterations, $N_{max}$, should be included in the simulation to avoid this.

The method described in this section is applied to the same example that was presented using perturbation theory (see Figure 6.6). As before, the dashed line is the original mode, the initializer in this example, while the solid is the output from the procedure (Figure 6.8).

As a final note regarding the use of these techniques, the initial modes were normalized functions. The results of either perturbation theory or relaxation, or any numerical technique, are not

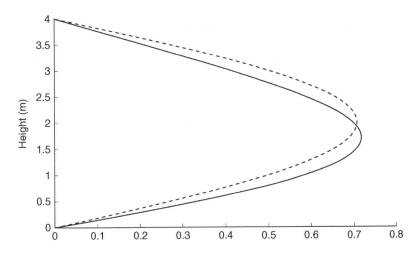

**Figure 6.8**   Relaxation applied to linear sound speed and soft boundary conditions

guaranteed to produce normalized results. Even if the equations have normalization imposed, roundoff error will result in an unnormalized function. The relaxation method presented here does not have normalization included in any constraint. Regardless of the method used to obtain the modes, results should be normalized before the results are used in series expansions of the propagating wave. Here a trapezoid rule was used on the output data and the result used for normalization.

## 6.8   Coupled Modes and the Pekeris Waveguide

This last subsection introduces two extensions of the use of normal modes presented previously. The first involves the introduction of nonideal boundary conditions and is used in underwater acoustics to model a penetrable ocean floor. The construction, attributed to Pekeris [6], is referred to as the Pekeris waveguide. The second is coupled mode theory introduced in underwater acoustic by Pierce [7]. A similar idea is the foundation of normal mode software called KRAKEN (Porter [8]). These two topics illustrate how the method can be extended beyond depth-dependent refraction in the presence of ideal boundary conditions.

### 6.8.1   Pekeris Waveguide

The waveguide has similar characteristics to the 1-dim boundary conditions introduced in Chapter 4, two parallel infinite planes with a fluid medium between them, with one of the surfaces being taken to be a pressure release surface. The extension involves two generalizations to the normal mode approach presented previously. The first is the application of appropriate boundary conditions at one of the planes to account for the presence of a penetrable medium on the other side. This follows the examples introduced in Chapter 5, boundary conditions for a fluid–fluid and fluid–solid interface. The introduction of this change means that waves will not experience an ideal hard or soft boundary condition at this interface, allowing waves to leak into

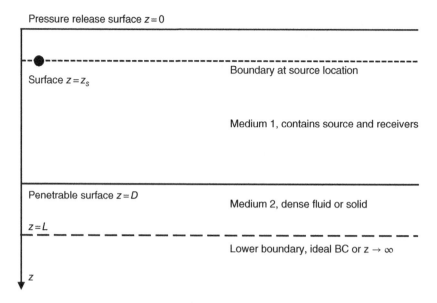

Pressure release surface $z = 0$

Surface $z = z_s$

Boundary at source location

Medium 1, contains source and receivers

Penetrable surface $z = D$

Medium 2, dense fluid or solid

$z = L$

Lower boundary, ideal BC or $z \rightarrow \infty$

$z$

**Figure 6.9**   Setup for the Pekeris waveguide

the second medium. The second medium can be considered to be an infinite half space. Some-times a pressure release or ideal hard boundary is placed at a second location some distance beyond the interface between the two media and ideal boundary conditions applied. At first glance this is a straightforward problem involving a fusion of two techniques introduced in earlier chapters. The second generalization is the modeling of a point source, that is, applying the boundary conditions to the vertical modes with source term included. This requires the introduction of additional conditions. Due to the nature of the inhomogeneous Helmholtz equa-tion for a point source, the contribution from the source will have a discontinuity in its deriva-tive at the location of the source. To account for this the wave in the medium containing the source is divided into two layers, divided by the plane containing the source. The standard setup is illustrated in Figure 6.9.

Following Pekeris' original work, it is assumed that the wave equation is separated in cylin-drical coordinates with a Bessel function accounting for the radial part of the wave function. It is further assumed that the properties of each medium are constant. Following the standard approach, plane wave solutions are assumed in both media, with the exception that to model the point source an additional linear combination of modes is added to medium 1:

$$p_1 = a_1 \sin(k_1 z) \tag{6.150}$$

$$p_2 = a_2 \sin(k_1 z) + a_3 \cos(k_1 z) \tag{6.151}$$

$$p_3 = a_4 \exp(i k_2 z) \tag{6.152}$$

The three pressure fields represent the following: $p_1$ is a solution for $0 \leq z \leq z_s$ to the homogeneous equation satisfying the boundary condition at $z = 0$, $p_2$ is a superposition of vertical modes designed to model the presence of the source, and $p_3$ is the pressure in the lower half space. The pressure in the lower medium must be chosen in such a way to ensure that the solution has the correct behavior depending on whether there is another boundary or if $z \to \infty$ is allowed, (6.152) corresponds to an unbounded region. The wavenumbers are

$$k_i = \sqrt{\left(\frac{\omega}{c_i}\right)^2 - k^2} \qquad (6.153)$$

The radial wavenumber is $k$. The vertical wavenumber can become imaginary, leading to evanescent modes. The total field in medium 1 is $p_1 + p_2$. Medium 2 is a dense fluid with boundary conditions at $z = D$, $p_2(D) = p_3(D)$ and $dp_2/dz|_D = dp_3/dz|_D$. At the source location, the fields must be continuous, $p_1(z_s) = p_2(z_s)$. Due to the nature of the point source, that is, being represented as a Dirac delta function, the boundary condition on the derivative at the source depth requires integrating the inhomogeneous equation over a small interval across the source position. This leads to a discontinuity in the first derivative $dp_1/dz|_{z_s} - dp_2/dz|_{z_s} = 2k$ [6]. These four boundary conditions specify the four unknown coefficients, $a_i$ $i = 1, 2, 3, 4$. The reader is referred to Ref. [6] for the solutions in this case. The point is to illustrate how concepts from more than one type of problem can be merged to build more sophisticated models. The same approach can be applied to a penetrable solid interface, and with an ideal boundary, for example, hard boundary conditions, at $z = L > D$. This type of waveguide can be extended to include a refractive sound speed profile in both media by use of known exact solutions or including the additional boundary conditions in a numerical approach.

## 6.8.2   Coupled Modes

The treatment of mode theory thus far has looked at ideal flat boundary conditions and environmental profiles that depend on depth only. The success of the approach rests in the fact that the wave equation is completely separable leaving an ODE for the modes. These one-dimensional models and the general approach to dealing with them can go a long way in modeling physical situations. It often happens that more complex refraction and geometrically irregular boundaries break this symmetry, creating a situation where separability is not possible. The general situation of an underwater environment with irregular boundary is depicted in Figure 6.10.

A classic approach to this problem in underwater acoustics is to replace the continuum by a discrete set of layers in the range direction. Then the sound speed and other environmental parameters are treated in each vertical layer as if they were only depth dependent. The problem is essentially solved in each layer by the methods described in previous sections. With a complete set of eigenfunctions in each layer, boundary conditions are imposed at the interface between the layers. These are continuity of pressure and particle velocity. The derivation presented here will follow very closely the original by Pierce as a starting point [7]. The idea is to assume a set of vertical normal modes satisfy the Helmholtz equation at each point, $\vec{x}_T$, in the horizontal plane, that is, they are solutions to equation $x$ with $c(z) = c(\vec{x}_T; z)$. The notation $\vec{x}_T$ refers to transverse coordinates, orthogonal to the $z$-axis. As a reminder that these modes are

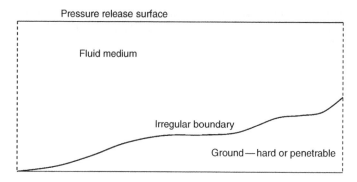

**Figure 6.10**  Waveguide with irregular boundary

distinct for each choice of $\vec{x}_T$, they are labeled $Z_n(\vec{x}_T;z)$. It is assumed that the modes at each point are normalized appropriately. A solution of the following form is assumed:

$$\psi(\vec{x}) = \sum_n f_n(\vec{x}_T)Z_n(\vec{x}_T;z) \tag{6.154}$$

Equation (6.154) states that the solution is a linear superposition of eigenmodes at each point. The coefficients of the expansion, $f_n(\vec{x}_T)$, are functions of the transverse position. Direct substitution of (6.154) into the Helmholtz equation and use of the eigenvalue equation yields the following reduced equation for the coefficients:

$$\sum_n \left\{ Z_n(\nabla_T^2 f_n + k_n^2 f_n) + 2\vec{\nabla}_T f_n \cdot \vec{\nabla}_T Z_n + f_n \nabla_T^2 Z_n \right\} = 0 \tag{6.155}$$

Since the eigenmodes are an orthonormal set at each horizontal location, (6.155) is multiplied by a different mode and integrated in the $z$ coordinates:

$$\nabla_T^2 f_n + k_n^2 f_n = -\sum_m \left\{ A_{mn} f_m + \vec{B}_{mn} \cdot \vec{\nabla}_T f_n \right\} \tag{6.156}$$

$$A_{mn} = \int Z_n \nabla_T^2 Z_m dz \tag{6.157}$$

$$\vec{B}_{mn} = 2 \int Z_n \vec{\nabla}_T Z_m dz \tag{6.158}$$

The coefficients obey the symmetry relations $A_{mn} = A_{nm}$ and $\vec{B}_{mn} = -\vec{B}_{nm}$, which can be derived from the orthogonality condition. The diagonal terms in $\vec{B}_{mn}$ vanish. Peirce's original treatment of the problem was meant to address slowly varying range-dependent changes. To simplify this problem the assumption of uncoupled modes was applied to (6.155). This amounts to setting $n = m$:

$$\nabla_T^2 f_n + (k_n^2 + A_{nn})f_n = 0 \tag{6.159}$$

In this approximation, the coefficients obey a two-dimensional wave equation, similar in form to the eigenvalue equation for $Z_n$ but with $A_{nn}$, taking place of the effective potential. In general, the coefficients (6.157) and (6.158) will depend on frequency and transverse coordinates. Since its introduction coupled modes have been elaborated on and applied to many problems in acoustics, the reader is directed to Refs. [8–13]. In theory, this procedure offers a solution to a problem that one would not normally consider solvable with normal mode theory. However, there are technical issues to overcome. There can be a very large, even infinite number of modes coupled together in (6.155). The integrals need to be worked out to produce the coefficients, and actual functions will not likely be available for these integrals. Equation (6.155) is a continuum equation. Implementations of the coupled mode paradigm divide the medium into vertical slices. By replacing the medium and boundary with thin vertical slices in which the boundaries are flat and the refraction locally dependent on only the $z$ coordinate, the standard methods presented here for evaluating modes in each column are valid and can be applied to the more general case. The rest of the effort required to calculating the acoustic field involves matching boundary conditions for the solution in each column at the boundary to neighboring columns. For modeling a point source in a cylindrically symmetric medium, the outgoing solution is chosen for the radial part of the wave function. In the slice approach to modeling propagation in range-dependent environments, a superposition of propagating modes in both directions is required in each column. Once again, the boundary conditions for the interface between two media are applied to the interface between each pair of columns, with appropriate boundary conditions in range for the ends of the environment. This procedure is implemented in the software package KRAKEN [8]. Figure 6.11 shows the setup for modeling range-dependent environments using vertical slices. So far density variations have been ignored, focusing attention on the sound speed profile. Much of the formalism developed in

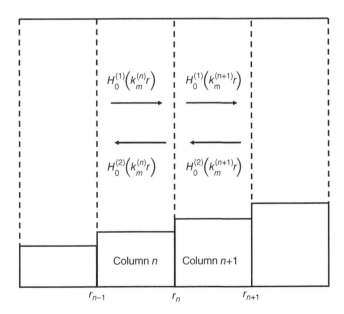

**Figure 6.11**   Setup for coupled mode

earlier chapters applied when the material density also varies with position. The reader is referred to Porter [8] and the references therein for more details on accounts of the derivation and discussion of the modes and dispersion. The Helmholtz equation with a point source in cylindrical coordinates is

$$\rho \vec{\nabla} \cdot \left(\frac{1}{\rho}\vec{\nabla}p\right) + \frac{\omega^2}{c^2}p = -\frac{S}{2\pi r}\delta(z-z_s)\delta(r) \tag{6.160}$$

The variable $r$ is the horizontal radius in cylindrical coordinates, and $S$ is the source strength. A solution of the following form can be found by separation of variables:

$$p = \frac{i}{4\rho(z_s)}\sum_{n=1}^{\infty}Z_n(z_s)Z_n(z)H_0^{(1,2)}(kr) \tag{6.161}$$

The choice of Hankel function for the outgoing wave depends on the convention chosen, and typically $H_0^{(1)}(r)$ is chosen as the outgoing solution. The modes $Z_n(z)$ obey an altered normalization condition with a weighting function:

$$\int\frac{1}{\rho}Z_nZ_mdz \tag{6.162}$$

For the range-dependent problem, the solution in each column is expressed as

$$p^{(n)} = \sum_{m=0}^{M}\left(a_m^{(n)}H_0^{(1)}\left(k_m^{(n)}r\right) + b_m^{(n)}H_0^{(2)}\left(k_m^{(n)}r\right)\right)Z_m(z) \tag{6.163}$$

All quantities are indexed by the column index, $n$, for example, $k_n$ is the wavenumber in column $n$. Applying boundary conditions to interface $r_n$ and projecting into an arbitrary vertical mode leads to a set of equations relating the coefficients for the $n$th and $(n+1)$th column:

$$a_l^{(n+1)} + b_l^{(n+1)} = \sum_{m=0}^{M}\left(a_m^{(n)}H_0^{(1)}\left(k_m^{(n)}r_n\right) + b_m^{(n)}H_0^{(2)}\left(k_m^{(n)}r_n\right)\right)c_{lm} \tag{6.164}$$

$$a_l^{(n+1)} - b_l^{(n+1)} = \sum_{m=0}^{M}\left(a_m^{(n)}H_0^{(1)}\left(k_m^{(n)}r_n\right) + b_m^{(n)}H_0^{(2)}\left(k_m^{(n)}r_n\right)\right)d_{lm} \tag{6.165}$$

where the constants $c_{lm}$ and $d_{lm}$ are

$$c_{lm} = \int\frac{1}{\rho^{(n+1)}}Z_l^{(n+1)}Z_m^{(n)}dz \tag{6.166}$$

$$d_{lm} = \frac{k_m^{(n)}}{k_l^{(n+1)}}\int\frac{1}{\rho^{(n)}}Z_l^{(n+1)}Z_m^{(n)}dz \tag{6.167}$$

With these relationships and boundaries at $r = 0$ and $r \to \infty$, a block matrix relating the coefficients in each column to the source is built. Solving this provides a solution to the problem.

## 6.9   Exercises

1. Combine the perturbation results for linear and exponential perturbations to develop the matrix $H'_{mn}$ for the Garrettt–Munk profile, (6.13).
   (a) Determine a frequency for which the perturbation parameter is greater than 1, or prove no such value exists.
   (b) Determine the first-order corrections for a low frequency.
2. Referring to the Helmholtz equation in cylindrical coordinates and its solution based on separation of variables, determine the discrete modes for a wedge boundary condition, that is, infinite planes at $\varphi = 0$ and $\varphi = \varphi_0 < \pi/2$.
3. Either from scratch or by searching the literature, develop the modes for a Pekeris waveguide with a solid medium for medium 2. You can assume a homogeneous isotropic solid with 2 Lamé parameters.
4. Develop a method to solve for modes in the Garrettt–Munk profile satisfying ideal boundary conditions, pressure release at one value of $z$ and hard boundary conditions at another value. Can the approximate results from Exercise 1 be used to fit the boundary conditions?

## References

[1] Arfkin, G. B. and Weber, H. J., Mathematical Method for Physicists, Fifth Edition, Harcourt, Academic Press, San Diego, 2001.

[2] Morse, H. and Feshbach, P. M., Methods of Theoretical Physics, Part II, International Series in Pure and Applied Physics, McGraw-Hill, New York, 1953.

[3] Gradshteyn, I. S. and Ryzhik, I. M., Table of Integrals, Series, and Products, Fifth Edition, Academic Press, Inc., San Diego, 1994.

[4] Liboff, R. L., Introductory Quantum Mechanics, Addison-Wesley, Reading, 1980.

[5] Press, W. H., Teukolsky, S. A., Vetterling, W. T., and Flannery, B. P., Numerical Recipes in C, Second Edition, Cambridge University Press, Cambridge, 1999.

[6] Pekeris, C. L., Theory of Propagation of Explosive Sound in Shallow Water, The Geological Society of America, Washington, DC, Memoir 27, October 15, 1948, Reprinted 1963.

[7] Pierce, A. D., Extension of the Method of Normal Modes to Source Propagation in an Almost-Stratified Medium, J. Acoust. Soc. Am., Vol. 37, No. 1, pp. 19–27, 1965.

[8] Porter, M. B., The KRAKEN Normal Mode Program, Naval Research Laboratory Report, NRL/MR/5120-92-6920, Available for public release, distribution unlimited, Naval Research Laboratory, Washington, DC, 1992.

[9] Boyles, C. A., Coupled mode solution for a cylindrically symmetric oceanic waveguide with a range and depth dependent refractive index and a time varying rough sea surface, J. Acoust. Soc. Am., Vol. 73, No. 3, pp. 800–805, 1983.

[10] Boyles, C. A., Dozier, L. B., and Joice, G. W., Application of the coupled mode theory to acoustic scattering from a rough sea surface overlying a surface duct, Johns Hopkins APL Tech. Dig., Vol. 6, No. 3, pp. 216–226, 1985.

[11] Evans, R. B., A coupled mode solutions for acoustic propagation in a waveguide with stepwise depth variation of a penetrable bottom, J. Acoust. Soc. Am., Vol. 74, No. 1, pp. 188–195, 1983.

[12] Evans, R. B. and Gilbert, K. E., Acoustic propagation in a refracting ocean waveguide with an irregular interface, Comput. Math. Appl., Vol. 11, No. 7/8, pp. 795–805, 1985.

[13] Porter, M. B. and Reiss, E. L., A numerical method for bottom interacting ocean acoustic normal modes, J. Acoust. Soc. Am., Vol. 77, No. 5, pp. 1760–1767, 1985.

# 7

# Ray Theory

## 7.1  Introduction

In Chapter 3 the reader encountered a discussion of waves divorced from any underlying theory connecting the phenomenon to the medium. It was suggested that under the right conditions one could arrive at a theory of propagating disturbances from a purely phenomenological point of view. The same could be said of ray theory. When trying to describe the behavior of a traveling wavefront, it is natural to divide it into segments and form a description of each segment in terms of particle motion. The difference is that in this case there is a continuum of these particles and the amalgamation of all their motion is the wavefront. In reality the situation is much more complicated, and this fails to be a useful description for diffraction and other wave phenomena. Rays are intimately connected to the bicharacteristics encountered in Chapter 4. The main difference is that bicharacteristics provide a general description of the evolution of a surface of initial data describing the PDE initial value problem (i.v.p.), while ray theory attempts to build up a description of the wave field using an expansion of the wave equation in the high frequency limit. This can give rise to some confusion. The usefulness of a ray-based calculation of a field may be limited, but the geometric structure of the rays is not. In this chapter a derivation of the acoustic ray equation for moving fluids is derived from the original wave equation. A good deal of effort will be spent providing a geometric description of the rays as paths in space and space-time based on the paradigm introduced in Chapters 3 and 4. Techniques for solving the ray equation to determine the precise nature of the paths for a variety of cases will be discussed, including analytic and numerical methods. After a complete description of rays, the equations for the amplitude factor will be introduced and their solution discussed. It is at this point that the restrictions of ray theory will become apparent. The amplitude predicted by ray theory can diverge when a finite result is expected, requiring separate methods to correct the infinity and provide more realistic results. This is one aspect that makes ray theory limited in its usefulness.

---

*Computational Acoustics: Theory and Implementation*, First Edition. David R. Bergman.
© 2018 John Wiley & Sons Ltd. Published 2018 by John Wiley & Sons Ltd.

In some sense, what starts out as a simple and elegant paradigm for describing propagation rapidly turns into a patchwork of correction factors. For direct propagation calculations from a source to receiver, rays are about the easiest way to predict the arrival times and leading-edge amplitudes of short pulses. Despite its limitations, ray theory is fairly easy to understand and implement and is a very popular paradigm for high frequency modeling.

## 7.2   High Frequency Expansion of the Wave Equation

Classic treatments of ray theory are usually restricted to the Helmholtz equation with a position-dependent refractive index. Since so much effort was invested in developing a wave equation for more generic circumstances, the derivation will go as far as possible without neglecting any terms. The only simplifying assumption made here is that the wave equation takes the following approximate form:

$$D^2\psi \approx 0 \tag{7.1}$$

The acoustic field is assumed to have the following form:

$$\psi = A\exp i\varphi \tag{7.2}$$

The amplitude, $A$, and phase, $\varphi$, are functions of position and time as well as frequency. This ansatz is inserted into (7.1), and the real and imaginary parts separated with an overall phase are factored out:

$$e^{i\varphi}\left\{\left(D^2A - Ag^{\mu\nu}\partial_\mu\varphi\partial_\nu\varphi\right) + i\left(2g^{\mu\nu}\partial_\mu A\partial_\nu\varphi + AD^2\varphi\right)\right\} = 0 \tag{7.3}$$

The phase depends linearly on frequency, and the amplitude is expanded in an infinite series in inverse powers of frequency:

$$\varphi = \omega\tau \tag{7.4}$$

$$A = \sum_{n=0}^{\infty}\left(\frac{i}{\omega}\right)^n A_n \tag{7.5}$$

Inserting (7.5) and (7.4) into (7.3) leads to an infinite series in the parameter $\omega^{-1}$:

$$\sum_{n=0}^{\infty}\left(\frac{i}{\omega}\right)^n\left\{\left(D^2A_n - \omega^2 A_n g^{\mu\nu}\partial_\mu\tau\partial_\nu\tau\right) + i\omega\left(2g^{\mu\nu}\partial_\mu A_n\partial_\nu\tau + A_n D^2\tau\right)\right\} = 0 \tag{7.6}$$

The coefficients of each term $\omega^{-n}$ must independently vanish, leading to a hierarchy of equations for the fields $\{A_n\}$. The lowest-order equation, that is, the coefficient of $\omega^2$, in the set is called the eikonal equation:

$$g^{\mu\nu}\partial_\mu\tau\partial_\nu\tau = 0 \tag{7.7}$$

The reader will recognize this as the characteristic equation. The next term in the series arises from the coefficient of $\omega^1$ and is referred to as the transport equation:

$$2g^{\mu\nu}\partial_\mu A_0 \partial_\nu \tau + A_0 D^2 \tau = \frac{1}{\sqrt{-g}}\partial_\mu\left(\sqrt{-g}g^{\mu\nu}A_0^2\partial_\nu\tau\right)/A_0 = 0 \tag{7.8}$$

For all other powers of $\omega^{-1}$, we get a series of equations relating the amplitudes of order $n$ and $n+1$:

$$D^2 A_n - \frac{1}{\sqrt{-g}}\partial_\mu\left(\sqrt{-g}g^{\mu\nu}A_{n+1}^2\partial_\nu\tau\right)/A_{n+1} = 0 \tag{7.9}$$

Equation (7.8) states that the covariant divergence of some vector field in space-time is zero:

$$\frac{1}{\sqrt{-g}}\partial_\mu\left(\sqrt{-g}g^{\mu\nu}A_0^2\partial_\nu\tau\right) = D_\mu\left(A_0^2 g^{\mu\nu}\partial_\nu\tau\right) = 0 \tag{7.10}$$

## 7.2.1  Eikonal Equation and Ray Paths

The eikonal is identical in form to the bicharacteristics, so the equations of Chapter 4 can be carried over and used here to develop ray tracing. Ray paths are defined as curves normal to the eikonal relative to the effective metric tensor:

$$g^{\mu\nu}\partial_\nu \tau = \frac{dx^\mu}{d\lambda} \tag{7.11}$$

The rays are null curves relative to the metric and the ray equation is the geodesic equation:

$$\frac{d^2 x^\mu}{d\lambda^2} + \Gamma^\mu_{\alpha\beta}\frac{dx^\alpha}{d\lambda}\frac{dx^\beta}{d\lambda} = 0 \tag{7.12}$$

The standard way of expressing the ray equation is as either a time or arclength parameterized curve in $\mathbf{R}^3$, $\vec{x}(t)$ or $\vec{x}(s)$. Any of the standard forms found in the literature can be derived from (7.12) and the explicit form of the Christoffel symbols. Using the geodesic equation may seem less intuitive than an equation for $\vec{x}(s)$, but there are advantages to understanding the theoretical consequences of (7.12). In particular, the ray path geometry is completely immune to multiplying the metric by any positive definite function of coordinates $f(x^\mu)$ along with an appropriate change in ray parameter $\lambda$. This allows terms to be gauged away, creating forms of the ray equation with less complexity (or potentially more depending on the choice) [1]. This also means that the ray geometry is completely immune to density variation since that factor appears equally in all entries of the metric. Another consequence of the geometry paradigm is that there is a simple method for generating conserved quantities along ray paths using isometry of the effective metric tensor. Suffice to say, the computational tools of differential geometry encompass all the same results as standard approaches. They also provide generalizations of ray theory for time-dependent environments. Experience with the techniques proves their value over time. From an implementation perspective, the geodesic equation is actually fairly simple. One can treat all the coefficients $\Gamma^\mu{}_{\alpha\beta}$ on equal footing. The procedures for developing the terms of the equation are all identical, and the complexity comes in writing a routine to calculate $\Gamma^\mu{}_{\alpha\beta}$. To this end, it is mentioned that MAPLE has a package called tensor, which is capable of generating all the fields of differential geometry from a given metric. What is lost is some of the three-dimensional geometric intuition gleaned from the traditional presentation of rays. If there is one take-away from this discussion, it might be that different conceptual paradigms are more appropriate for different types of analysis.

## 7.2.2 Paraxial Rays

This section introduces another advantage to the novel paradigm used to express the ray paths. Paraxial ray tracing techniques are used to trace neighboring rays in close proximity to a particular ray. There is an advantage to doing this in that the paraxial equation is linear in the field variables and is transported along the ray path. The paraxial technique is widely used in optics and acoustics to describe the deformation of a ray bundle as it propagates through the medium. The paraxial vector can be used to estimate the area of a cross section of a ray bundle centered on a ray, thus providing a direct calculation of geometric spread. The reader is directed to references for alternate treatments of the paraxial and dynamic ray tracing procedures for specialized cases [2, 3]. The medium is assumed to depend on spatial coordinates and time, and the ray paths are treated as curves in the four-coordinate space-time system.

A small deviation is added to the ray coordinate, and the geodesic equation expanded about the ray path to derive an ordinary differential equation (ODE) for the deviation [4–7]:

$$x^\mu = x^\mu_\lambda + Y^\mu \tag{7.13}$$

In (7.13), $x^\mu_\lambda$ represents the coordinates of a specific geodesic path, indicated by the subscript $\lambda$. The subscript is not a tensor index in this case. The coordinates $x^\mu$ are constrained to lie on another geodesic in close proximity to $x^\mu_\lambda$. Equation (7.13) represents neighboring geodesic points relative to those on the initial geodesic, $x^\mu_\lambda$. The deviation $Y^\mu$ is not a coordinate in the same sense as $x^\mu_\lambda$. Rather it is a local vector based at the point $x^\mu_\lambda$ and pointing away from $x^\mu_\lambda$ to nearby points. To distinguish the geodesic $x^\mu_\lambda$ from its neighbors, it will be referred to as the base geodesic, since it acts as a base point for this expansion. Because the neighboring points lie on another geodesic, they will also obey the null constraint. The deviation vector is likewise constrained to be orthogonal to the base geodesic at every point along it:

$$\frac{dx^\mu}{d\lambda} Y^\nu g_{\mu\nu} = 0 \tag{7.14}$$

With this constraint in place, the deviation vector tracks neighboring paths for equal value of affine parameter, $\lambda$. The constraints placed on the deviation vector reduce the number of degrees of freedom to two. Even though there are four coordinates in this approach, recall that only a subspace, the space made of null curves, is being used in this process. One is not looking for any nearby point but only those nearby points that lie in the three-dimensional subspace.

Due to the reduced number of degrees of freedom, it is more convenient to describe the deviation in a local set of coordinates on the ray path, $x_\lambda^\mu$. This can be thought of as a four-dimensional analog of the Frenet frame used in the theory of curves embedded in $\mathbf{R}^3$. To make this frame centered on the base geodesic its velocity vector, $dx^\mu/d\lambda$, is chosen as one of the local coordinate directions. The second basis vector will be another null vector, $L^\mu$, $L^\mu L^\nu g_{\mu\nu} = 0$, that obeys the following constraint:

$$\frac{dx^\mu}{d\lambda} L^\nu g_{\mu\nu} = -1 \tag{7.15}$$

The final elements of the basis are two space-like vectors, $\hat{e}_1^\mu$ and $\hat{e}_2^\mu$, satisfying the following constraints:

$$\frac{dx^\mu}{d\lambda} \hat{e}_I^\mu g_{\mu\nu} = 0, \quad I = 1,2 \tag{7.16}$$

$$L^\mu \hat{e}_I^\mu g_{\mu\nu} = 0, \quad I = 1, \tag{7.17}$$

$$\hat{e}_I^\mu \hat{e}_J^\nu g_{\mu\nu} = \delta_{IJ}, \quad I,J = 1,2 \tag{7.18}$$

An example of this construction is illustrated in Figure 7.1. The example shows the base geodesic and all the vectors required for the paraxial procedure in both Cartesian coordinates and relative to the local basis vectors. Due to obvious spatial constraints, this example contains two spatial dimensions and time. There are enough constraints in place to specify the vector $L^\nu$, making it superfluous.

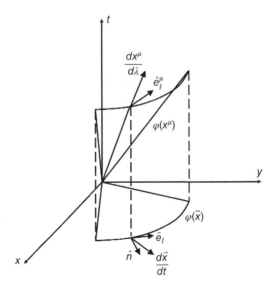

**Figure 7.1**   Illustration showing quantities defined in the paraxial ray trace

A basis needs to be defined at every point along the base geodesic. This is accomplished not by building the frame from scratch at each new point but by propagating the frame from some initial point to every other point by the process of parallel transport. It turns out that only the space-like vectors are required to describe the neighboring rays. The space-like basis vectors obey the following differential equation:

$$\frac{d\hat{e}_I^\mu}{d\lambda} + \Gamma^\mu{}_{\alpha\beta} \frac{dx^\alpha}{d\lambda} \hat{e}_I^\beta = 0 \tag{7.19}$$

The components of the deviation vector can be projected into the internal frame:

$$Y_I = Y^\mu \hat{e}_I^\nu g_{\mu\nu}, \quad I = 1,2 \tag{7.20}$$

This equation can be inverted to give an expression for $Y^\nu$ in terms of the two internal components:

$$Y^\nu = Y_I \hat{e}_I^\nu \tag{7.21}$$

The sum over the index $I$ is implied.

Before continuing to the derivation of the deviation equation, some comments are in order. The internal basis and their constraints may seem sort of abstract at first glance. The geodesic tangent vector is seen to contain the three-dimensional ray path:

$$\frac{dx^\mu}{d\lambda} = \begin{bmatrix} dt/d\lambda \\ d\vec{x}/d\lambda \end{bmatrix} = \frac{dt}{d\lambda} \begin{bmatrix} 1 \\ d\vec{x}/dt \end{bmatrix} = \frac{dt}{d\lambda} \begin{bmatrix} 1 \\ c\hat{n} + \vec{v}_0 \end{bmatrix} \tag{7.22}$$

The space-like basis vectors can be expressed in terms of the time component and a three-dimensional spatial vector:

$$\hat{e}_I^\nu = \begin{bmatrix} e_I^0 \\ \hat{e}_I \end{bmatrix} \tag{7.23}$$

The space-like basis also contains familiar information. It is a straightforward task to show using the two constraints, $\hat{e}_I^\mu \hat{e}_I^\nu g_{\mu\nu} = 1$ and $(dx^\mu/d\lambda)\hat{e}_I^\nu g_{\mu\nu} = 0$, that the following constraints are obeyed by each internal space-like basis vector:

$$\hat{n} \cdot \left( \hat{e}_I - e_I^0 \frac{d\vec{x}}{dt} \right) = 0 \tag{7.24}$$

$$\left| \hat{e}_I - e_I^0 \frac{d\vec{x}}{dt} \right| = 1 \tag{7.25}$$

This motivates the definition of a new vector, referred to here as an auxiliary basis:

$$\tilde{e}_I = \hat{e}_I - e_I^0 \frac{d\vec{x}}{dt} \tag{7.26}$$

Equation (7.24) states that $\tilde{e}_I$ is orthogonal to the wavefront normal in three-dimensional space with respect to the ordinary dot product, and (7.25) states that $\tilde{e}_I$ is a unit vector. The two new basis vectors, $\tilde{e}_I$, $I = 1, 2$, are tangent to the wavefront in space. Furthermore, the parallel transport of all these vectors preserves this relationship, and the set of three unit vectors, $\{\hat{n}, \tilde{e}_1, \tilde{e}_2\}$, provide a three-dimensional frame of the wavefront in space. It was stated that $\hat{e}_I^\mu$ was "space-like." It is possible for this vector to develop a time component value as it is propagated, but it will always maintain a positive definite magnitude. The new vector, $\tilde{e}_I$, however, is purely spatial and remains so as it evolves.

An equation for $Y^\nu$ is derived by applying this variation to the original ray equation, (7.12). In the process of applying the deviation to the original equation, the Christoffel symbols must be Taylor expanded:

$$\Gamma^\mu{}_{\alpha\beta}(x_\lambda^\mu + Y^\mu) = \Gamma^\mu{}_{\alpha\beta}(x_\lambda^\mu) + \left.\frac{\partial \Gamma^\mu{}_{\alpha\beta}}{\partial x^\nu}\right|_{x_\lambda^\mu} Y^\nu + \cdots \qquad (7.27)$$

Direct substitution of (7.13) in (7.12) gives the geodesic equation for the neighboring ray:

$$\frac{d^2\left(x_\lambda^\mu + Y^\mu\right)}{d\lambda^2} + \Gamma^\mu{}_{\alpha\beta}\left(x_\lambda^\mu + Y^\mu\right)\frac{d\left(x_\lambda^\alpha + Y^\alpha\right)}{d\lambda}\frac{d\left(x_\lambda^\beta + Y^\beta\right)}{d\lambda} = 0 \qquad (7.28)$$

Applying (7.27) to first order and ignoring nonlinear terms in $Y^\mu$ simplifies (7.28) to

$$\frac{d^2 x_\lambda^\mu}{d\lambda^2} + \Gamma^\mu{}_{\alpha\beta}\frac{dx_\lambda^\alpha}{d\lambda}\frac{dx_\lambda^\beta}{d\lambda} + \frac{d^2 Y^\mu}{d\lambda^2} + \frac{\partial \Gamma^\mu{}_{\alpha\beta}}{\partial x^\nu}Y^\nu\frac{dx_\lambda^\alpha}{d\lambda}\frac{dx_\lambda^\beta}{d\lambda} + 2\Gamma^\mu{}_{\alpha\beta}\frac{dx_\lambda^\alpha}{d\lambda}\frac{dY^\beta}{d\lambda} = 0 \qquad (7.29)$$

It is understood that the Christoffel symbols and their derivatives are evaluated at $x_\lambda^\mu$. The first two terms are the geodesic equation for the base geodesic and hence vanish, leaving the following for the deviation vector:

$$\frac{d^2 Y^\mu}{d\lambda^2} + \frac{\partial \Gamma^\mu{}_{\alpha\beta}}{\partial x^\nu}Y^\nu\frac{dx_\lambda^\alpha}{d\lambda}\frac{dx_\lambda^\beta}{d\lambda} + 2\Gamma^\mu{}_{\alpha\beta}\frac{dx_\lambda^\alpha}{d\lambda}\frac{dY^\beta}{d\lambda} = 0 \qquad (7.30)$$

This is a perfectly fine stopping point. But this would track four degrees of freedom and does not explicitly include all the constraints on the degrees of freedom. The goal is to develop an equation for the internal field $Y_I$. Differentiation of the expression $Y^\mu = Y_I \hat{e}_I^\mu$ and use of the transport equation, (7.19), provides the following expression:

$$\frac{dY^\mu}{d\lambda} = \frac{dY_I}{d\lambda}\hat{e}_I^\mu + Y_I\frac{d\hat{e}_I^\mu}{d\lambda} = \frac{dY_I}{d\lambda}\hat{e}_I^\mu - Y_I\Gamma^\mu{}_{\alpha\beta}\frac{dx_\lambda^\alpha}{d\lambda}\hat{e}_I^\beta \qquad (7.31)$$

Inserting this into (7.30) leads to the following expression:

$$\frac{d^2 Y^\mu}{d\lambda^2} = \frac{d^2 Y_I}{d\lambda^2}\hat{e}_I^\mu - 2\Gamma^\mu{}_{\alpha\beta}\frac{dY_I}{d\lambda}\frac{dx_\lambda^\alpha}{d\lambda}\hat{e}_I^\beta - \frac{\partial \Gamma^\mu{}_{\alpha\beta}}{\partial x^\nu}\frac{dx_\lambda^\nu}{d\lambda}\frac{dx_\lambda^\alpha}{d\lambda}\hat{e}_I^\beta Y_I + \Gamma^\mu{}_{\alpha\beta}\Gamma^\alpha{}_{\sigma\rho}\frac{dx_\lambda^\beta}{d\lambda}\frac{dx_\lambda^\alpha}{d\lambda}\hat{e}_I^\sigma Y_I$$

$$+ \Gamma^\mu{}_{\alpha\beta}\Gamma^\beta{}_{\sigma\rho}\frac{dx_\lambda^\sigma}{d\lambda}\frac{dx_\lambda^\beta}{d\lambda}\hat{e}_I^\alpha Y_I \qquad (7.32)$$

Inserting the aforementioned results into (7.30), canceling and rearranging terms, and multiplying through by $g_{\mu\nu}\hat{e}_J^\nu$, the following ODE is derived for the field $Y_J$:

$$\frac{d^2 Y_J}{d\lambda^2} + \left\{ g_{\mu\rho} R^\rho{}_{\alpha\nu\beta} \frac{dx^\alpha}{d\lambda} \frac{dx^\beta}{d\lambda} \hat{e}_I^\mu \hat{e}_J^\nu \right\} Y_I = 0 \tag{7.33}$$

This equation is defined on the base geodesic and along with (7.19) can be added to the geodesic equation and considered one set of fields for a larger system of ODE's. In the process a new tensor is defined, the Riemann curvature tensor:

$$R^\rho{}_{\alpha\nu\beta} = \frac{\partial \Gamma^\rho{}_{\alpha\beta}}{\partial x^\nu} - \frac{\partial \Gamma^\rho{}_{\alpha\nu}}{\partial x^\beta} + \Gamma^\rho{}_{\alpha\nu}\Gamma^\sigma{}_{\alpha\beta} - \Gamma^\rho{}_{\alpha\beta}\Gamma^\sigma{}_{\alpha\nu} \tag{7.34}$$

As the name suggests, this tensor contains information regarding the curvature associated with the metric tensor. One can construct other measures of curvature from this four-index tensor, the Ricci curvature, $R_{\alpha\beta} = R^\nu{}_{\alpha\nu\beta}$, and the scalar curvature, $R = g^{\alpha\beta} R_{\alpha\beta}$. From the point of view of this discussion, this is a computational tool for accounting for the effects of fluid motion and refraction on ray trajectories without appealing to any approximations. The enhanced system of equations consisting of Equations (7.12), (7.19), and (7.33) comprises a paraxial, or dynamic, ray trace procedure for rays in a fluid medium with background quantities, $c(\vec{x},t)$ and $\vec{v}_0(\vec{x},t)$.

The deviation equation traces two linearly independent deviation components, $Y_1$ and $Y_2$. Based on the discussion regarding the nature of the deviation vector, the internal basis, and their geometric interpretation, these components can be used to generate a measure of the cross-sectional area of a ray bundle. The cross product of the two independent deviation vectors provides an estimate of the infinitesimal area element of the wavefront. To develop this estimate in 3-dimensional (3-dim) space, the deviation vectors are projected in the auxiliary basis $\tilde{Y}_I = Y^\mu \tilde{e}_{I\mu}$. Based on the discussion regarding this vector, the cross product of the two independent vectors will be normal to the wavefront, that is, in the direction of $\hat{n}$. To get the cross-sectional area of the ray bundle, this is projected onto a unit vector in the direction of the 3-dim ray path tangent, $d\vec{x}/d\lambda$. Denoting this vector $\hat{t}$, the area is

$$dA_\lambda = |\hat{t} \cdot (Y_1 \times Y_2)| \tag{7.35}$$

## 7.3 Amplitude

For a large majority of applications, the medium can be considered time independent. The approach taken thus far carries over but with the introduction of additional mathematical machinery. Blokhintzev [8] provides a derivation of the pressure amplitude based on the geometric spread of a ray bundle for a generic inhomogeneous, time-independent moving medium. The results were used by Boone and Vermaas [9], and the author, in the development of a ray trace algorithm for moving inhomogeneous media. The result of this derivation, known as the Blokhintzev invariant, is presented here [8] as

$$\frac{p^2 \Lambda}{\rho c^2} dA_\lambda = \text{constant} \tag{7.36}$$

In (7.36) $p$ is the pressure perturbation. The ray theoretic approach in the higher frequency limit applies to pressure and the scalar field, $\psi$. The factor in the numerator, $\Lambda$, is a function of the ray path geometry. In this presentation, the factor is expressed explicitly in terms of the coordinate derivative along the ray path, $d\vec{x}/d\lambda$, but is equivalent to that found in [8]:

$$\Lambda = \frac{\left|\frac{d\vec{x}}{d\lambda}\right|\left(\hat{n}.\frac{d\vec{x}}{d\lambda}\right)}{\left(\frac{dt}{d\lambda}\right)} \tag{7.37}$$

The factor expressed in (7.37) can be written in terms of the local environmental fields, $c$ and $\vec{v}_0$, and the wavefront normal, $\hat{n}$, if desired. For low Mach number $\Lambda \sim c^2$, and this term cancels the same factor in the denominator. The last factor is the cross-sectional ray tube area measured on the ray, hence the subscript $\lambda$. Equation (7.36) relates the amplitude of the pressure field along the ray to the ray geometry and environmental parameters. Comparing values at two points allows the amplitude to be determined at an arbitrary point given the value at some initial point. Labeling all quantities at a point by the same index,

$$\frac{p_1^2 \Lambda_1}{\rho_1 c_1^2} dA_1 = \frac{p_2^2 \Lambda_2}{\rho_2 c_2^2} dA_2 \tag{7.38}$$

The environmental parameters are known and can be evaluated at each point along the ray. The output from a ray trace procedure can be used to evaluate (7.37) and the cross-sectional area of a ray bundle. With this information, the pressure at any point on a ray path can be determined from (7.38) given an initial value at some point on the same ray. In Ref. [9] the relation is simplified for horizontal propagation by assuming that $\rho$, $c$, and $\Lambda$ are approximately constant for the source and receiver locations.

As a final note to the section, it is pointed out that in the most general case, the cross-sectional area of a ray bundle can shrink to zero. Such an occurrence is called a caustic and represents a breakdown of the ray approximation. Using a pure ray theoretic calculation results in an infinite value for the acoustic pressure. The caustic also indicates that the wavefront is folding through itself. There are two ways that the area can shrink to zero, either in a single direction, leading to a line for the cross section, or in both independent directions, resulting in a point. This folding process in one direction introduces a discrete phase shift of $\pi/2$. Near a caustic, the field amplitude is not reliable and needs a better method of estimation. The method for evaluating the field at or near a caustic involves expanding the wave equation in a neighborhood in the vicinity of the caustic. For details the reader is referred to the literature, specifically Ludwig [10] for a detailed treatment and Foreman [11] for a clear review and applied examples. Another treatment for caustic analysis can be found in Brekhovskikh [12]. Lastly a detailed account of the types of caustics that can develop and their classification and evaluation can be found in Kravtsov and Orlov [13].

## 7.4   Ray Path Integrals

It is frequently the case in underwater and atmospheric acoustics that the background fields describing the medium vary only in one coordinate and are independent of time [14–19].

In Chapter 4 an equivalent Lagrangian and Hamiltonian were introduced for the bicharacteristics, along with the concept of an effective momentum. When the background fields are independent of one of the coordinates, the momentum conjugate to that coordinate is conserved. From this very simple idea, a set of first-order equations can be derived from this momentum conservation law, which are essentially an integrated form of the geodesic equation. For the purpose of this introduction, the background fields are assumed to depend only on one Cartesian coordinate, $z$, which represents the vertical direction. The horizontal plane will be referred to as the transverse plane, and projections of vectors into that plane will be referred to as the transverse component of that vector, for example, $\vec{V}_T$. No time dependence implies that $p_0 = \kappa_0$, where $\kappa_0$ is a constant. This can be stated in terms of the ray geometry:

$$-c\hat{n}\cdot\frac{d\vec{x}}{d\lambda} = \kappa_0 \tag{7.39}$$

No dependence on the transverse coordinates implies that $\vec{p}_T = \vec{\kappa}$, where $\vec{\kappa}$ is a constant vector in the transverse plane. In terms of ray path geometry, this conservation law implies the following:

$$c\hat{n}_T\frac{dt}{d\lambda} = \vec{\kappa} \tag{7.40}$$

A minus sign can be absorbed into the definition of $\kappa_0$. Then taking the ratio of these two equations yields the following:

$$\frac{\hat{n}_T}{c+\hat{n}\cdot\vec{v}_0} = \frac{\vec{\kappa}}{\kappa_0} \equiv \vec{\alpha} \tag{7.41}$$

In (7.41) the ray path coordinates have been eliminated using the equation $d\vec{x}/dt = c\hat{n} + \vec{v}_0$. Defining the polar angle, $\theta$, relative to the transverse plane rather than the $z$-axis, the normal vector can be described at any point using spherical coordinates:

$$\hat{n} = [\sin\theta\cos\varphi \quad \sin\theta\sin\varphi \quad \cos\theta]^T \tag{7.42}$$

Taking the ratio of the components of (7.41) gives a conservation law for the azimuth angle of the wavefront normal:

$$\tan\varphi = \frac{\kappa_y}{\kappa_x} \tag{7.43}$$

Equation (7.43) states that the wavefront normal vector propagates along the ray path with a fixed azimuth. The magnitude of (7.41) provides an equation for the polar angle of the wavefront normal:

$$\frac{\sin\theta}{c+w\sin\theta+u\cos\theta} = |\vec{\alpha}| \tag{7.44}$$

The following quantities are defined in (7.44) to simplify the form $w = v_{0x} \cos\varphi + v_{0y} \sin\varphi$ and $u = v_{0z}$. Defining $s = |\vec{a}|^{-1}$, (7.44) can be solved to give an equation for $\cos\theta$:

$$\cos\theta = \frac{-cu \pm (s-w)\sqrt{(s-w)^2 + (u^2-c^2)}}{u^2 + (s-w)^2} \tag{7.45}$$

Equation (7.45) is a version of Snell's law. It can be converted to an equation for the ray path polar angle using the relation between the ray velocity and the wavefront normal (Kornhauser [20]). Although interesting from a historical perspective, this generalization will not be as useful as other approaches. The conservation laws for the ray path can be expressed in terms of the ray velocity:

$$\left(c^2 - v_0^2\right)\frac{dt}{d\lambda} + \vec{v}_0 \cdot \frac{d\vec{x}}{d\lambda} = \kappa_0 \tag{7.46}$$

$$\left(\frac{d\vec{x}}{d\lambda} - \vec{v}_0\frac{dt}{d\lambda}\right)_T = \vec{\kappa} \tag{7.47}$$

For the remainder of this derivation and discussion, it is assumed that the background fluid flow is horizontal, that is, $u = 0$. With some algebraic manipulation, (7.46) and (7.47) can be used to derive the following:

$$\frac{dt}{d\lambda} = \frac{\kappa_0}{c^2}\left(1 - \vec{v}_0 \cdot \vec{a}\right) \tag{7.48}$$

$$\frac{d\vec{x}_T}{d\lambda} = \kappa_0\left(\vec{a} + \frac{\vec{v}_0}{c^2}\left(1 - \vec{v}_0 \cdot \vec{x}\right)\right) \tag{7.49}$$

The last equation needed is that for the $z$ coordinate of the ray path. This is derived by inserting (7.48) and (7.49) into the null constraint and solving for $dz/d\lambda$. The result is presented in (7.50) and left to the reader to derive:

$$\frac{dz}{d\lambda} = \pm\frac{\kappa_0}{c}\sqrt{\left(1 - \vec{v}_0 \cdot \vec{a}\right)^2 - c^2\alpha^2} \tag{7.50}$$

The term $\kappa_0$ has been factored out of each equation and $\alpha \equiv |\vec{a}|$. Equations (7.48)–(7.50) are a first-order nonlinear system for solving the ray paths. The vertical component of the ray velocity requires some care. Using the null condition, it can be shown that the quantity under the radical in (7.50) is always positive. This doesn't guarantee that it will be positive when implemented on a computer using finite precision. In fact, this quantity can easily become negative, leading to imaginary values for $dz/d\lambda$. The $\pm 1$ factor accounts for the direction of travel of the ray path along the $z$-axis. If the rays do not experience turning points, the sign can be chosen to match the initial conditions. Otherwise the sign needs to be changed after a turning point is encountered. Turning points can be solved for by setting (7.50) and the components of (7.49) equal to zero and solving for $z$. For complicated $c$ and $\vec{v}_0$, or for simulations using data tables for these quantities, it may not be possible to determine turning points *a priori*. These equations can be used to develop a solution for the ray paths in terms of a set of integrals. Rather than treat $\lambda$ as the independent variable, it is assumed that the $z$ coordinate can serve this purpose. This is valid along segments of the ray path where $z(\lambda)$ is invertible, that is, between turning points. Equation (7.50) is divided into (7.48) and (7.49), and the results are integrated from a starting value $z_0$ to some arbitrary value:

$$t(z) - t_0 = \int_{z_0}^{z} \frac{1 - \vec{v}_0 \cdot \vec{\alpha}}{c \sqrt{\left(1 - \vec{v}_0 \cdot \vec{\alpha}\right)^2 - c^2 \alpha^2}} \, dz' \tag{7.51}$$

$$\vec{x}_T(z) - \vec{x}_{T0} = \int_{z_0}^{z} \frac{c^2 \vec{\alpha} + \vec{v}_0 \left(1 - \vec{v}_0 \cdot \vec{\alpha}\right)}{c \sqrt{\left(1 - \vec{v}_0 \cdot \vec{\alpha}\right)^2 - c^2 \alpha^2}} \, dz' \tag{7.52}$$

The final results are presented here for the plus sign on (7.50). A similar result holds for the minus sign. These equations are most useful when the environmental parameters are described by functions that allow them to be integrated in closed form. The integrals define time and range along segments of the path between turning points. To extend them beyond a turning point requires resetting the parameter of the integral and the initial conditions and adding the result to the previous segment. For example, if one is calculating the travel time for a ray path from a particular initial position to a desired final location down range, the process might be as follows:

1. Calculate the travel time from $z_0$ to $z_{tp}$ and call it $T_0$.
2. Calculate the travel from $z_{tp}$ to the next point. If another turning point is encountered before the end range is reached, calculate the travel time between turning points and call it $T_{tp}$.
3. Continue accumulating turning point to turning point times until the desired final range is reached and then calculate the travel time from the last turning point to the end point $z_f$ and call it $T_f$.
4. Calculate the total travel time by adding together the results from steps 1 to 3.

The aforementioned description is only a high-level notional process. The astute reader will realize that the process isn't feasible in general. It essentially solves an i.v.p. and we are really describing a type of two-point boundary value problem (b.v.p.) in range. To determine how many turning points exist between the initial and final range, one needs solutions to the range integral for comparison, which may not be possible, and a value for the turning points based on Snell's law, which again may not be possible. There are a few additional points to be made that will help make use of these expressions. Depending on the symmetry present in the problem, it is sometimes possible to demonstrate the following:

1. That the turning points always occur at the same value or set of values for $z_{tp}$.
2. That the quantity calculated in step 2 of the process, $T_{tp}$, is translationally invariant and the same value for positive and negative values of $dz/d\lambda$.

Step 2 only needs to be calculated one time and the result multiplied by $N$, the number of complete trips between turning points encountered between the end points. The same statement holds for range integrals as well.

One case where the previous statements are true is the case when $\vec{v}_0 = 0$ and all refraction is due to the sound speed. The ray integrals simplify, but more importantly ray paths do not bend out of the vertical plane defined by the initial azimuth angle. This result has a far-reaching consequence in that it means the propagation problem is truly a two-dimensional problem. The entire ray trace of one plane can be rotated to produce the exact result in any other plane. When the flow vector, $\vec{v}_0$, is nonzero, ray paths have a tendency to bend in the direction of the flow. If the flow is one-dimensional, say, $v_{0x}$ is the only nonzero component, then ray paths that are initially in the plane containing the flow vector will only turn in that plane and never leave. However, ray paths initially propagating out of that plane will start to turn out of their initial osculating plane. This tendency to bend in the direction of flow alters the ray tube geometry and the number and type of multiple paths that may intersect a sensor or field point in a calculation. The turning points of the ray paths in the coordinate directions can be determined by setting the appropriate term equal to zero. Since this section deals with cases where there is only $z$ dependence, turning points in the $z$ direction are likely. From (7.50) the vertical turning points will be a solution to

$$1 - \vec{v}_0 \cdot \vec{\alpha} = c\alpha \qquad (7.53)$$

The horizontal turning points are determined from (7.49):

$$c^2 \vec{\alpha} = -\vec{v}_0 \left(1 - \vec{v}_0 \cdot \vec{\alpha}\right) \qquad (7.54)$$

It is a straightforward matter to show that $\left(1 - \vec{v}_0 \cdot \vec{\alpha}\right) > 0$. Due to the nature of these integrals, it is clear that the integrand will diverge at the turning points. This makes their use potentially difficult. There exist a fairly good number of choices for the functions $c$ and $\vec{v}_0$ for which these integrals can be evaluated in closed form in terms of ordinary, transcendental, or special functions. This makes them a valuable tool for conceptual understanding of ray theory. To make the equations and the discussion regarding their use more concrete, two examples with exact solutions will be worked out. A few classic cases are considered next, five to be precise. These are chosen due to their having exact closed form solutions to the range and time integrals in terms of ordinary functions. They also illustrate specific features related to the effect of the environment on ray path geometry. The first example is a linear sound speed profile and no fluid motion, and the second example is of a linear wind profile and a constant sound speed. These cases share a common feature in that they do not have caustics. The third example is known as the Kormilitsin profile. The sound speed goes as $\sim 1/\sqrt{z}$. This is a fairly simple example that contains a caustic. The fourth example is of a refractive waveguide. This is presented to illustrate the presence of multiple turning points and the ability for the environment to trap ray paths. The final example combines the first two, linear sound speed and linear wind profiles. This example does have a closed-form solution to the ray path but is a bit more involved in terms of explaining all the details of the ray paths. In the presentation of the general theory, the standard definition of spherical coordinates was used. For problems in underwater and aero-acoustics in layered refractive media, it is customary to define the polar angle, $\theta$, with respect to the horizontal plane, for example, $z = r \sin \theta$.

## Example 7.1   Linear SSP

This example involves one of the simplest nontrivial sound speed profiles:

$$c(z) = c_0(1 + \xi z) \tag{7.55}$$

The sound speed increases with increasing $z$. For $z = -\xi^{-1}$ the sound speed is zero, and for $z < -\xi^{-1}$ it is negative. These last two cases are unphysical, yet this is still a good model for refraction in air and water if reflective boundaries are included. This is a case when the ray system has azimuthal symmetry and can be treated as a two-dimensional system. And since the flow is zero, the wavefront normal will align with the ray velocity. Snell's law then gives

$$\frac{\cos \theta}{c(z)} = \frac{\cos \theta_0}{c(z_0)} = \alpha \tag{7.56}$$

The turning point(s) occur when $\theta = 0$. Setting this value in (7.56) provides an equation for the value of $z$ at the turning point. This is true in general for any problem with an arbitrary $c(z)$ and $\vec{v}_0 = 0$. For this example there is only one turning point:

$$z_{tp} = z_0 \sec \theta_0 + \xi^{-1}(\sec \theta_0 - 1) \tag{7.57}$$

For the initial condition $\theta_0 = 0$, $z_0$ is the turning point. To solve the ray path integrals, a change of variables is made from $z$ to $c$, and the differential in the integral, $c_0 \xi dz = dc$. Taking the horizontal direction to be the $x$-axis, (7.51) and (7.52) become

$$t = \frac{1}{c_0 \xi} \int \frac{d(\alpha c)}{(\alpha c)\sqrt{1 - (\alpha c)^2}} \tag{7.58}$$

$$x - x_0 = \frac{1}{\alpha c_0 \xi} \int \frac{(\alpha c) d(\alpha c)}{\sqrt{1 - (\alpha c)^2}} \tag{7.59}$$

An obvious identification of the variable $\alpha c$ can be made, $t_0 = 0$ is assumed, and integration limits have been suppressed. Integrating (7.59) gives the geometry of the ray paths in space:

$$\pm \alpha c_0 \xi \Delta x = -\sqrt{1 - (\alpha c)^2} + \sqrt{1 - \alpha^2 c^2(z_0)} \tag{7.60}$$

With a little algebra, this can be turned into the equation of a circle, defined by the following equation and parameters:

$$(\Delta x - \Delta x_c)^2 + (z - z_c) = r^2 \tag{7.61}$$

$$z_c = -\xi^{-1} \tag{7.62}$$

$$\Delta x_c = \pm(z_0 - z_c)\tan\theta_0 \tag{7.63}$$

$$r = (z_0 - z_c)\sec\theta_0 \tag{7.64}$$

Equation (7.62) states that all circles, regardless of initial conditions, have their center at the same vertical location, where $c = 0$. This makes sense from Snell's law as the solution for the angle would be $\pi/2$. The negative case is accounted for by the minus sign in the ray path equations. This is the solution for $-\pi/2 < \theta_0 < \pi/2$. By reflection symmetry, this covers all ray paths. Typically, the special case $\theta_0 = \pm\pi/2$ needs to be handled separately. Snell's law evaluates to zero for these angles, implying that these rays will not turn in the $z$ direction and continue to propagate along a straight line, $\Delta x = 0$.

The travel time is determined by integrating (7.51):

$$\pm c_0\xi t = \ln\left(\frac{\alpha c}{\sqrt{1-(\alpha c)^2}+1}\right) - \ln\left(\frac{\alpha c(z_0)}{\sqrt{1-\alpha^2 c^2(z_0)}+1}\right) \tag{7.65}$$

With a bit of algebra and using the turning point formula, this can be expressed as

$$\pm c_0\xi t = \ln\left(\frac{(z-z_c)}{\sqrt{(z_{tp}-z_c)^2-(z-z_c)^2}+(z_{tp}-z_c)}\right) - \ln\left(\frac{\cos\theta_0}{|\sin\theta_0|+1}\right) \tag{7.66}$$

For $\theta_0 = \pm\pi/2$, the original integral gives

$$c_0\xi t = \pm\ln\left(\frac{z-z_c}{z_0-z_c}\right), \quad \theta_0 = \pm\frac{\pi}{2} \tag{7.67}$$

The minus sign on the left-hand side (l.h.s.) covered the case of ray traveling downward. For the positive solution, the time to reach a turning point is

$$c_0\xi t = \frac{1}{c_0\xi}\ln\left(\frac{\sin\theta_0+1}{\cos\theta_0}\right) \tag{7.68}$$

For downward traveling rays, the time to reach $z_c$ is infinite. Not all problems yield exact solutions and even some that do will not lend themselves to such easy analysis.

A sample ray trace is shown in Figure 7.2 for a source at $z = 10$ m, $c_0 = 343$ m/s, and $\xi = 0.05$ m$^{-1}$. Left–right symmetry is clearly present; only half the rays in the plane are required, a feature shared by all 1-dim sound speed profiles. This profile shares a feature with

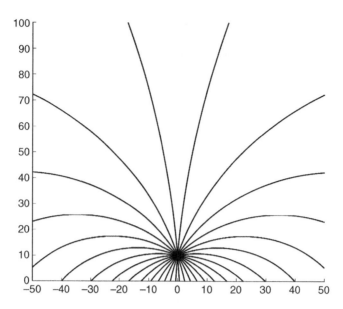

**Figure 7.2**    Ray trace for the linear sound speed profile

the trivial case in that there are no multiple paths produced by the refractive properties of the medium. Each pair of points, for $z > -20$m, is connected by a single path. Interference and caustics will not occur unless boundaries are present.

**Example 7.2    Linear Wind**

This example illustrates the effects of a nontrivial wind profile on ray propagation as well as the additional analysis required to fully determine the geometry of the ray paths. The background sound speed profile is assumed to be constant, $c_0$. The fluid motion is assumed to flow in the $x$ direction and vary linearly with the $z$ coordinate:

$$\vec{v}_0 = v_0 \xi z \hat{e}_x \tag{7.69}$$

The ray path integrals for range and time take the following forms:

$$t = \int_{z_0}^{z} \frac{1 - \alpha v_0 \xi z'}{c_0 \sqrt{\left(1 - \alpha v_0 \xi z'\right)^2 - c_0^2 \alpha^2}} \, dz' \tag{7.70}$$

$$x = \pm \int_{z_0}^{z} \frac{c_0^2 \alpha + v_0 \xi z' \left(1 - \alpha v_0 \xi z'\right)}{c_0 \sqrt{\left(1 - \alpha v_0 \xi z'\right)^2 - c_0^2 \alpha^2}} \, dz' \tag{7.71}$$

These are converted into a standard form as follows:

$$t = \frac{1}{c_0} \left\{ I_0 - \alpha I_1 \right\} \tag{7.72}$$

$$x = \frac{1}{c_0 \alpha} \left\{ c_0^2 \alpha^2 I_0 + a I_1 - a^2 I_2 \right\} \tag{7.73}$$

The following integral is defined:

$$I_n = \int \frac{z^n dz}{R} \tag{7.74}$$

For $n = 0$, (7.74) evaluates to the following:

$$I_0 = \frac{1}{|a|} \ln(-\text{sign}(a) R + 1 - az) \tag{7.75}$$

The results for $n = 1, 2$ are related to this result:

$$I_1 = \frac{1}{a^2} R + \frac{1}{a} I_0 \tag{7.76}$$

$$I_2 = \left( \frac{3 + az}{2a^3} \right) R + \left( \frac{3 - b}{2a^2} \right) I_0 \tag{7.77}$$

The constants appearing in these equations are defined as follows: $a = \alpha v_0 \xi$, $b = 1 - c_0^2 \alpha^2$. For convenience, the following definition is made:

$$R(z) = \sqrt{b - 2az + a^2 z^2} \tag{7.78}$$

With these results the equations for travel time and range along a segment of the ray path are determined:

$$c_0 a \Delta t = \pm \{ R(z_0) - R(z) \} \tag{7.79}$$

$$2 c_0 \alpha a \Delta x = \pm \left\{ c_0^2 \alpha^2 a (I_0(z) - I_0(z_0)) - (1 + az) R(z) + (1 + az_0) R(z_0) \right\} \tag{7.80}$$

The plus and minus signs correspond to solutions evaluated along paths that have positive and negative $dz/dt$, respectively. The analysis of turning points is a little more involved in this case as compared with that of a linear sound speed. There no longer is a left–right symmetry in this problem. More generally there is no azimuthal symmetry. As a result of this, one cannot rotate a plane of ray paths to fill all of space. For the purpose of illustrating this technique and the general behavior of ray paths in the presence of wind, only those rays in the plane containing the wind vector will be solved for. In this case there will be horizontal as well as vertical turning points. Before delving into turning point analysis, the ray paths are determined for the special case, $\alpha = 0$. Rather than attempting to take the limit of the previous results, the original integrals are solved:

$$t = \pm \int_{z_0}^{z} \frac{1}{c_0} dz' = \pm \frac{\Delta z}{c_0} \tag{7.81}$$

$$x = \pm \int_{z_0}^{z} \frac{v_0 \xi z'}{c_0} dz' = \pm \frac{v_0 \xi}{2c_0} \left( z^2 - z_0^2 \right) \tag{7.82}$$

These curves represent a limiting case. These paths are parabolas opening in the horizontal direction, each having its critical point at the $\vec{v} = 0$ line, in this case at $z=0$. The specific nature of these curves depends not only on the initial direction but also on the placement of the source. If the source is on $x$-axis, these curves will be reflected mirror images of each other. For a source placed above the $x$-axis, the upward traveling ray path will not encounter a horizontal turning point, while the downward traveling ray path will encounter a horizontal turning point, specifically at $z=0$. More generally ray paths will encounter horizontal turning points when

$$\alpha + \frac{v}{c^2}(1 - v\alpha) = 0 \tag{7.83}$$

For the vertically launched rays with $\alpha = 0$, (7.83) gives horizontal turning points at $v = 0$. Now the turning points are determined for all other rays in this system. The vertical turning points will occur when $\cos\theta_0 = 1$ and can be determined by Snell's law as before. The starting point is the equation

$$\alpha = \frac{\cos\theta_0 \cos\varphi_0}{c_0 + v_0 \xi z_0 \cos\theta_0 \cos\varphi_0} = \frac{\cos\varphi_0}{c_0 + v_0 \xi z_{vtp} \cos\varphi_0} \tag{7.84}$$

Solving this gives the $z$ coordinate of the vertical turning points for each ray, labeled $z_{vtp}$:

$$z_{vtp} = z_0 + \frac{c_0}{v_0 \xi} \left( \frac{1 - \cos\theta_0}{\cos\theta_0 \cos\varphi_0} \right) \tag{7.85}$$

The $\cos\varphi_0$ factor is evaluated at $\varphi_0 = 0$ for rays launched to the right and $\varphi_0 = \pi$ for rays launched to the left. Since the profile is linear, each ray will have one and only one vertical turning point. Rays launched up and to the right will have a single turning point above $z_0$, while those launched down and to the left will have turning points below $z_0$ by the same distance, as indicated by (7.85). Ray paths launched down and to the right or up and to the left will not encounter vertical turning points. The horizontal turning points come from (7.83). The factor $(1 - v\alpha)$ is positive definite. This means that the only way to have a horizontal turning point is if $\alpha$ and $v$ have opposite sign. For $\alpha > 0$ and $\dot{z} < 0$, rays will encounter horizontal turning points when $v < 0$, somewhere below the $x$-axis, while for $\alpha < 0$ and $\dot{z} < 0$, the turning points will occur above the $x$-axis. Rearranging terms in Equation 7.83 and solving for $v$ gives

$$v = \frac{1}{2\alpha} \left( 1 \pm \sqrt{1 + 4c_0^2 \alpha^2} \right) \tag{7.86}$$

where the negative root is taken for the right-hand side (r.h.s.). The range and time integrals required care when evaluated at the turning point. To derive an exact expression for the indefinite integrals at the turning points, note that $R(z_{vtp}) = 0$ by definition:

$$t(z_{vtp}) = 0 \tag{7.87}$$

$$x\left(z_{vtp}\right) = \text{sign}(\alpha)\frac{c_0}{2v_0\xi}\ln(c_0|\alpha|) \tag{7.88}$$

When evaluating the range for the turning points, (7.88) should be used instead of (7.80). Attempting to evaluate the individual terms in the indefinite integral at or near the turning point results is anomalous complex values.

A sample ray trace is shown in Figure 7.3 for a point source at the origin, $c_0 = 343$ m/s, $v_0 = 10$ m/s, and $\xi = 0.5$ m$^{-1}$. The 180° rotational symmetry is clear from the trace. In this environment ray paths from a point source will not cross and caustics will not form unless boundaries are present.

### Example 7.3    Kormilitsin Profile
The sound speed profile is

$$c = \frac{c_0}{\sqrt{\xi z}} \tag{7.89}$$

In this representation, the profile has the following properties, $c \to \infty$ as $z \to 0$, and $c$ becomes imaginary when $z < 0$. The physical space for this problem is $z > 0$. The turning points are determined by Snell's law:

$$z_{tp} = z_0 \cos^2\theta_0 \tag{7.90}$$

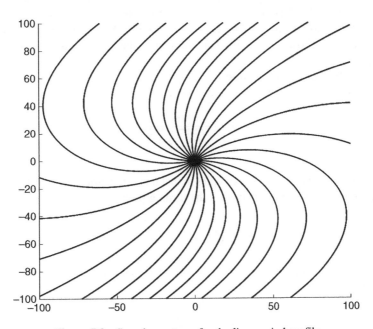

**Figure 7.3**    Sample ray trace for the linear wind profile

A more convenient form of the ray integrals is

$$t = \int \frac{c^{-2} dz'}{\sqrt{c^{-2} - \alpha^2}} \tag{7.91}$$

$$x - x_0 = \alpha \int \frac{dz'}{\sqrt{c^{-2} - \alpha^2}} \tag{7.92}$$

From Snell's law it follows that $c_0^2 \alpha^2 = \xi z_{tp}$. The range integral becomes

$$\frac{\xi \Delta x}{c_0 \alpha} = \int \frac{d(\xi z')}{\sqrt{\xi z' - \xi z_{tp}}} = 2\sqrt{\xi(z - z_{tp})} - 2\sqrt{\xi(z_0 - z_{tp})} \tag{7.93}$$

Solving for $z$ gives the ray path as a curve in space and is identified as a parabola:

$$z = z_{tp} + \xi^{-1} \left( \frac{\xi \Delta x}{2 c_0 \alpha} + \sqrt{\xi(z_0 - z_{tp})} \right)^2 \tag{7.94}$$

The dependence on initial conditions is hidden in $z_{tp}$. The travel time integral becomes

$$c_0 \xi t = \int \frac{(\xi z')d(\xi z')}{\sqrt{(\xi z') - \alpha^2}} = \frac{2}{3} \left\{ \sqrt{\xi(z - z_{tp})} \left( 2\xi z_{tp} + \xi z \right) - \sqrt{\xi(z_0 - z_{tp})} \left( 2\xi z_{tp} + \xi z_0 \right) \right\} \tag{7.95}$$

It is clear from the solution and by inspection of the sound speed function that these ray paths initially traveling toward $z = 0$ will have one and only one turning point after which they travel to infinity. Ray initially traveling upward will never encounter a turning point.

A sample ray trace for this profile is provided in Figure 7.4 for a source at $z = 25$ m. The parameters for the profile are $c_0 = 343$ m/s and $\xi = 0.05$. A salient feature of this profile, and the reason it was chosen as a particular example, is that it contains a caustic.

## Example 7.4   Refractive Waveguide

This example is of a sound speed profile that creates a natural waveguide:

$$c = \frac{c_0}{\sqrt{\left( 1 - (\xi z)^2 \right)}} \tag{7.96}$$

This profile differs from the other examples in that it is symmetric. The profile will approach infinity as $z \to \pm \xi$. Snell's law determines the turning points:

$$z_{tp} = \pm \xi^{-1} \sqrt{1 - \cos^2 \theta_0 \left( 1 - (\xi z_0)^2 \right)} \tag{7.97}$$

The same form of the ray integrals is useful here. The range integral gives the following ray path geometry:

$$z = z_{tp} \sin \left( \frac{\xi \Delta x}{\alpha c_0} + \varphi_0 \right) \tag{7.98}$$

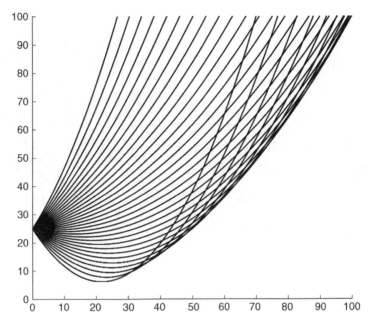

**Figure 7.4**  Sample ray trace for the Kormilitsin profile

where the initial phase of the path is

$$\varphi_0 = \arcsin\left(\frac{z_0}{z_{tp}}\right) \tag{7.99}$$

The travel time integral evaluates to the following:

$$
\begin{aligned}
c_0 \xi t = {} & \left(1 - \frac{(\xi z_{tp})^2}{2}\right)\left(\arcsin\left(\frac{z}{z_{tp}}\right) - \arcsin\left(\frac{z_0}{z_{tp}}\right)\right) \\
& + \frac{(\xi z_{tp})^2}{2}\left(\frac{z}{z_{tp}}\sqrt{1 - \left(\frac{z}{z_{tp}}\right)^2} - \frac{z_0}{z_{tp}}\sqrt{1 - \left(\frac{z_0}{z_{tp}}\right)^2}\right)
\end{aligned}
\tag{7.100}
$$

This example introduces a new feature that the other profile examples do not have, periodicity. The ray paths oscillate between a pair of turning points in the $z$ direction. This property will cause interference and caustic formation. An example is shown in Figure 7.5. The source is at the origin and the environmental parameters are $c_0 = 343$ m/s and $\xi = 0.5$ m$^{-1}$. For these parameters the turning points occur at $\pm 2|\sin\theta_0|$.

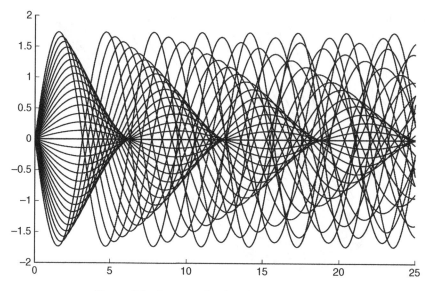

**Figure 7.5**  Ray trace for the refractive waveguide

### Example 7.5  Linear SSP and Wind

This example combines two of the previous environments. The sound speed is (7.55) and the wind is (7.69). The time and range integrals have a similar form as that of the previous example with more complex coefficients. The following are defined to make expressions easier to deal with:

$$b_0 = 1 - c_0^2 \alpha^2 \tag{7.101}$$

$$b_1 = -2\left(v_0 \xi_2 \alpha + c_0^2 \xi_1 \alpha^2\right) \tag{7.102}$$

$$b_2 = \left(v_0^2 \xi_2^2 - c_0^2 \xi_1^2\right) \tag{7.103}$$

$$R = \sqrt{b_0 + b_1 z + b_2 z^2} \tag{7.104}$$

$$I_n = \pm \int_{z_0}^{z} \frac{z'^n dz'}{(1 + \xi_1 z') R(z')} \tag{7.105}$$

The constants $\xi_1$ and $\xi_2$ measure the sound speed and wind gradients respectively.

Due to the nature of the integrand, it will be more convenient to define the variable $y = (1 + \xi_1 z)$. In terms of this variable, (7.104) becomes

$$R = \frac{1}{\xi_1} \sqrt{d_0 + d_1 y + d_2 y^2} = \frac{1}{\xi_1} \tilde{R} \tag{7.106}$$

with the new set of constants, $d_n$, defined as follows, with $d_2 = b_2$.

$$d_0 = (\xi_1 + a)^2 \tag{7.107}$$

$$d_1 = -2a(\xi_1 + a) \tag{7.108}$$

The goal now is to express the time and range integrals in terms of the new variables:

$$t = \frac{1}{c_0 \xi_1} \int \frac{(\xi_1 + a) - ay}{y \widetilde{R}(y)} dy \tag{7.109}$$

$$x = \frac{1}{c_0 \alpha \xi_1^2} \int \frac{-a(\xi_1 + a) + a(\xi_1 + 2a)y - d_2 y^2}{y \widetilde{R}(y)} dy \tag{7.110}$$

The following integrals are defined:

$$I_n = \int \frac{y^{n-1} dy}{\widetilde{R}(y)} \tag{7.111}$$

With these definitions the travel time and range integrals become

$$c_0 \xi_1 t = (\xi_1 + a) I_0 - a I_1 \tag{7.112}$$

$$c_0 \alpha \xi_1^2 x = -a(\xi_1 + a) I_0 + a(\xi_1 + 2a) I_1 - d_2 I_2 \tag{7.113}$$

The factor $a$ has the same definition as in the previous example. The vertical turning points are determined, as before, from Snell's law:

$$z_{vpt} = z_0 \left( \frac{c_0 \xi_1 \pm v_0 \xi_2 \cos\theta_0}{c_0 \xi_1 \pm v_0 \xi_2} \right) \sec\theta_0 + \left( \frac{c_0(1 - \cos\theta_0)}{c_0 \xi_1 \pm v_0 \xi_2} \right) \sec\theta_0 \tag{7.114}$$

The plus (minus) sign provides turning points for rays launched to the right (left). Before discussing the solution for arbitrary launch angle, the solution for $\alpha = 0$ is provided:

$$\Delta t = \pm \frac{1}{c_0 \xi_1} \ln\left( \frac{1 + \xi_1 z}{1 + \xi_1 z_0} \right) \tag{7.115}$$

$$\Delta x = \pm \frac{v_0 \xi_2}{c_0 \xi_1^2} \left( \xi_1 \Delta z - \ln\left( \frac{1 + \xi_1 z}{1 + \xi_1 z_0} \right) \right) \tag{7.116}$$

To determine the general solution, the indefinite form of the integrals in (7.111) is provided for $n = 0, 1, 2$. In terms of the new variables, $c_0 > 0$, but $c_{1,2}$ may be either sign depending on the environment and the ray parameter, $a$. In the evaluation of these integrals, the following quantity is needed:

$$\Delta = 4 c_2 c_0 - c_1^2 = -4 c_0^2 \alpha^2 \xi_1^2 (\xi_1 + a)^2 \tag{7.117}$$

The indefinite integrals in (7.111) for $n = 0, 1, 2$ are as follows:

$$I_0 = -\frac{1}{\sqrt{c_0}}\left\{\ln\left(2\sqrt{c_0}\widetilde{R}+c_1 y+2c_0\right)-\left(\ln y\right)\right\} \tag{7.118}$$

$$I_1 = \frac{1}{\sqrt{|c_2|}}\begin{cases} \ln\left(2\sqrt{c_2}\widetilde{R}+2c_2 y+c_1\right), & c_2>0 \\ -\arcsin\left(\dfrac{2c_2 y+c_1}{\sqrt{|\Delta|}}\right), & c_2>0 \end{cases} \tag{7.119}$$

$$I_2 = \frac{1}{c_2}\widetilde{R}-\frac{c_1}{2c_2}I_1 \tag{7.120}$$

Constants have been absorbed into the arbitrary constant of integration. While the last several cases were more straightforward in this example, the nature of the ray paths depends on the relative strength of the sound speed and wind gradients. This is discussed in more detail in Ref. [21], which also provides a sample wavefront and ray trace in space-time. In particular, for weak wind gradients, caustics will not form, but stronger wind gradients will begin to produce caustics in regions above and below the source.

## 7.5   Building a Field from Rays

Given solutions to the ray equation for the ray path and its geometric spread, a full description of the pressure field at an arbitrary point requires determining all rays that connect the sources to that point. In a few cases presented previously, the trivial case, a linear sound speed profile, and a linear wind profile, there will be a unique ray path connecting the source to any arbitrary point in physical space. In general, this is not the case. The refractive properties of the medium, presence of boundaries, or both will cause multiple paths with different initial conditions to cross paths at one or more points in space. In the frequency domain with a monochromatic c.w. source, the crossing of paths will create an interference pattern due to the different phases along each path. The phase differences will be due in part to the different travel times along each ray and the interaction with boundaries. Assume that all the ingredients needed are known and that the ray paths connecting source and field point are known. These ray paths are indexed by $n = 1,...,N$, which could be a function of the initial conditions of the ray path, and the following quantities are defined on each of these rays: travel time, $\tau_n$, amplitude, $A_n$, and the net phase shift due to boundary interactions and other discrete phase effects, $\delta_n$. Then the total field at the field point is

$$p = \sum_{n=1}^{N} A_n \exp[i(\omega\tau_n + \delta_n)] \tag{7.121}$$

The solutions presented thus far provide a method for calculating range and travel time as a function of depth given an initial position and ray direction, $(\theta_0, \varphi_0)$. This is an initial value problem, i.v.p. Locating the multiple paths required to evaluate (7.121) is a b.v.p. Using the exact solution for a model profile, an ideal approach would be to specify the desired final location, $(x_f, z_f)$, and solve the ray path equation for $(\theta_0)$. In practice this approach doesn't lead to a useful solution except in the most trivial cases. Another approach is to trace the rays out to the range of the desired field point, for example, a sensor location, and then start binning the results in depth. For the rays that are close to the desired point, the initial conditions are known, or knowable. Using this subset of initial conditions, each path is refined by tracing new ray paths with small deviations in initial conditions and testing the final location to judge whether the result is closer to the desired field point. When a ray path is traced, and meets the desired final location, the search is stopped. This type of approach is referred to as shooting and correcting, and while it may seem like a lot of work for many types of problems, it can yield results quickly and is amenable to parallelization. This type of search can be done without an initial ray trace by binning the initial conditions in small intervals, for example, $[\theta_n, \theta_{n+1}]$, $\Delta\theta \sim 5°$, and setting a stop condition when the location is good or a maximum number of trials have been reached with no result. This type of approach works well for layered media, but starts to get costlier for full 3-dim ray traces with strongly varying environments. As the simple case of a linear wind profile illustrates, ray paths can have turning points in both the $x$ and $z$ directions. This situation is more complex for a full 3-dim ray trace. When conditions permit, it is sometimes possible to exclude large portions of the initial conditions from the start. This is the case for a depth-dependent sound speed profile and no background flow. For this type of environment, turning points in $x$ will not occur, and the ray trace will be identical in both range directions. A half plane search is sufficient to find the multipaths. What this discussion illustrates is that it is a good idea in practice to tailor the approach to finding multipaths to the type of problem considered.

The amplitude factor appearing in (7.121) is not meant to represent the geometric spread on the ray bundle alone, but a complete amplitude estimate. The presence of caustics injects more than one effect on the ray path parameters. If the field point is on or near a caustic, the amplitude derived from geometric spread is not correct and requires a more refined estimate. Correcting a ray trace near a caustic involves performing an expansion of the field equation in the vicinity of the caustic and estimating a solution to this equation. The approach is similar to the WKB expansion of a wave function near a turning point in quantum mechanics. The caustic occurs when the rays with infinitesimally close initial conditions meet. This is not the same situation as occurs with multipaths. Multiple paths approach the field point from completely different directions, whereas when a caustic occurs, the cross-sectional area of the bundle of rays about the ray path shrinks to zero. This leads to an infinite result for the ray amplitude based on geometric spread alone. Caustics can occur along any and all rays in the system and form a family of parameterized curves or surfaces, depending on the dimensionality of the problem. The full caustic represents a boundary in space where a subset of ray paths appears to be reflected off this boundary. Expanding the field in a small neighborhood about a ray path essentially produces a paraxial approximation of the wave equation in ray path coordinates. The other effect produced by the presence of a caustic is a phase shift of $\pi/2$ for each time a caustic occurs along a path and for each direction along which the ray bundle shrinks. The field correction is required when a calculation is being performed at a point on or near a caustic, but these phase shifts get carried along the path, altering its phase, regardless of the final location. When the ray experiences a caustic, the wavefront is being folded through itself. This can occur in either or both of the independent directions in the wavefront surface. In theory, this type of analysis requires knowing what a family of ray paths, each being slightly different initial conditions, in a small neighborhood of a given ray path are doing at all points along the ray. This is exactly the information provided in the deviation vector discussed in the paraxial ray trace procedure. A good amount of this information can be discovered by simply checking for changes in sign of

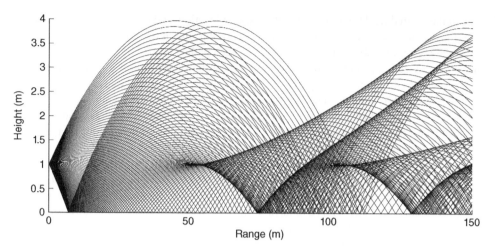

**Figure 7.6**  Ray trace for a linear refractive sound speed profile and hard reflective ground as $z = 0$

the two independent deviation vectors at each point along the ray. And for each zero encountered, for each of the two vectors, factor of $\pi/2$ is included in $\delta_n$. If a total of $m$ caustics are encountered along the ray path, the phase is $m\pi/2$.

Lastly, a point that was not mentioned in the previous sections is that there will be hard and soft boundaries present in the environment. When a ray path encounters a boundary, it must be reflected. If the normal to the reflective boundary is $\hat{N}$, then $\hat{N} \cdot \hat{n}_R = -\hat{N} \cdot \hat{n}_I$, where the indices $I, R$ stand for incident and reflection. This resets the propagation of the ray path in the medium. In addition to reflecting the ray path, the field will pick up a phase when reflected from a pressure release surface. One can account for the total phase shift due to idea the boundary by keeping track of the number of reflections, $m\pi$. When reflecting from a nonideal absorbing boundary described by a scalar absorption coefficient, $\alpha$, the amplitude will be affected by $\alpha^m$ where $m$ is the number of interactions with the surface. As mentioned in the examples, reflection from a boundary can introduce caustics in the ray trace that need to be dealt with in the calculation of the acoustic field. An example is shown in Figure 7.6 for a linear sound speed profile (Example 7.1) and a hard ground at $z = 0$. This example illustrates the difference between interference between multiple paths and the development of caustic curves (surfaces) in the trace.

## 7.6   Numerical Approach to Ray Tracing

Numerical techniques for solving the ray equation are needed to model full three-dimensional propagation. They are also needed for problems that can be reduced to ray path integrals since closed-form solutions are rare, and the form of the integrals is such that the integrand is singular near the turning points. As illustrated by the exact solutions presented in this chapter, care is required when evaluating expressions at turning points. Though numerical techniques for integration exist, when confronted with the type of integrals in (7.51) and (7.52), it is usually best to go back to the original second-order ODE and implement one of the standard solvers. For purposes of illustration, the basics of numerical solutions to ODE systems will be presented here.

From the discussion of the preceding sections, the ray equation is a system of second-order quasi-linear homogeneous ODE (a mouth full). It is quasi-linear because the equation is linear in the highest-order derivative and nonlinear in the lower-order derivatives. Reduction of order is applied to the ray equations to produce an enlarged first-order system, as was done for the relaxation method. Deviating from the specifics of ray theory for a moment, the goal is to solve an ODE in the following form:

$$\frac{dy_k}{d\lambda} = f_k(y_j, \lambda) \tag{7.122}$$

Equation (7.122) describes the evolution of a state vector $y_k$, $k = 1, \ldots, K$, with $K$ indicating the total number of variables, with respect to an arbitrary independent variable, $\lambda$ (not wavelength). The r.h.s. is an arbitrary, but well behaved, function of the state $y_k$ and possibly the independent variable. A second-order ODE in the variables $x_k$ can be cast in this form by the following substitution:

$$h(x_j)\frac{dx_k}{d\lambda} + u_k(x_j) = p_k \tag{7.123}$$

The new independent variable $\vec{y} = [\vec{x}, \vec{p}]^T$ is defined. The dependence of the r.h.s. on the independent variable can be eliminated by introducing a new parameter and adding an equation to the system:

$$\frac{d\sigma}{d\lambda} = 1 \tag{7.124}$$

When the ray equation is expressed as a geodesic equation, (7.12), there will never be explicit dependence on the independent variable, $\lambda$.

Integrating (7.122) will produce a family of solutions due to the arbitrary constant(s) produced in the integration process. There will be one arbitrary constant per degree of freedom present in (7.122). A particular solution is specified by supplying data for enough points to specify each arbitrary constant. Two ways to do this are (1) to specify the complete initial state of $\vec{y} = [\vec{x}, \vec{p}]^T$, which in most cases is equivalent to specifying $\vec{x}$, and $d\vec{x}/d\lambda$, or (2) to specify the part of the solution, $\vec{x}$, at two points in physical space, $\vec{x}_1$ and $\vec{x}_2$. The first choice specifies the initial conditions, and solving (7.122) with this choice is referred to as the i.v.p. The second choice requires the solution to pass through two points in space but leaves the initial velocity unspecified. Solving (7.122) with the second choice is referred to as a two-point b.v.p. Methods for solving the i.v.p. are presented in detail, and the b.v.p. is addressed in later discussion. To develop the technique, the value of the dependent variables is estimated at $\lambda = \lambda'$ given the values at $\lambda = \lambda_0 < \lambda'$ by a Taylor series expansion:

$$y_k(\lambda) = y_k(\lambda_0) + \frac{dy_k}{d\lambda}\bigg|_{\lambda=\lambda_0}(\lambda-\lambda_0) + \frac{1}{2!}\frac{d^2y_k}{d\lambda^2}\bigg|_{\lambda=\lambda_0}(\lambda-\lambda_0)^2 + \cdots + \frac{1}{n!}\frac{d^ny_k}{d\lambda^n}\bigg|_{\lambda=\lambda_0}(\lambda-\lambda_0)^n + \cdots$$

The step size $h = (\lambda' - \lambda_0)$ is defined as well as the aforementioned expression at the final value to yield (7.125):

$$y_k(\lambda') = y_k(\lambda_0) + \left( \frac{dy_k}{d\lambda} \bigg|_{\lambda = \lambda_0} \right) h + \cdots + \frac{1}{n!} \left( \frac{d^n y_k}{d\lambda^n} \bigg|_{\lambda = \lambda_0} \right) h^n + \cdots \tag{7.125}$$

The coefficients of $h^n$ are evaluated. The first term is simply the initial condition, and the second term is the r.h.s. of (7.122) evaluated at the initial conditions, that is, $f_k(y_k(\lambda_0), \lambda_0)$. Higher-order terms are developed by explicitly differentiating the original equation. The quadratic term is provided as follows as an example:

$$\frac{d^2 y_k}{d\lambda^2} = \frac{df_k}{d\lambda} = \frac{\partial f_k}{\partial \lambda} + \frac{dy_j}{d\lambda} \frac{\partial f_k}{\partial y_j} \tag{7.126}$$

Equation (7.126) includes explicit differentiation of the force with respect to $\lambda$ for completeness, but will not be present in this development of the ray equation. Since the force vector is a function of the state vector, a gradient operator with respect to the state variables, $\vec{\nabla}_{\vec{y}}$, is defined. The gradient operator and the original ODE are used to define the operator in (7.127):

$$M \equiv \vec{f} \cdot \vec{\nabla}_{\vec{y}} \tag{7.127}$$

The coefficients in (7.125) can all be determined by applying this operator to the force vector:

$$y_k(\lambda') = y_{k0} + f_{k0}h + \frac{1}{2!} M(f_{k0})h^2 + \cdots + \frac{1}{n!} M^{n-1}(f_{k0})h^n + \cdots \tag{7.128}$$

Now each coefficient is expressed solely in terms of known quantities or quantities that can be determined analytically given an explicit form of the force vector. The explicit form of these operators up to second order is presented:

$$\frac{dy_k}{d\lambda} = f_k \tag{7.129}$$

$$\frac{d^2 y_k}{d\lambda^2} = M(f_k) = \vec{f} \cdot \vec{\nabla}_{\vec{y}}(f_k) \tag{7.130}$$

$$\frac{d^3 y_k}{d\lambda^3} = M^2(f_k) = \vec{f} \cdot \vec{\nabla}_{\vec{y}} \left[ \vec{f} \cdot \vec{\nabla}_{\vec{y}}(f_k) \right] = f_i \frac{\partial f_j}{\partial y_i} \frac{\partial f_k}{\partial y_j} + f_i f_j \frac{\partial^2 f_k}{\partial y_i \partial y_j} \tag{7.131}$$

Using these results a truncated Taylor expansion can be used to approximate the state for a small change in $\lambda > \lambda_0$. For small $h$ a fairly low-order expansion, for example, linear or quadratic, may suffice to produce a good estimate for a single step.

The first-order approximation is a starting point for a class of methods known as the Runge–Kutta methods that are now discussed. Using the first-order approximation only to estimate a solution to the ODE is referred to as Euler's method:

$$y_k' = y_{k0} + f_{k0}h \tag{7.132}$$

For any reasonably well-behaved function, this will always either under- or overestimate the final value. A solution for a large sequence of points, $\{\lambda_0, \lambda_1, \ldots, \lambda_{n-1}, \lambda_n, \lambda_{n+1}, \ldots, \lambda_{N-1}, \lambda_N\}$, is sought. Starting from the initial point, the state at the next point in the sequence is approximated using (7.132). This new value for the state is then used as the initial conditions for the next step. It is easy to imagine how, after just a few steps, the solution could drift considerably from the true solution. Euler's method has the feature that it does not depend on history. Each time the system is evaluated for a step, the new values are used to initialize the next step. Other than that no additional data is saved.

The distance between consecutive points need not be identical, but for many applications it is common to start out with a fixed step size throughout the entire process. Step sizes are denoted $h_n = \lambda_n - \lambda_{n-1}$. For a fixed step size evaluation, $h_n = h$ for all $n$. One argument in favor of this choice is convenience. This will be discussed again in more detail later, but in a nut shell if a fixed step is defined at the beginning of the process, the rest of the algorithm is simply looping through the steps in sequence. The Euler step approximates the solution to first order in the step size, $h$; hence we call this a first-order routine. The estimated error is on the order of $O(h^2)$. More generally a routine that approximates the solution to order $h^n$ will have an error on the order of $O(h^{n+1})$. With fixed step size routines, there is no way of checking the results until the end of the routine, and at that point results may be unreliable. A way to regulate the error in the results by developing a reasonable estimate for the error in the function values at each step is desired. With an error estimate the result can be checked at each step, and if it is found to be above a desired threshold, the step can be evaluated again with a smaller step size. Application of this process in the loop is called step size control, or regulation. The simplest step size control method is to cut the step size in half if a result fails to meet the error threshold. Once a step size is changed, a decision needs to be made as to whether or not to keep the new step size as a default or use the original step size. The latter choice could mean that at least twice as many calculations are performed in the worst-case scenario because the default step is too big. However, the former approach could force the procedure down the rabbit hole, choosing smaller and smaller step sizes, resulting in even more unnecessary evaluations and longer run time. Ideally it would be nice to be able to make a smart choice that changes the default step size so that it is either increased or decreased when needed. This is the ideal behind variable step size.

Euler's method is extended to something more sophisticated, the Runge–Kutta method. From a high-level perspective, a Runge–Kutta method uses a weighted average of several Euler-type steps within the step size, $h$. These algorithms fall into two broad categories, explicit (ERK) and implicit (IRK). The RK step is defined as follows:

$$\vec{y}_{n+1} = \vec{y}_n + h \sum_{i=1}^{s} b_i \vec{k}_i \tag{7.133}$$

$$\vec{k}_i = \vec{f}\left(\lambda_n + c_i h, \ \vec{y}_n + h \sum_{j=1}^{S'} a_{ij} \vec{k}_j\right) \tag{7.134}$$

where $S' = i - 1$ for ERK and $S' = s$ for IRK. For ERK we have $c_1 = 0$ and $a_{1j} = 0$ for all $j$. The $\{b_i\}$ are referred to as weights and the $\{c_i\}$ as nodes. The independent variable is included in (7.134) for completeness. The coefficients $\{a_{ij}\}$ are referred to collectively as the RK matrix.

The collection of coefficients is usually presented in a table referred to as Butcher *tableaux*. The standard forms for ERK and IRK are presented in (7.135) and (7.136), respectively:

$$
\begin{array}{c|ccccc}
0 & & & & & \\
c_2 & a_{21} & & & & \\
c_3 & a_{31} & a_{32} & & & \\
\vdots & \vdots & \vdots & & & \\
c_s & a_{s1} & a_{s2} & \cdots & a_{s,s-1} & \\
\hline
 & b_1 & b_2 & \cdots & b_{s-1} & b_s
\end{array}
\tag{7.135}
$$

$$
\begin{array}{c|cccc}
c_1 & a_{11} & a_{12} & \cdots & a_{1s} \\
c_2 & a_{21} & a_{22} & \cdots & a_{2s} \\
\vdots & \vdots & \vdots & & \\
c_s & a_{s1} & a_{s2} & \cdots & a_{s,s} \\
\hline
 & b_1 & b_2 & \cdots & b_s
\end{array}
\tag{7.136}
$$

The value of $s$ counts the number of stages, or intermediate steps; the RK routine will take and determine the order of the technique. Even though the RK step is a series of Euler steps inside the interval $h$, a higher order of accuracy in the solution at the end point is produced. The order is determined by explicitly Taylor expanding each intermediate step and comparing terms. The coefficients are not all free parameters, but are chosen to satisfy a set of constraints on to ensure consistency. The constraints can be derived for low-order RK methods by explicitly expanding the $\vec{k}_i$, building up an explicit form for the RK as a series, and comparing the result to the Taylor expansion of the exact solution. Requiring a term-by-term match up to the order of the method produces a set of algebraic equations connecting the sets of coefficients. This is a lengthy process and usually not presented beyond order 3 or 4. Research in this area has led to the development of techniques using graph theory to extract sets of RK coefficients. Deriving RK coefficients is beyond the scope of this discussion, and the reader is referred to the literature [22] for more details. There are a few items worth mentioning about these procedures that the reader should be aware of. First, there is an equation of consistency for ERK that will always be present:

$$
\sum_{j=1}^{i-1} a_{ij} = c_i, \quad i = 2, \ldots, s
\tag{7.137}
$$

An ERK method is referred to as consistent when this relation holds. The l.h.s. is the sum of a row of the RM matrix and implies, for example, that $a_{21} = c_2$. Next, there can be many choices of RK coefficients that satisfy the constraints, but not all of them will be stable. Stability analysis of RK methods is another specialized area of research (see Refs. [23, 24]). One of the most common RK methods is the "classical" fourth-order RK method. Its *tableaux* is provided in (7.138) as an example:

$$
\begin{array}{c|cccc}
0 & & & & \\
\dfrac{1}{2} & \dfrac{1}{2} & & & \\[2mm]
\dfrac{1}{2} & & \dfrac{1}{2} & & \\[2mm]
1 & & & 1 & \\
\hline
& \dfrac{1}{6} & \dfrac{1}{3} & \dfrac{1}{3} & \dfrac{1}{6}
\end{array}
\tag{7.138}
$$

Lastly there is the issue of order. To achieve accuracy to a desired order, $n$, will require a certain number of stages, $s$. In general, for an ERK method, one must choose $s \geq n$ if $n < 5$ and $s > n$ for higher orders. Hence, many of the common routines appearing in texts for orders up to and including 4 have $s = n$, which can lead the reader to conclude that this is a general trend when it is not. These are just some of the things to be aware of when implementing RK routines.

Now that a step algorithm has been presented, error estimation is discussed in the context of the RK method. Error in ODE solvers is estimated by taking the difference between an $n$-th-order and an $(n+1)$-th-order estimate. At first glance it would seem like one needs to take two independent steps inside the loop to develop an error estimate. Luckily there is a special class of RK methods, the embedded RK, that provide two consecutive order estimates with the same set of $\{c_i\}$ and $\{a_{ij}\}$ but with different $\{b_i\}$. An embedded RK of orders $n$ and $n-1$ is one for which the coefficients of the lower-order RK method are identical to a subset of the higher-order method, referred to as RK$n(n-1)$. A generic tableau for an embedded ERK is provided in (7.139):

$$
\begin{array}{c|ccccc}
0 & & & & & \\
c_2 & a_{21} & & & & \\
c_3 & a_{31} & a_{32} & & & \\
\vdots & \vdots & \vdots & & & \\
c_s & a_{s1} & a_{s2} & \cdots & a_{s,s-1} & \\
\hline
& b_1 & b_2 & \cdots & b_{s-1} & b_s \\
& b_1^* & b_2^* & \cdots & b_{s-1}^* & b_s^*
\end{array}
\tag{7.139}
$$

A specific example of this is the tableau for the Cash–Karp parameters for an embedded RK5 (4) method [25]. Using any embedded RK scheme allows the error at the end point of the step to be estimated:

$$
\vec{\varepsilon} = \vec{y}^{(n+1)} - \vec{y}^{(n)} = \sum_{j=1}^{s} \left( b_j - b_j^* \right) \vec{k}_j
\tag{7.140}
$$

Take note that the costliest part of the routine is evaluating the force vector, r.h.s. of (7.134), $s$ times. This depends only on the $\{c_i\}$ and $\{a_{ij}\}$; therefore once these are evaluated, each of the two order solutions is just a matter of taking the weighted sum. If output data from two independent propagation algorithms was used for the error estimate, the processing requirement would be driven by the number of force evaluations, the most expensive part of the algorithm.

The error estimate can be used to judge whether to accept a result or shrink the step size and try again. The process of changing the step size based on an error threshold is called adaptive step size control. It is possible to arrive at an "ideal" step by taking advantage of the definition of the error as the next term in the Taylor expansion. Given an $n$-th-order algorithm, the error in the estimate scales as $h^{n+1}$ based on the series expansion of the solution. The coefficients depend only on the initial data and are therefore the same regardless of the step size. Assume two errors, $\varepsilon_0$ and $\varepsilon_1$, from the same order propagator with different step sizes, $h_0$ and $h_1$, respectively. Then the ratio of the errors is related to the ratio of the step sizes:

$$\left(\frac{h_0}{h_1}\right)^{n+1} = \frac{\varepsilon_0}{\varepsilon_1} \tag{7.141}$$

This expression is for a scalar function, or a single variable; in general this ratio is formed for each variable in the vector $\vec{y}$. One of the parameter sets $(h_0, \varepsilon_0)$ represents the desired accuracy (or limit on error) and the step size necessary to achieve it, while the other set $(h_1, \varepsilon_1)$ represents the current step size and the error associated with it. Equation (7.141) provides a way to choose the next step appropriately should the current step fail to achieve our desired accuracy:

$$h_0 = h_1 \left(\frac{\varepsilon_0}{\varepsilon_1}\right)^{\frac{1}{n+1}} \tag{7.142}$$

For an array of dependent solutions, an accuracy requirement is set for each element of $\vec{\varepsilon}_0$. For multivariable problems a common strategy is to regulate the step size using the variable with the worst error, a choice that can vary from one step to the next. If the error from the output of a step is bigger than desired, (7.142) will give a smaller new step, while if the output error is less than needed, the same will give a larger choice for the next step. This type of step size control avoids the problem of the step size getting smaller and smaller by providing a prescription for tightening and loosening the step as the equation is propagated. Rather than setting an error tolerance on the forward propagated step, another technique is to back propagation the output to recover the original input. Then a threshold is applied to the back propagated result.

## 7.7 Complete Paraxial Ray Trace

Accumulating the results of the previous sections a complete first-order paraxial system is built suitable for implementing numerical techniques:

$$\frac{d^2 x^\mu}{d\lambda^2} + \Gamma^\mu{}_{\alpha\beta} \frac{dx^\alpha}{d\lambda} \frac{dx^\beta}{d\lambda} = 0 \tag{7.143}$$

$$\frac{d\hat{e}_I^\mu}{d\lambda} + \Gamma^\mu{}_{\alpha\beta} \frac{dx^\alpha}{d\lambda} \hat{e}_I^\beta = 0 \tag{7.144}$$

$$\frac{d^2 Y_J}{d\lambda^2} + \left\{ g_{\mu\rho} R^\rho{}_{\alpha\nu\beta} \frac{dx^\alpha}{d\lambda} \frac{dx^\beta}{d\lambda} \hat{e}^\mu_I \hat{e}^\nu_J \right\} Y_I = 0 \qquad (7.145)$$

The starting point is the geodesic equation, parallel transport equation for the basis vectors, and the geodesic deviation equation. Reduction of order is applied to the geodesic and geodesic deviation equations:

$$\frac{dx^\mu}{d\lambda} = p^\mu \qquad (7.146)$$

$$\frac{dp^\mu}{d\lambda} = -\Gamma^\mu{}_{\alpha\beta} p^\alpha p^\beta \qquad (7.147)$$

$$\frac{d\hat{e}^\mu_I}{d\lambda} = -\Gamma^\mu{}_{\alpha\beta} p^\alpha \hat{e}^\beta_I \qquad (7.148)$$

$$\frac{dY_J}{d\lambda} = P_J \qquad (7.149)$$

$$\frac{dP_J}{d\lambda} = -\left\{ g_{\mu\rho} R^\rho{}_{\alpha\nu\beta} \right\} p^\alpha p^\beta \hat{e}^\mu_I \hat{e}^\nu_J Y_I \qquad (7.150)$$

The vector of unknowns is defined by stacking together all the variables defined in the aforementioned equations, while the r.h.s. defines the "force" vector:

$$\boldsymbol{u} = \left[ x^\alpha, p^\beta, \hat{e}^\mu_I, Y_J, P_J \right]^T \qquad (7.151)$$

$$\boldsymbol{F} = \left[ p^\alpha, -\Gamma^\mu{}_{\alpha\beta} p^\alpha p^\beta, -\Gamma^\mu{}_{\alpha\beta} p^\alpha \hat{e}^\beta_I, P_J, -\left\{ g_{\mu\rho} R^\rho{}_{\alpha\nu\beta} \right\} p^\alpha p^\beta \hat{e}^\mu_I \hat{e}^\nu_J Y_I \right]^T \qquad (7.152)$$

The ODE solver is applied to the equation

$$\frac{d\boldsymbol{u}}{d\lambda} = \boldsymbol{F} \qquad (7.153)$$

Assuming a profile for the environmental parameters, $c$ and $\vec{v}_0$, is provided, the quantities $\Gamma^\mu{}_{\alpha\beta}$ and $g_{\mu\rho} R^\rho{}_{\alpha\nu\beta}$ are determined by the appropriate formulas. Given these parameters as functions of position and time, the derivative can be worked out by hand and code written to update the connection and tensors as a function of coordinates with each step. The full first-order system in 4-dim space-time contains 20 degrees of freedom—4 each for $x^\alpha$ and $p^\beta$, 4 for each basis vector $\hat{e}^\mu_I$, and 2 each for $Y_J$ and $P_J$. However, the system is subject to several constraints: the null constraint on ray velocity and the orthonormality conditions on the basis vectors. Applying all constraints reduces the number of independent variables to 12—four coordinates $x^\alpha$, three momentum components $p^k$, the deviation and its momentum $(Y_J, P_J)$, and a single degree of freedom for the wavefront basis set—which can be interpreted as a rotation in the plane defined by $\hat{n}$. Reducing the number of components in $\boldsymbol{u}$ can lead to issues since some of the constraints are nonlinear. Experience indicates that tracing the larger system and using the constraints to help regulate error works. To get the true geometric spread of the ray bundle in space, the output

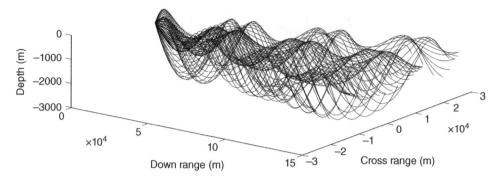

**Figure 7.7**   Three-dimensional ray trace for the Garrett–Munk profile

$Y_I$ is projected onto the auxiliary basis, $\tilde{e}_1$, defined in (7.26). From this the area of a cross section of the ray bundle centered on the ray can be determined.

Figure 7.7 shows an example of a 3-dim ray trace for a point source in the deep underwater sound channel described by the Garrett–Munk profile.

## 7.8   Implementation Notes

Designing a ray trace procedure that accounts for all the effects listed in this chapter requires a bit of planning. There are several ways to utilize the various results presented in this chapter. The most generic ray trace routine will involve a large number of degrees of freedom and require more steps in post-processing data to evaluate the geometry of caustic surfaces. It is generally a good idea to limit the scope of development to suite a specific type of problem. For depth-dependent sound speed profiles and no fluid flow, a common model of underwater environments, the full ray system reduces to two degrees of freedom in spatial coordinates, there is only one paraxial vector to trace, and the wavefront basis is determined completely by constraints. The reduced order system will contain twice as many degrees of freedom. It is possible to evaluate the ray spreading by numerically differencing the output of the ray trace data, but this will inject error in the results and become less reliable as rays spread farther apart. This requires filling in the ray trace with extra data to increase fidelity. The paraxial approach always yields a measure of the instantaneous geometric spread in the neighborhood of the ray, and the extra degree(s) of freedom will be checked for precision along with the ray coordinates. In theory, if one can locate the path connecting the source and receiver using a paraxial approach, the information to find the amplitude without additional finite differencing is part of the output. Again, for depth-dependent profiles, it may be tempting to use the first-order ODE system derived from the conservation laws. One drawback with this is that these equations involve square roots for the initial conditions and for evaluating $dz/dt$. This leads to additional work to ensure complex values are not accidentally propagated. The second-order equations are the safest approach to avoiding extra work in this regard. Then the expected conserved quantities can be used to regulate the steps.

   It is worth noting that the exact solutions for simple background profiles can be used to develop models of more sophisticated environments by replacing the sound speed and wind profiles by piecewise linear profiles and applying boundary conditions on the ray path variables at the interface. This type of approach reduces the problem of tracing rays to piecing together segments of ray paths by finding appropriate initial conditions and evaluating the integrals. The appropriate boundary conditions can be derived from the original ray equations and requirements of continuity.

Including boundaries introduces the need to check whether a ray path has encountered a boundary at each step and applying to appropriate ray path reflection and phase shift with each bounce. For environments with refractive media and boundaries, it is sometimes possible to predict which rays will interact with boundaries and which will not prior to propagation, but in general this will not be the case. All aspects to the paraxial ray trace are local, including a count of the number of caustics encountered along the path, allowing for parallelization methods to be applied. The only part of a ray-based field analysis that will require nonlocal analysis is the evaluation of caustic corrections should a receiver be placed on or near a caustic.

## 7.9   Gaussian Beam Tracing

This chapter closes with a brief description of a novel approach to using ray traces for field calculations that avoids some of the issues encountered with caustics. The method, known as Gaussian beam tracing, can be thought of as a hybrid technique. The idea is to develop the wave equation in ray-centric coordinates initially, rather than as an afterthought to fix the amplitude near a caustic. To illustrate the process highlights from Porter and Bucker [26] are presented here. A detailed presentation of 3-dim Gaussian beam tracing can be found in Cerveny [2]. The starting point is the Helmholtz equation with a depth-dependent sound speed profile, that is, a cylindrically symmetric medium. A coordinate transformation is performed in which a specific ray path is used as one curvilinear coordinate and a set of normal directions used for the other two. In the case of a cylindrically symmetric medium, one of the normal directions is the azimuthal coordinate, $\phi$. For a point source in this type of medium, the solution will not depend on the azimuthal coordinate. This is included to properly account for the scale factor in the coordinate transformation of the Laplacian operator. Derivations usually assume arclength parameterization of the 3-dim ray paths, but the derivation can be performed using travel time or any other parameterization. It should be noted that if other parameterizations are used, slightly different forms of the final equation are obtained. The new ray-centric coordinates are $(s, y, \varphi)$, where $s$ is the arclength along the ray and $y$ the coordinate in the direction perpendicular to the ray at each point $s$, borrowing notation from the paraxial ray trace procedure presented in this chapter. The transformation from cylindrical coordinates, $(\rho, \varphi, z)$, to ray-centric coordinates will involve derivatives of one system with respect to the other, for example, $\partial \rho / \partial s$ and $\partial \rho / \partial y$. It is assumed that the solution has the form

$$p(s,y) = P(s,y)\exp(-i\omega\tau(s)) \tag{7.154}$$

The travel time along each ray expressed as a function of the ray parameter is

$$\tau(s) = \int \frac{ds}{c(s)} \tag{7.155}$$

The phase in (7.154) involves the local sound speed on a specific ray path, whereas the Helmholtz equation involves the sound speed as function of all coordinates. Both depend on $s$, so the latter will be denoted as $c(y)$ and the former simply as $c$, the dependence on arclength being understood. In addition to this ansatz, a coordinate scaling is introduced, $\eta = \sqrt{\omega} y$. The combination of these three actions—coordinate transformation, introduction of (7.154), and coordinate scaling—leads to the following form of the Helmholtz equation:

$$\omega^2 h f_0 P + \frac{\omega}{hc}\left(if_1 P - 2i\frac{\partial P}{\partial s} + h^2 c\frac{\partial^2 P}{\partial \eta^2}\right) + \sqrt{\omega}hf_2\frac{\partial P}{\partial \eta} + \frac{1}{h}\frac{\partial^2 P}{\partial s^2} - \frac{1}{h^3}f_3\frac{\partial P}{\partial s} \qquad (7.156)$$

The following functions are defined in (7.156):

$$f_0 = \left(\frac{1}{c^2(y)} - \frac{1}{h^2 c^2}\right) \qquad (7.157)$$

$$f_1 = \left(\frac{1}{c}\frac{\partial c}{\partial s} + \frac{1}{h}\frac{\partial h}{\partial s} - \frac{1}{r}\frac{\partial r}{\partial s}\right) \qquad (7.158)$$

$$f_2 = \left(\frac{1}{h}\frac{\partial h}{\partial y} + \frac{1}{r}\frac{\partial r}{\partial y}\right) \qquad (7.159)$$

$$f_3 = \left(\frac{1}{h}\frac{\partial h}{\partial s} - \frac{1}{r}\frac{\partial r}{\partial s}\right) \qquad (7.160)$$

$$h = 1 + \frac{1}{yc}\frac{\partial c}{\partial y} \qquad (7.161)$$

In (7.156) terms are ordered in decreasing powers of $\omega$. The functions appearing in (7.157)–(7.160) are understood to be evaluated at $(s, \eta)$. At this point terms containing $c(y)$ and $h$ are Taylor expanded in the variable $\eta$. The field amplitude, $P(s, y)$, is then assumed to have an expansion in powers of $\omega^{-1/2}$:

$$P = \sum_{n=0}^{\infty} P^{(n)} \omega^{-n/2} \qquad (7.162)$$

Combining these results leads to the following equation to first order in $\omega$:

$$\left(-\frac{1}{c^3}\frac{\partial^2 c}{\partial y^2}\eta^2 + \frac{i}{c^2}\frac{\partial c}{\partial s} - \frac{i}{rc}\frac{\partial r}{\partial s}\right)P^{(0)} - \frac{2i}{c}\frac{\partial P^{(0)}}{\partial s} + \frac{\partial^2 P^{(0)}}{\partial \eta^2} = 0 \qquad (7.163)$$

This can be thought of as a paraxial approximation to the wave equation, or parabolic equation, along a specific ray path. A new field is introduced $P^{(0)} = W(s,\eta)\sqrt{c/r}$ and the final form of the equation becomes

$$-\frac{2i}{c}\frac{\partial W}{\partial s} + \frac{\partial^2 W}{\partial \eta^2} - \eta^2\left(\frac{1}{c^3}\frac{\partial^2 c}{\partial y^2}\right)W = 0 \qquad (7.164)$$

Two more substitutions are made, $W(s,\eta) = w(s)\exp(-i\eta^2\gamma(s)/2)$ and $\gamma(s) = \alpha(s)/\beta(s)$. This substation leads to the following set of equations for $\alpha(s)$, $\beta(s)$:

$$\frac{\partial\beta}{\partial s} = c\alpha \tag{7.165}$$

$$\frac{\partial\alpha}{\partial s} = -\frac{1}{c^2}\frac{\partial^2 c}{\partial y^2}\beta \tag{7.166}$$

The amplitude is then found to be $w = w_0\sqrt{\beta}$. With solutions to these last two equations, the field is developed:

$$p(s,y) = w_0\sqrt{\frac{c}{r\beta}}\exp\left(-i\omega\left(\tau + \frac{y^2\alpha/\beta}{2}\right)\right) \tag{7.167}$$

All quantities in (7.167) are evaluated on the ray path, that is, are functions of $s$ only. It was stated previously that these equations can be derived for any ray parameter; in particular one can take advantage of the nature of the ray paths to scale them by any function of coordinates. In particular, using the parameterization introduced in the first section for describing rays as geodesics leads to a form of (7.165) and (7.166) that is identical to the equation for geodesic deviation used in the paraxial ray trace. Clearly the meaning of the parameters is different when used for a Gaussian beam calculation, but the form of the equations already developed can be reused to evaluate these parameters.

## 7.10   Exercises

1.  Given the following sound speed profile

$$c(z) = A\exp(\alpha z) + B\exp(\beta z)$$

where $A$, $B$, $\alpha$, and $\beta$ are constants,
    (a) Determine the ray paths and travel times using the ray path integrals.
    (b) Evaluate for $A = B$ and $\alpha = -\beta$.
    (c) Either prove or demonstrate by example the ideal focusing feature of this profile.
2.  Generalize the ray path integrals for the case when the background fluid velocity has all three components: $v_{0x}$, $v_{0y}$, and $v_{0z}$.
    (a) Analyze the special case when the flow is perpendicular to the initial ray path, for example, the ray starts in the $x$–$z$ plane and $v_{0y}$ is the only nonzero component.
    (b) Derive the ray path integrals for 2-dim rays in the $x$–$z$ plane with $c$, $v_{0y}$, and $v_{0z}$, all functions of $z$.
    (c) Starting from the conservation laws along the ray path, derive a result for the case where the environmental parameters $c$, $\vec{v}_0$ are independent of coordinates but functions of time.
3.  Assuming a purely 2-dim spatial environment with sound speed profile $c(x, z)$,
    (a) Derive the Christoffel symbols using the definition from Chapter 4.
    (b) Derive the components of the Riemann tensor (7.34).

4. Using MATLAB or an open-source equivalent, write a routine that will apply the RK algorithm to the ray equations and test out on the ideal waveguide example presented in this chapter. Compare results with the exact solution.

# References

[1] Bergman, D. R., Internal symmetry in acoustical ray theory, Wave Motion, Vol. 43, pp. 508–516, 2006.

[2] Cerveny, V., Seismic Ray Theory, Cambridge University Press, Cambridge, 2001.

[3] Guenther, R., Modern Optics, John Wiley & Sons, Inc., New York, 1990, (specifically Chapter 5, see section entitled "Propagation in a Graded Index Optical Fiber" for a wave guide paraxial ray approximation).

[4] Abromowicz, A. and Kluzniak, W., Epicyclic orbital oscillations in Newton's and Einstein's dynamics, Gen. Relativ. Gravit., Vol. 35, No. 1, pp. 69–77, 2003.

[5] Bazanski, S. L., Kinematics of relative motion of test particles in general relativity, Ann. Inst. Henri Poincare A, Vol. 27, No. 2, pp. 115–144, 1977.

[6] Hawking, S. W. and Ellis, G. F. R., The Large Scale Structure of Space: Time, Cambridge University Press, Cambridge, 1973.

[7] Schneider, P., Ehlers, J., and Falco, E. E., Gravitational Lenses, Astronomy and Astrophysics Library, Springer-Verlag, Berlin, 1992.

[8] Blokhintzev, D., The propagation of sound in an inhomogeneous and moving medium, I, J. Acoust. Soc. Am., Vol. 18, No. 2, pp. 322–328, 1945.

[9] Boone, M. M. and Vermaas, E. A., A new ray-tracing algorithm for arbitrary inhomogeneous and moving media, including caustics, J. Acoust. Soc. Am., Vol. 90, No. 4 (Part 1), pp. 2109–2117, 1991.

[10] Ludwig, D., Uniform asymptotic expansion at a caustic, Commun. Pure Appl. Math., Vol. 19, pp. 215–250, 1966.

[11] Foreman, T., A Frequency Dependant Ray Theory, Technical Report, ARL-TR-88-17, Applied Research Laboratories, The University of Texas at Austin, Austin, 1988.

[12] Brekhovskikh, L. M., Waves in Layered Media, Academic Press, New York, 1960.

[13] Kravtsov, Yu. A. and Orlov, Yu. I., Caustics, Catastrophes, and Wave Fields, Second Edition, Springer Series on Wave Phenomena, Lotsch, H. K. V., Managing Editor, Springer, Heidelberg, 1999.

[14] Franchi, E. and Jacobson, M., Ray propagation in a channel with depth: Variable sound speed and current, J. Acoust. Soc. Am., Vol. 52, No. 1 (Part 2), pp. 316–331, 1972.

[15] Heller, G. S., Propagation of acoustic discontinuities in an inhomogeneous moving liquid medium, J. Acoust. Soc. Am., Vol. 25, No. 5, p. 950, 1953.

[16] Rudenko, O. V., Sukhorukova, A. K., and Sukhorukov, A. P., Full solutions to the equations of geometrical acoustics in stratified moving media, Acoust. Phys., Vol. 43, No. 3, pp. 339–343, 1997.

[17] Thompson, R. J., Ray theory for an inhomogeneous moving medium, J. Acoust. Soc. Am., Vol. 51, No. 5 (Part 2), p. 1675, 1972.

[18] Ugincius, P., Acoustic-ray equations for a moving, inhomogeneous medium, J. Acoust. Soc. Am., Vol. 37, No. 3, pp. 476–479, 1965.

[19] Ugincius, P., Ray acoustics and Fermat's principle in a moving inhomogeneous medium, J. Acoust. Soc. Am., Vol. 51, No. 5 (Part 2), pp. 1759–1763, 1972.

[20] Kornhauser, E. T., Ray theory for moving fluids, J. Acoust. Soc. Am., Vol. 25, No. 5, pp. 945–949, 1953.

[21] Bergman, D. R., Generalized space-time paraxial acoustic ray tracing, Waves Random Complex Media, Vol. 15, No. 4, pp. 417–435, 2005.

[22] Iserles, A., A First Course in the Numerical Analysis of Differential Equations, Cambridge University Press, Cambridge, 2002.

[23] Butcher, J. C., The Numerical Analysis of Ordinary Differential Equations, Runge-Kutta and General Linear Methods, John Wiley & Sons, Ltd, Chichester, 1987.

[24] Gottlieb, S., Ketcheson, D., and Shu, C., Strong Stability Preserving Runge–Kutta and Multistep Time Discretizations, World Scientific, River Edge, 2011.

[25] Press, W. H., Teukolsky, S. A., Vetterling, W. T., and Flannery, B. P., Numerical Recipes in C++: The Art of Scientific Computing, Second Edition, Cambridge University Press, Cambridge, 2005.

[26] Porter, M. and Bucker, H., Gaussian beam tracing for computing ocean acoustic fields, J. Acoust. Soc. Am., Vol. 82, No. 4, pp. 1349–1359, 1987.

# 8

# Finite Difference and Finite Difference Time Domain

## 8.1 Introduction

This chapter presents the finite difference (FD) technique in the frequency domain and the finite difference time domain (FDTD) method. The former is applied in the frequency domain primarily to the second-order scalar wave equation with spatial inhomogeneities. The FD method involves discretizing the differential equation by creating finite difference representations of the derivatives in the spatial variable on a discrete lattice of points in space. The continuum is replaced by points, and the continuous operators replace with couplings between the discrete points designed to provide estimates for those operators to some desired accuracy. The resulting equation is a matrix equation $\mathbf{A} \cdot \mathbf{x} = \mathbf{b}$, where the components of $\mathbf{x}$ are the values of the unknown scalar field at the lattice points, $\mathbf{b}$ is the source term evaluated on the lattice, and the matrix $\mathbf{A}$ is a sparse matrix containing terms related to the discretized differential operators and evaluations of the sound speed, density, and other environmental fields evaluated on the lattice.

The application of finite difference techniques to the time derivative introduces a new set of issues to deal with. In this case a first-order system is preferable. The basic equations required for propagating an initial field configuration forward in time are presented for the first-order linear equations of acoustics, with pressure and particle velocity as the field variables. The primary FD scheme is still valid, but splitting of the variables is required. A method borrowed from computational electrodynamics, known as Yee's method, is presented as a procedure for an FDTD approach to acoustics. For the FDTD approach an initial field is propagated forward in time in tiny steps. In this case processing time is spent on the numerous applications, possibly tens of thousands, of the matrix to the updated source terms in a feedback loop. This last statement depends on the method being explicit in time. For implicit methods matrix inversion is required at each step. This chapter presents the basic theory behind developing FD and FDTD operators and the building up of FD/FDTD simulations.

*Computational Acoustics: Theory and Implementation*, First Edition. David R. Bergman.
© 2018 John Wiley & Sons Ltd. Published 2018 by John Wiley & Sons Ltd.

## 8.2 Finite Difference

This section introduces finite differencing applied to the wave equation in the frequency domain. This will serve as a foundation for discussing approaches to modeling time evolution using finite differencing schemes. The starting point is expressing first and second derivatives of an unknown function of a single variable, which will be generalized to two and three dimensions. Consider a scalar function of a single spatial coordinate, $f(x)$. It is assumed that $f$ obeys a second-order linear differential equation in the interval $I = (x_1, x_2)$ with boundary conditions at the end points. A finite set of sampling points, typically equally spaced, is defined in $I$, $x_n \in I$, and $n = 1, ..., N$, and the unknown function is evaluated on these sample points:

$$f_n = f(x_n) \tag{8.1}$$

This set of sample points will be called a lattice or grid. For the time being it will not be assumed that the samples are uniformly spaced. What is needed is an evaluation of the derivatives of $f$, at least to the second order, at the grid points. Focusing attention on a specific point $x_n$, the value of the function near this point can be expressed by a Taylor expansion. Evaluating the function at a point close to $x_n$, $x_n + h$ where $h > 0$ is a small displacement from $x_n$, and applying Taylor's theorem gives the following:

$$f(x_n + h) = f(x_n) + f'(x_n)h + \frac{1}{2}f''(x_n)h^2 + \cdots \tag{8.2}$$

Equation (8.2) is solved for the first derivative:

$$f'(x_n) = \frac{f(x_n + h) - f(x_n)}{h} - \frac{1}{2}f''(x_n)h - \cdots \tag{8.3}$$

The first term is taken as an approximation to the derivative. The leading term in the error, or the difference between the derivative and this approximation, is of the order $h$. This would be referred to as a first-order approximation. This is also called a forward difference since $x + h > x$. A similar approximation can be made using a backward difference. Evaluating the function at $x - l$ and $l > 0$ and performing the same steps gives the backward difference:

$$f'(x_n) = \frac{f(x_n) - f(x_n - l)}{l} + \frac{1}{2}f''(x_n)l - \cdots \tag{8.4}$$

Going back to the original Taylor expansion for both and taking the difference gives the following:

$$f(x_n + h) - f(x_n - l) = f'(x_n)(h + l) + \frac{1}{2}f''(x_n)(h^2 - l^2) + \cdots \tag{8.5}$$

Again, solving for $f'(x_n)$:

$$f'(x_n) = \frac{f(x_n + h) - f(x_n - l)}{(h + l)} + \frac{1}{2}f''(x_n)(h - l) + \frac{1}{3!}f'''(x_n)\frac{(h^3 + l^3)}{(h + l)} + \cdots \tag{8.6}$$

Choosing $l = h$ leads to the central difference approximation for the derivative:

$$f'(x_n) = \frac{f(x_n + h) - f(x_n - h)}{2h} + \frac{1}{3!}f'''(x_n)h^2 + \cdots \tag{8.7}$$

Notice that the leading order for the error is now $h^2$. By taking the average over two equal-sized intervals, a better estimate is achieved. Second derivatives are approximated by adding the expansion for $x + h$ and $x - h$:

$$f(x_n + h) + f(x_n - h) = 2f(x_n) + f''(x_n)h^2 + \frac{1}{12}f^{iv}(x_n)h^4 + \cdots \tag{8.8}$$

Notice that in the process the third-order term has been cancelled. Solving for $f''(x_n)$,

$$f''(x_n) = \frac{f(x_n + h) + f(x_n - h) - 2f(x_n)}{2h^2} + O(h^2) \tag{8.9}$$

Equation (8.9) is a second-order approximation to the second derivative at $x_n$. This discussion started out with arbitrary spacing between points, but it becomes clear that the finite difference formulae are more accurate in this representation for equally spaced points. From this point forward, with the exception of handling boundary conditions, it will be assumed that the points are described by the sequence, $x_n = nh$, $n = 1, \ldots, N$, and $h = x_{n+1} - x_n$ for all $n$ up to $N - 1$. Based on the definitions of these finite difference derivatives, it is clear that there will only be enough data to form derivatives at $N - 2$ points. There will not be derivatives at $n = 1, N$. These will be boundary points and boundary conditions will be imposed on them. The points $n = 2, \ldots, N - 1$ are called interior points.

To extend this approach to two and three dimensions, a lattice of points is defined within the region where the differential equation is defined. It is assumed that each dimension is sampled at uniform intervals. Furthermore, it is assumed that the sampling size is the same for all dimensions. An index is used for each direction in space, $(i, j, k)$ for $(x, y, z)$, with ranges, $i, j, k = 1, \ldots, N_1, N_2, N_3$. A point in the space is described as follows, where a reference point, $(x_0, y_0, z_0)$, is included:

$$[x_i, y_j, z_k] = [x_0, y_0, z_0] + h[(i-1), (j-1), (k-1)] \tag{8.10}$$

To convert a PDE into a discrete matrix equation, it will be convenient to map a double or triple indexed point into a single index. This can be accomplished, for example, by the following function:

$$m(i, j, k) = i + N_1(j-1) + N_1 N_2(k-1) \tag{8.11}$$

For two dimensions set, $N_2 = 0$. The values of the function, for the time being, will be referenced by the multi index set.

$$f_{i,j,k} = f(x_i, y_j, z_k) \tag{8.12}$$

A second-order estimate for the Laplacian operator using finite differences is developed by using the 1-dimensional (1-dim) central difference as many times as needed. In two and three dimensions, this gives the following:

$$\left(\nabla^2 f\right)_{i,j,k} = \frac{f_{i+1,j,k} + f_{i-1,j,k} + f_{i,j+1,k} + f_{i,j-1,k} - 4f_{i,j,k}}{h^2} \tag{8.13}$$

$$\left(\nabla^2 f\right)_{i,j,k} = \frac{f_{i+1,j,k} + f_{i-1,j,k} + f_{i,j+1,k} + f_{i,j-1,k} + f_{i,j,k+1} + f_{i,j,k-1} - 6f_{i,j,k}}{h^2} \tag{8.14}$$

respectively. Mixed partials require a little more thought. To get an estimate for the mixed partials at a point, one needs the values of the first derivative at neighboring points. For example, consider the mixed partial $\partial^2 f/\partial x \partial y$ in three dimensions:

$$\left(\frac{\partial^2 f}{\partial y \partial x}\right)_{i,j,k} = \frac{1}{2h}\left(\left(\frac{\partial f}{\partial x}\right)_{j+1} - \left(\frac{\partial f}{\partial x}\right)_{j-1}\right) \tag{8.15}$$

Equation (8.7) is used to arrive at the previous equation, treating $\partial f/\partial x$ as a function. Applying (8.7) again to each term in (8.15) provides a finite difference for the mixed partial:

$$\left(\frac{\partial^2 f}{\partial y \partial x}\right)_{i,j,k} = \frac{f_{i+1,j+1,k} + f_{i-1,j-1,k} - f_{i+1,j-1,k} - f_{i-1,j+1,k}}{4h^2} \tag{8.16}$$

The simplest approximations have been made to illustrate the process of converting differential operators into finite difference operators.

With these definitions in place, the wave equation is converted to a linear system in the unknown variables $f_{i,j,k}$. For illustrative purposes consider the Helmholtz equation with a position-dependent sound speed. In finite difference form, this equation becomes

$$f_{i+1,j,k} + f_{i-1,j,k} + f_{i,j+1,k} + f_{i,j-1,k} + f_{i,j,k+1} + f_{i,j,k-1} - 6f_{i,j,k} + \frac{h^2\omega^2}{c_{i,j,k}}f_{i,j,k} = -h^2 S_{i,j,k} \tag{8.17}$$

The notation for $c_{i,j,k}$ and $S_{i,j,k}$ follows that of $f_{i,j,k}$. In the l.h.s. of (8.17), the coefficients of $f_{i,j,k}$ are the entries of a matrix. What is required is knowing the correct placement for these coefficients. It is here that the use of the index mapping in (8.11) is useful. Focusing attention on interior points (8.14) couples seven field variables. These are the nearest neighbors given by the following set of indices in Table 8.1.

A convenient representation of the finite difference formulae derived previously is the use of a stencil, a visual representation made from a piece of the lattice that represents the points used in the derivative operator along with the corresponding weights. Stencils for the first and second derivatives accurate to second order are presented in Table 8.2 for 1-, 2-, and 3-dim spaces. A solid dot represents a point used in the derivative operator, while hollow points do not contribute, for example, the first derivative stencils representing a central difference. The "weights" are the coefficients of the function values used in definition of each finite difference operator, for example, the seven-point difference in (8.14).

**Table 8.1** Indices for neighboring points contributing to Laplacian in 3 dim

| Three-index set | Equivalent $m$ | Shorthand |
|---|---|---|
| $i,j,k$ | $m$ | $m_0$ |
| $i+1,j,k$ | $m+1$ | $m_2$ |
| $i-1,j,k$ | $m-1$ | $m_1$ |
| $i,j+1,k$ | $m+N_1$ | $m_4$ |
| $i,j-1,k$ | $m-N_1$ | $m_3$ |
| $i,j,k+1$ | $m+N_1N_2$ | $m_6$ |
| $i,j,k-1$ | $m-N_1N_2$ | $m_5$ |

**Table 8.2** Stencils for spatial derivative

| | Stencils for first and second derivative to second order | |
|---|---|---|
| Dim | First derivative | Second derivative |

Since boundary conditions are placed on the faces of the region, the matrix will not be filled for every value of the indices using these stencils. Implementing boundary conditions in a finite difference scheme requires identifying boundary points on the grid. This is straightforward when the boundaries coincide with the Cartesian coordinate planes. For a rectangular enclosure in three dimensions, there are six such boundary planes. Boundaries are located at $i = 1, N_1$, $j = 1, N_2$, and $k = 1, N_3$. The index sets for each face are presented in Table 8.3.

Figure 8.1 provides an example of a grid for a two-dimensional problem with interior points and boundary points identified with filled circles and asterisks, respectively. It is clear by the nature of the finite difference scheme that a boundary will be required; otherwise the grid would extend to infinity. This is not an issue for interior problems, but for using the finite difference method on free problems, an artificial boundary condition that simulates this is required. Problems with curved boundaries require additional work. For a curved or irregularly shaped boundary, the intersection of the coordinate lines with the boundary is determined. These serve as boundary points. Far enough in the interior region, the uniform grid structure will suffice to

**Table 8.3**   Index sets for boundary points

| Plane | $(i, j, k)$ index | $m$ index |
|---|---|---|
| $y$–$z$ plane at $x_1$ | $(1, j, k)$ | $1 + N_1(j-1) + N_1 N_2(k-1)$ |
| $y$–$z$ plane at $x_{N_1}$ | $(N_1, j, k)$ | $N_1 j + N_1 N_2(k-1)$ |
| $x$–$z$ plane at $y_1$ | $(i, 1, k)$ | $i + N_1 N_2(k-1)$ |
| $x$–$z$ plane at $y_{N_2}$ | $(i, N_2, k)$ | $i - N_1 + N_1 N_2 k$ |
| $x$–$y$ plane at $z_1$ | $(i, j, 1)$ | $i + N_1(j-1)$ |
| $x$–$y$ plane at $z_{N_3}$ | $(i, j, N_3)$ | $i + N_1(j-1) + N_1 N_2(N_3-1)$ |

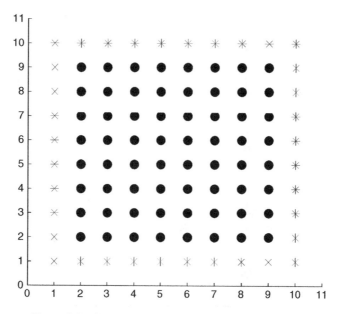

**Figure 8.1**   Sample 2-dim grid with rectangular boundary

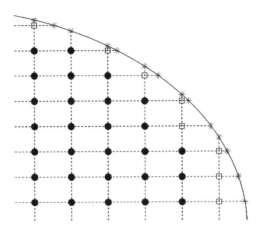

**Figure 8.2** Example of irregular boundary illustrating interior, boundary, and near points

discretize the equation. For points that are close to the boundary, there will be irregular distances in the grid for the finite difference approximation to the derivatives. This situation is illustrated in Figure 8.2. A cutaway of a small portion of a two-dimensional grid with a circular boundary is shown. Interior points are indicated by solid black dots, boundary points by asterisks, and near-boundary points (or near points for short [1]) by squares. For any interior point the finite difference equation for the second derivatives will work. At the points indicated by squares, a new finite difference operator will be needed.

As a first guess one might just use the nearest four points on the grid and boundary to develop a finite difference. The problem with this is that, if one attempts to use the five-point formula in two dimensions for near points in this case, the order of the operator will be reduced. Rather, one would desire finite difference operators for these points that are the same order as those for the interior. This can be fixed by using a six-point formula that includes an additional interior point. For an internal grid spacing of $h$, consider a boundary point that is a distance $l < h$ away from the grid point where the operator is to be applied. For the one-dimensional case, this amounts to taking the points $x-h$, $x-2h$, and $x+l$ and Taylor expanding the function about $x$ for each:

$$f(x-h) = f(x) - f'(x)h + \frac{1}{2}f''(x)h^2 - \frac{1}{6}f'''(x)h^3 + \cdots \tag{8.18}$$

$$f(x-2h) = f(x) - f'(x)2h + \frac{1}{2}f''(x)4h^2 - \frac{1}{6}f'''(x)8h^3 + \cdots \tag{8.19}$$

$$f(x+l) = f(x) + f'(x)l + \frac{1}{2}f''(x)l^2 + \frac{1}{6}f'''(x)l^3 + \cdots \tag{8.20}$$

To develop a suitable nonsymmetric formula for the second derivative, take a linear superposition of the three previous equations and require the coefficients of $f'(x)$ and $f'''$ to vanish:

$$c_1 f_{-h} + c_2 f_{-2h} + c_3 f_l = a_1 f - a_2 f' + \frac{1}{2}a_3 f'' - \frac{1}{6}a_4 f''' + \cdots \tag{8.21}$$

In (8.21) the following have been defined: $a_1 = (c_1 + c_2 + c_3)$, $a_2 = ((c_1 + 2c_2)h - c_3 l)$, $a_3 = ((c_1 + 4c_2)h^2 + c_3 l^2)$, and $a_4 = ((c_1 + 8c_2)h^3 - c_3 l^3)$. Setting $a_2 = a_4 = 0$ gives $c_2$ and $c_3$ in terms of $c_1$. With $c_1 = 1$ the other two coefficients are

$$c_2 = -\frac{1}{2}\frac{h^2 - l^2}{4h^2 - l^2} \tag{8.22}$$

$$c_3 = \frac{1}{l}\frac{3h^3}{4h^2 - l^2} \tag{8.23}$$

and a second-order expression for the second derivative near a boundary point is obtained:

$$f'' = 2\frac{c_1 f_{-h} + c_2 f_{-2h} + c_3 f_l - a_1 f}{a_3} \tag{8.24}$$

Although this formula was obtained in one dimension, the technique applies equally to any of the three directions. If a near point is close enough to boundary points in multiple directions to destroy the uniformity of the grid spacing, this formula is applied to each direction to repair the loss of order in the usual stencil.

There is still one issue related to the application of boundary conditions. For Dirichlet boundary conditions the unknown field will be set equal to constant at the boundary points. In terms of the matrix equation, this is represented by the inclusion of a 1 in the diagonal whose index corresponds to the boundary point. The source vector will then have 0 and the same index value. The effect is trivial and one could just strike the boundary points from the problem as long as the effect on the remaining equations is accounted for correctly. Neumann conditions constrain the normal derivative at the boundary. To implement Neumann boundary conditions, additional equations need to be added to the system. One approach is to add a fictitious point outside the boundary. Then the full wave equation is evaluated at the boundary point with the new point included. To illustrate this, consider a one-dimensional system with a Neumann boundary at $m = 1$. Let the fictitious point be indexed $m = 0$. Then the second-order operator evaluated at boundary is

$$\left.\frac{d^2 f}{dx^2}\right|_{m=1} = \frac{f_2 + f_0 - 2f_1}{h^2} \tag{8.25}$$

The Neumann boundary condition can be expressed as

$$\frac{f_2 - f_0}{2h} = \beta \tag{8.26}$$

where $\beta$ is the value of the derivative at the boundary. Equation (8.26) can be solved for $f_0$ and the result included in (8.25):

$$\left.\frac{d^2 f}{dx^2}\right|_{m=1} = \frac{2f_2 - 2f_1 - 2h\beta}{h^2} \tag{8.27}$$

This mechanism is necessary to maintain consistency with respect to the order within the system. Applying a forward difference method to the equation would inject a first-order equation into a second-order system. The inclusion of fictitious points was just a device to get a second-order accurate boundary condition for implementing Neumann boundary conditions, but in the end that point is not part of the system. The effect is present in the modified wave equation at boundary points using (8.27) for the second derivative operator. It is possible to avoid this mechanism by simply using second-order forward difference for the first derivative at the boundary. In the end the result of boundary conditions will be the inclusion of additional constraint equations to the finite difference matrix. For an ideal hard boundary, $\beta = 0$.

When the boundary is curved, the situation is a little more complicated. The Neumann boundary conditions constrain the normal derivative of the field, $\hat{n} \cdot \overline{\nabla} f = \beta$, at the boundary. Given that the entire domain of the problem, including the boundary, is a set of points, one now has the task of providing both, a second-order version of the gradient operator at the boundary and an estimate of the unit normal to the boundary curve or surface. Given the previous discussions on dealing with operators on irregularly spaced lattices and the use of fictitious points and extended point operators, the reader can combine this information to develop a suitable second-order finite difference operator for the gradient at boundary points. The local normal will have to be estimated given the discrete data. To construct an irregular boundary for a lattice requires some model of the boundary, either a parameterized equation for some regular geometry such as a sphere or cylinder or a facet file model of the boundary. This model is used to apply a data cut to the uniform spaced lattice, and the intersection of the coordinate lines with the surface model used to generate boundary points. This process ruins the three-index to one-index map used for filling the matrix, but a new map can be generated once the set of points is finalized. The surface model used to cut the data contains enough information to estimate local normal vectors. For example, when the boundary is described by a parameterized curve or surface, normal vectors for each lattice point on the boundary can be determined ahead of time and used as input data for a simulation. When the boundary is described as a tiled surface, the data points associated with each tile can be used to generate the normal and mapped to the boundary points that intersect that tile. These are just a couple of ideas for producing the required information for a Neumann boundary condition in three dimensions. An illustrative example is provided in Figure 8.3. Shown are (1) a cubic grid of evenly spaced points, (2) a representation of a spherical surface used as a boundary, and (3) the remaining points in the volume after boundary is applied. The reader may notice that for a sphere the triangular

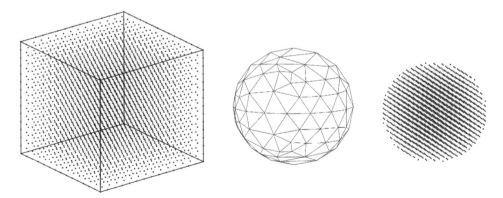

**Figure 8.3**  Cutting of a cubic point lattice by a sphere

tiles version is not necessary, but it is used to illustrate what might occur in more realistic situations without unnecessary complexity. The image of the sphere is intentionally undersampled to make the figure clear. Figure 8.4 shows an enlargement of one of the triangles. Depending on how the surface was made, *e.g.* what geometric software package was used, and the file format of the data, the outward pointing normal vectors for each tile may be provided. In case they are not, the cross product of two edges in the correct order will provide this.

At the risk of belaboring the point, another example in two dimensions is shown in Figure 8.5. Illustrated in this figure is the cutting process for one line of lattice points in the $x$ direction by an irregular boundary represented by a polygonization. Having identified where the lattice line intersects, the boundary provides locations for specific boundary points as well as a normal direction given the slope of the intersecting line on the curve. This process is repeated for all lattice lines creating a full set of interior points, boundary points, and normal direction for each.

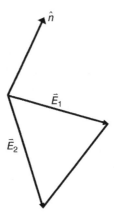

**Figure 8.4**   Local normal derived from edge vectors of a triangular patch

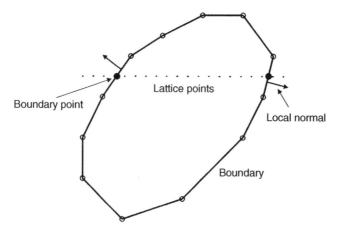

**Figure 8.5**   Identification of interior and boundary points and local normal vectors for implementing Neumann boundary conditions in 2 dim

The finite difference operators and normal vectors for boundary evaluation have been discussed in the context of ideal Dirichlet and Neumann boundary conditions. This can be extended to more general cases when either the field, its normal derivative, or a linear combination is equated to a constant, fairly easily. Modeling waves free to travel to infinity, that is, without boundary, requires the introduction of special boundary conditions to which the reader is referred to the literature for more information [2, 3].

Before continuing to other topics, an example is presented to illustrate the exact form of the operator matrix for the two-dimensional Laplacian with Dirichlet boundary conditions. When the full wave equation is set up as a finite difference equation, the various operators take on special forms. The inhomogeneous sound speed is evaluated only at interior lattice points resulting in a diagonal operator. The matrix for the second derivative, Laplacian, will depend on the stencil used. Finally, the Dirichlet boundary conditions will be represented by diagonal elements, while Neumann boundary conditions will couple different lattice points together. The entire process of building up an FD representation of the wave equation in 2 dim with Dirichlet boundary conditions is presented here for the special case of a $5 \times 5$ gird. The numbering scheme is shown in Figure 8.6.

Only the exterior indices are shown explicitly. The first block of the matrix will be a $6 \times 6$ identity matrix, as will be the last block. The next major block will be the bottom three interior nodes, $m = 7, 8, 9$. Using the five-point stencil and the indexing map leads to the following blocks for these three points:

$$D_y^- = \begin{bmatrix} 1 & 0 & 0 & 0 \\ 0 & 1 & 0 & 0 \\ 0 & 0 & 1 & 0 \end{bmatrix} \tag{8.28}$$

$$D_y^+ = \begin{bmatrix} 0 & 1 & 0 & 0 \\ 0 & 0 & 1 & 0 \\ 0 & 0 & 0 & 1 \end{bmatrix} \tag{8.29}$$

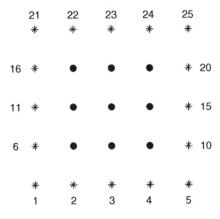

**Figure 8.6**  Example $5 \times 5$ grid

$$D_x = \begin{bmatrix} 1 & -4 & 1 & 0 & 0 \\ 0 & 1 & -4 & 1 & 0 \\ 0 & 0 & 1 & -4 & 1 \end{bmatrix} \tag{8.30}$$

The matrix $D_y^-$ starts at column 2, row 7, while $D_y^+$ ends at column 14, row 9, the rest of that row block being zeros. Rows 7–9 can be expressed as $\left[0_{3\times1}, D_y^-, D_x, D_y^+, 0_{3\times11}\right]$. The grouping of nontrivial elements, $D = \left[D_y^-, D_x, D_y^+\right]$, will repeat for rows 12–14 and 17–19. In between these sets of three rows, there will be $2\times2$ identity matrices for the boundary conditions at $m = 10, 11$ and $m = 15, 16$. The diagonal terms from the effective potential will occur in the same rows and $D$, adding to the center point evaluation. The source vector on the right-hand side of the wave equation will contain whatever profile is chosen to model the source, but all entries corresponding to boundary points will be zero. For this example, the operators were filled in the order of the index $m$, but these indices can be permuted such that all boundary conditions are grouped together in one block and likewise for the interior points with the same order applied to the source term.

## 8.3 Time Domain

Adding time to the equation introduces one more variable to discretize. However, unique issues arise in dealing with propagation of an initial solution in time that requires specific techniques not needed in finite differencing in the frequency domain. Though the main goal is the introduction of techniques suitable for modeling the acoustic field in three dimensions of space and time, some of the basic techniques are best understood by starting with a less complex equation. Consider the ordinary wave equation in 1, 2, or 3 dimensions with a constant sound speed:

$$-\frac{\partial^2 p}{\partial t^2} + c^2 \nabla^2 p = 0 \tag{8.31}$$

A spatial grid of uniformly spaced points along Cartesian directions was introduced to discretize the spatial derivative. In the same manner, the time axis is sampled at a set of discrete points $t_n = (n-1)\Delta t$, $n = 1, \ldots, N$. The function evaluated at a discrete space–time point is indicated by the following notation:

$$p(x_i, y_j, z_k, t_n) = p_{i,j,k}^{(n)} = p_{m(i,j,k)}^{(n)} \tag{8.32}$$

Applying the same second-order finite difference method introduced for space variables to time leads to the following representation of (8.31):

$$\left(p_{m_0}^{(n+1)} + p_{m_0}^{(n-1)} - 2p_{m_0}^{(n)}\right) = \frac{c^2 \Delta t}{h} \left\{\left(p_{m_1}^{(n)} + p_{m_2}^{(n)} - 2p_{m_0}^{(n)}\right) + \cdots\right\} \tag{8.33}$$

The ellipsis in (8.33) is to indicate the possible inclusion of a second or third spatial second-derivative operator. In this discretization three points in time are required to express the second derivative, as one would expect. The spatial derivatives are evaluated at the center point in time. This approach makes sense following similar treatments of function evaluations in the finite difference approach. Think of the right-hand side of (8.33) as a function of time and it is clear that following that same convention, this would be evaluated at the center point of the set used to estimate the second derivative. Equation (8.33) can be augmented to include a source distribution with time behavior, but a common type of problem for which the FDTD method is used involves modeling the propagation of an initial field profile through a medium with boundaries and inhomogeneities. Rather than a model of a source, what is needed to drive the simulation is a model of an initial pressure distribution on the spatial lattice. Looking at the form of (8.33), it is clear that with this differencing scheme in time, two profiles are needed to initialize the propagation; for example, one needs $p_m^{(1)}$ and $p_m^{(2)}$ to obtain $p_m^{(3)}$. Reduction of order is accomplished by introducing a new set of variables as follows:

$$s = \frac{\partial p}{\partial t} \tag{8.34}$$

$$\vec{w} = c \vec{\nabla} p \tag{8.35}$$

The wave equation is then expressed in terms of the four-dimensional array. A system of differential equation is then derived by taking the time derivatives of these new variables and using the definition of $s$ and the wave equation:

$$\frac{\partial s}{\partial t} = c \vec{\nabla} \cdot \vec{w} \tag{8.36}$$

$$\frac{\partial \vec{w}}{\partial t} = c \vec{\nabla} s \tag{8.37}$$

By defining a four-dimensional vector $\mathbf{Y} = [s, \vec{w}]$, this set of equations is cast in operator matrix form:

$$\frac{\partial \mathbf{Y}}{\partial t} = -c \begin{bmatrix} 0 & \vec{\nabla}^T \\ \vec{\nabla} & 0_{3 \times 3} \end{bmatrix} \mathbf{Y} \tag{8.38}$$

First-order difference operators can now be applied to both space and time derivatives. In the first attempt, one would likely be inclined to use a time difference in which the initial time coincided with the same time at which the spatial derivatives are evaluated and a later time one step forward, with the fields evaluated at a given lattice point, $m$:

$$\frac{\partial \mathbf{Y}}{\partial t} \rightarrow \frac{\mathbf{Y}_m^{(n+1)} - \mathbf{Y}_m^{(n)}}{\Delta t} \tag{8.39}$$

By definition this is an estimate at the time index $n + 1/2$. The spatial derivatives on the right-hand side are usually taken using central difference with all fields referenced to the same time, $n$. This type of scheme is referred to as forward time centered space (FTCS). While it looks easy to implement and understand, it is unstable. There are a variety of fixes to this problem [4]. One is the Lax method, and another is called the leapfrog method. Each of these will be described here, followed by a brief introduction to stability analysis. In the Lax method, the term $\mathbf{Y}_m^{(n)}$ in the time difference equation is replaced by a two-point spatial average:

$$\mathbf{Y}_m^{(n)} \rightarrow \frac{\mathbf{Y}_{m_{k+1}}^{(n)} + \mathbf{Y}_{m_k}^{(n)}}{2} \tag{8.40}$$

Another approach to creating a finite difference of the time derivative is the staggered leap-frog method. In a nut shell, it employs a central difference for the time derivative, centered at the current time, $n$:

$$\frac{\partial \mathbf{Y}}{\partial t} \rightarrow \frac{\mathbf{Y}_m^{(n+1)} - \mathbf{Y}_m^{(n-1)}}{\Delta t} \tag{8.41}$$

Spatial derivatives are the same as before. The nature of the name should be clear. To predict the value of the field at time $n + 1$, the spatial operator is evaluated at time $n$, but the initial time for the propagation is $n - 1$. The time operator has in essence "leapt over" the time when the right-hand side is evaluated. The staggering nature of the algorithm can be understood by realizing that, since the algorithm uses central difference for both time and spatial variables, it propagates subsets of the grid in a manner that completely decouples them. The leapfrog method is depicted in Figure 8.7. Just as stencils provide a good visual representation of the spatial derivatives, stencils are used to represent time evolution. The stencil for each of the three-time operator presented here is provided in Table 8.4.

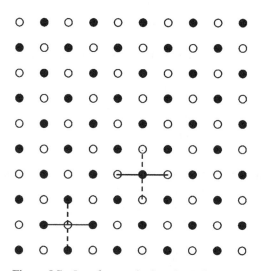

**Figure 8.7**  Leapfrog method on $1 + 1$ dim lattice

**Table 8.4**   Stencils for finite differences in time

| | Time stencils |
|---|---|
| Method | Stencil |

FTCS

Lax

Leapfrog

To better understand the virtues of some of these methods, and the reason why FTCS cannot work, the concept of stability is introduced. For purposes of illustration this is done in the case of one spatial dimension for a scalar field. In other words, the canonical equation is

$$\frac{\partial f}{\partial t} = -a\frac{\partial f}{\partial x} \tag{8.42}$$

To assess the stability of the finite difference method, a technique called von Neumann stability analysis is introduced. The discretized fields are expressed as a Fourier series in spatial degrees of freedom:

$$f_m^{(n)} = \sum_k \delta^{(n)} \ \exp(ikx_m) \tag{8.43}$$

In (8.43) $\delta^{(n)}$ is an amplitude factor, $k$ is the wave number, and $x_m$ is a lattice point. A single mode of the expansion is inserted into the discretized equation. This will involve the ratio of the amplitude factor at consecutive times:

$$\eta = \frac{\delta^{(n+1)}}{\delta^{(n)}} \tag{8.44}$$

The magnitude of this ratio is called the growth factor associated with the method. The growth factor will, in general, be a function of the wave number. If it is greater than 1, the method is unstable. What is desired is that the growth factor be less than or equal 1 for all modes. A method is conditionally stable if the requirement $|\eta| \leq 1$ places a restriction on the choice of step sizes. If no such condition exists, then the method is unconditionally stable. As an example, this is applied to (8.42) using the FTCS, Lax, and leapfrog methods

$$f_m^{(n+1)} = f_m^{(n)} - a\frac{\Delta t}{2h}\left(f_{m+1}^{(n)} - f_{m-1}^{(n)}\right) \tag{8.45}$$

$$f_m^{(n+1)} = \frac{1}{2}\left(f_{m+1}^{(n)} + f_{m-1}^{(n)}\right) - a\frac{\Delta t}{2h}\left(f_{m+1}^{(n)} - f_{m-1}^{(n)}\right) \tag{8.46}$$

$$f_m^{(n+1)} = f_m^{(n-1)} - a\frac{\Delta t}{h}\left(f_{m+1}^{(n)} - f_{m-1}^{(n)}\right) \tag{8.47}$$

respectively. Plugging in the expression for a single Fourier series term leads to the following expressions for each method:

$$\delta^{(n+1)} = \delta^{(n)}\left(1 - i\frac{a\Delta t}{h}\sin(kh)\right) \tag{8.48}$$

$$\delta^{(n+1)} = \delta^{(n)}\left(\cos(kh) - i\frac{a\Delta t}{h}\sin(kh)\right) \tag{8.49}$$

$$\delta^{(n+1)} = \delta^{(n-1)} - \delta^{(n)}\left(2i\frac{a\Delta t}{h}\sin(kh)\right) \tag{8.50}$$

The lowest index growth rate is divided from each equation, leading to a single equation for the amplitude ratio for each method:

$$\eta = 1 - i\frac{a\Delta t}{h}\sin(kh) \tag{8.51}$$

$$\eta = \cos(kh) - i\frac{a\Delta t}{h}\sin(kh) \tag{8.52}$$

$$\eta^2 = 1 - \eta\left(2i\frac{a\Delta t}{h}\sin(kh)\right) \tag{8.53}$$

The magnitude of the first expression will clearly be greater than 1 for all choices of parameters. Therefore, the FTCS method is unconditionally unstable and not useful. For the Lax method, consider $|\eta|^2$. Taking the magnitude of (8.52), and applying a little algebra and trig identities, gives the following stability condition:

$$\frac{a\Delta t}{h} \leq 1 \tag{8.54}$$

This is known as the Courant–Friedrichs–Lewy stability condition. The nature of this condition places an upper limit on the time step given the parameter $a$ and a spatial grid step size of $h$. To analyze the stability of the leapfrog equation, first solve (8.53):

$$\eta = -ia\sin(kh) \pm \sqrt{1-\alpha^2\sin^2(kh)} \tag{8.55}$$

The parameter $\alpha = a\Delta t/h$ is introduced to simplify the equation. In this case $|\eta|^2 = 1$ and the method is unconditionally stable.

## 8.4   FDTD Representation of the Linear Wave Equation

The previous section introduced time evolution operators in the finite difference scheme and introduced stability analysis. In this section these ideas are applied to develop a full FDTD version of the first-order linear acoustics equations:

$$\frac{\partial p}{\partial t} + \rho_0 c^2 \,\vec{\nabla}\cdot\vec{v} + \vec{v}\cdot\vec{\nabla}p_0 + p\vec{\nabla}\cdot\vec{v}_0 = 0 \tag{8.56}$$

$$\frac{\partial\vec{v}}{\partial t} + \frac{1}{\rho_0}\vec{\nabla}p + \vec{v}\cdot\vec{\nabla}\vec{v}_0 + \frac{p}{\rho_0 c^2}\frac{\partial\vec{v}_0}{\partial t} = 0 \tag{8.57}$$

To help make the variables clear, the coefficients that depend on the background fields are redefined in an obvious way:

$$\frac{\partial p}{\partial t} + a_1\,\vec{\nabla}\cdot\vec{v} + \vec{v}\cdot\vec{b} + a_2 p = 0 \tag{8.58}$$

$$\frac{\partial\vec{v}}{\partial t} + a_3\,\vec{\nabla}p + \vec{v}\cdot S + a_4 p = 0 \tag{8.59}$$

The coefficient $S$ is a dyadic, left multiplied by $\vec{v}$. In most realistic situations, it is reasonable to ignore the time dependence of the background fields, so it will be assumed that the coefficients depend on position only or are constant. The goal now is to discretize this set of equations. To illustrate this, for the time being, the background fluid motion is ignored, resulting in $\vec{b} = 0$, $S = 0$, and $a_2 = a_4 = 0$. The primary task is in estimating the derivatives on a spatial grid and propagating the values forward in time. A method by K. Yee originally applied to Maxwell's equations is well suited for this [5]. The approach evaluates different components of the vector field at intermediate points within the lattice, that is, at center points between grid points. In addition to this splitting of the variables, the pressure field and the velocity fields are updated at different times. The process is described now in some detail. The starting point is the finite difference grid defined in the FD method. The end points define boundaries in space. Attention will be focused only on interior points for defining the finite difference operators and the time step. The pressure field is defined at the grid points $(i, j, k)$. The three components of the velocity field are defined at points shifted away from this point in each direction:

$$v_x \to v_x\left(i+\frac{1}{2}, j, k\right) \tag{8.60}$$

$$v_y \to v_y\left(i, j+\frac{1}{2}, k\right) \tag{8.61}$$

$$v_z \to v_z\left(i, j, k+\frac{1}{2}\right) \tag{8.62}$$

The velocity field is updated at the time lattice points, $n$, while the values of the pressure field are updated at the center points between consecutive time points, $n+1/2$. The following replacement is made to avoid too many indices, $v_x = u$, $v_y = v$, and $v_z = w$. With this convention in place, the spatial derivatives of the pressure and velocity fields are

$$\frac{\partial p}{\partial x} \to \frac{1}{2h}\left(p_{i+1,j,k}^{(n+1/2)} - p_{i-1,j,k}^{(n+1/2)}\right) \tag{8.63}$$

$$\frac{\partial u}{\partial x} = \frac{1}{2h}\left(u_{i+1/2,j,k}^{(n)} - u_{i-1/2,j,k}^{(n)}\right) \tag{8.64}$$

and similarly for the other spatial derivatives and components of $\vec{v}$. The time derivative of the pressure and velocity fields become

$$\frac{\partial p}{\partial t} \to \frac{1}{2\Delta t}\left(p_{i,j,k}^{(n+1/2)} - p_{i,j,k}^{(n+1/2)}\right) \tag{8.65}$$

$$\frac{\partial u}{\partial t} \to \frac{1}{2\Delta t}\left(u_{i+1/2,j,k}^{(n+1)} - u_{i+1/2,j,k}^{(n)}\right) \tag{8.66}$$

and similarly for the other components of $\vec{v}$. When the coefficients depend on position, they are evaluated at the lattice points and at the same time as the fields used in the derivatives. The fully discretized first-order wave equation takes on the following forms:

$$p_{i,j,k}^{(n+1/2)} = p_{i,j,k}^{(n+1/2)} - \frac{\Delta t}{h}a_{1\,i,j,k}\left(u_{i+1/2,j,k}^{(n)} - u_{i-1/2,j,k}^{(n)} + \cdots\right) \tag{8.67}$$

$$u_{i+1/2,j,k}^{(n+1)} = u_{i+1/2,j,k}^{(n)} - \frac{\Delta t}{h}a_{3\,i,j,k}\left(p_{i+1,j,k}^{(n+1/2)} - p_{i-1,j,k}^{(n+1/2)}\right) \tag{8.68}$$

$$v_{i+1/2,j,k}^{(n+1)} = v_{i+1/2,j,k}^{(n)} - \frac{\Delta t}{h}a_{3\,i,j,k}\left(p_{i,j+1,k}^{(n+1/2)} - p_{i,j-1,k}^{(n+1/2)}\right) \tag{8.69}$$

$$w_{i+1/2,j,k}^{(n+1)} + w_{i+1/2,j,k}^{(n)} - \frac{\Delta t}{h}a_{3\,i,j,k}\left(p_{i,j,k+1}^{(n+1/2)} - p_{i,j,k-1}^{(n+1/2)}\right) \tag{8.70}$$

Dots in (8.67) are used as placeholders for indices in the pressure equation to make it less cluttered and easier to read. It is clear that the time propagation of each field is leapfrog, but it is also the case that different fields are being leapfrogged at different time, one after the other. The earliest times will be 1 for the velocity field and 3/2 for the pressure field. Given this data the first step taken updates the velocity fields at 2, followed by a step to update the pressure field at 5/2. Now with data at 5/2, a step can be taken to update the velocity at 3, followed by a pressure update at 7/3, and so on. The placement of the fields is illustrated in three dimensions in Figure 8.8. The propagation from one time to the next is illustrated for a two-dimensional grid in Figure 8.9. Shown are three times: $n$, $n + 1/2$, and $n + 1$. The figure illustrates by example that field variables on each time slice contribute to the evaluation of the variables in the next time slice. In the derivation, the external force field was ignored but is clearly easy to put back in. Any term in the equation that is not being propagated is just a known function of position and possibly time. Therefore, these can be evaluated at all lattice points, and their respective center points, and all times. The same holds for the coefficients that were assumed to depend on position only; time dependence may be included. The stepping forward in time represented in all

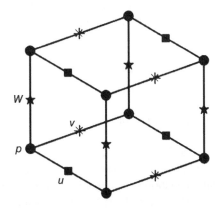

**Figure 8.8**   Location on lattice where variables are evaluated

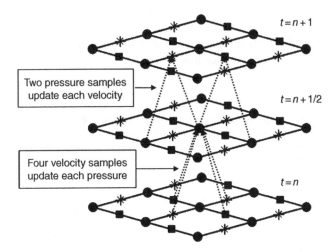

**Figure 8.9**   Time steps for a $2 + 1$ dim implementation

these equations is explicit. The update in the variables depends only on previous data. These methods do not require matrix inversion, only multiplication, since the initial conditions are simply being used to evaluate the spatial part of the equation, which in turn provides an estimate for the distribution at the next time. When the effects of background flow are not ignored, terms linear in the field variables need to be included and inserted into the equations.

To get an idea of the amount of data that would be required for an FD calculation for a typical room, the following estimates are provided. Case 1 is an office with dimensions $4 \text{ m} \times 7 \text{ m} \times 4 \text{ m}$ with a total volume of $\sim 112 \text{ m}^3$. To model sound with a max frequency of 220 Hz in air, $c \sim 343$ m/s with a requirement of $n$ samples per wavelength gives

$$N \sim n^3 \frac{112}{3.8} \sim 29n^3$$

The second example is a small theater with 200 seats and an estimated volume of $3000 \text{ m}^3$. Estimating the upper register of a violin to be $\sim 1300$ Hz, modeling this frequency in a theater of this size would require

$$N \sim n^3 \frac{3,000}{0.018} \sim 17,000n^3$$

Using a seven-point stencil and ignoring boundary conditions for the moment leads to a system of $\sim 7 \times N$, to leading order (Table 8.5).

Due to the nature of the FD algorithm, the resulting matrices will be sparse, on the order of $n_s \times N_g$ where $n_s$ is the number of points in the stencil and $N_g$ is the number of grid points. Even though this seems to be a matrix equation in $N_g$ degrees of freedom, it is not really and $N_g \times N_g$ matrix of coefficients, $n_s \ll N_g$. The size of the system really scales as $O(N_g)$. Sparse matrices can be handled differently compared with full matrices. There will generally be a large number of zeros in the full matrix. A storage technique uses an array of indices and matrix values instead of simply filling the entire matrix. Rather than $M(i,j) = m$ for all $i, j$, one stores the collection $(i, j, m)$ only for the values $m \neq 0$. MATLAB has a function that will create a sparse matrix given two index arrays and an array of numbers for the entries of $M(i, j)$ and $M = \text{sparse}(i,j,m)$. The "back slash" operator works on sparse matrices as well, so no additional effort is required to solve a system once it is set up.

**Table 8.5**  Example estimates of the number of sample points required for setting up an FD grid

| Sample number | Number of points required | |
|---|---|---|
| | Case 1 | Case 2 |
| 3 | 783 | 459,000 |
| 5 | 3,625 | 2,125,000 |
| 7 | 9,947 | 5,831,000 |
| 10 | 29,000 | 17,000,000 |

The method for evolving the equations in time presented here is an example of an explicit method. The value of the fields at a future time is determined from earlier data. Other methods include the lattice points at the time being solved for, termed implicit methods, where matrix inversion is required to determine the fields at the next time step. In deriving the time step equations, the terms containing background fluid flow were tacitly dropped to make the presentation easier to follow. The staggered leapfrog method is known to work when applied to the first-order acoustic equations with no background flow but suffer defect when applied to the same equation with $\vec{v}_0 \neq 0$. The general equations are investigated in detail by Ostashev *et al.* [6] where a time evolution scheme called "nonstaggered leapfrog" is discussed.

## 8.5 Exercises

1. Develop a fourth-order stencil for first and second derivatives in 1 dim, on a uniform lattice.
2. With an understanding of how ideal boundary conditions constrain the degrees of freedom, eliminate these from the full system to the degree possibly leaving their effect on the rest of the system. How does the indexing change in the process?
3. Assume the pressure field in the presence of a background with position-dependent sound speed and flow vector obeys $D^2 p = 0$; see Chapter 3.
   (a) Develop an FD method for this equation in the FD in 2 dim, with ideally hard rectangular boundary conditions.
4. Write a software routine to solve the equations developed in Exercise 3.

## References

[1] Iserles, A., A First Course in the Numerical Analysis of Differential Equations, Cambridge University Press, Cambridge, 2002.

[2] Bradley, A., Greengard, L., and Hagstrom, T., Nonreflecting boundary conditions for the time-dependent wave equation, J. Comput. Phys., Vol. 180, pp. 270–296, 2002.

[3] Düz, B., Huijsmans, H. M., Veldman, E. P., Borsboom, M. J. A., and Wellens, P. R., An Absorbing Boundary Condition for Regular and Irregular Wave Simulations, in MARINE 2011, IV International Conference on Computational Methods in Marine Engineering, Selected Papers, Eca, L., Onate, E., García-Espinosa, J., Kvamsdal, T., and Bergan, P. (Eds.), Springer, Dordrecht, 2013.

[4] Press, W. H., Teukolsky, S. A., Vetterling, W. T., and Flannery, B. P., Numerical Recipes in C++: The Art of Scientific Computing, Second Edition, Cambridge University Press, Cambridge, 2002.

[5] Yee, K. S., Numerical solution of initial boundary value problems involving Maxwell's equations in isotropic media, IEEE Trans. Antennas Propag., Vol. 14, No. 3, pp. 302–307, 1966.

[6] Ostashev, V. E., Wilson, D. K., Liu, L., Aldridge, D. F., and Symons, N. P., Equations for finite-difference, time-domain simulation of sound propagation in moving inhomogeneous media and numerical implementation, J. Acoust. Soc. Am., Vol. 117, No. 2, pp. 503–517, 2005.

# 9

# Parabolic Equation

## 9.1 Introduction

The parabolic equation (PE) typically arises from applying an approximation to the full wave equation, or via operator factorization. A common example from physics is the derivation of Schrödinger's equation as a low velocity approximation of the Dirac equation, or Klein–Gordon equation. In acoustics one encounters the PE when modeling propagation in ducted environments where the dominant contribution is expected to be from low grazing angles. This approach is popular in underwater acoustics where propagation is confined in the vertical direction, due to the boundaries and refractive gradients, and encounters weak range-dependent variations. The PE can be developed along any direction in space, or in time, or along more general stationary curves in space, for example, coordinate curves in generalized coordinates. In the case of beam tracing, one develops the PE approximation along solutions to the ray equation, that is, along the bicharacteristics of the wave equation. The takeaway is that the approximation is developed in consideration of the environment or other factors that are expected to drive a particular behavior. Along these lines specialized techniques have been developed for evaluating operators and implementing numerical solutions to the PE. Several methods of factorization are discussed followed by an introduction to the methods for representing the pseudo-differential operator that emerges from the factorization process. The last section introduces the Crank–Nicolson method for numerically marching the wave equation in the range direction.

## 9.2 The Paraxial Approximation

There is more than one way to arrive at the PE. The starting point is the Helmholtz equation for the pressure field with a position-dependent sound speed and no fluid motion in cylindrical coordinates.

---

*Computational Acoustics: Theory and Implementation*, First Edition. David R. Bergman.
© 2018 John Wiley & Sons Ltd. Published 2018 by John Wiley & Sons Ltd.

$$\frac{\partial^2 p}{\partial \rho^2} + \frac{1}{\rho}\frac{\partial p}{\partial \rho} + \frac{1}{\rho^2}\frac{\partial^2 p}{\partial \phi^2} + \frac{\partial^2 p}{\partial z^2} + k_0^2 n^2 p = 0 \tag{9.1}$$

Some new notation has been introduced. Assuming the sound speed to be $c(\vec{x}) = c_0 C(\vec{x})$, where $c_0$ is a reference sound speed, and a reference wave number and refractive index are defined as follows:

$$k_0 = \frac{\omega}{c_0} \tag{9.2}$$

$$n(\vec{x}) = \frac{1}{C(\vec{x})} \tag{9.3}$$

Classic derivations typically proceed in one of two ways. The first is to assume that the field and the refractive index are independent of azimuth (referred to an angle coupling) and range, respectively. This essentially assumes that propagation in different vertical planes can be considered a two-dimensional problem with the effect of a three-dimensional propagation factored into the radial operator, which contributes to the geometric spread of the waves as they propagate downrange. Under these assumptions, a solution of the following form is assumed:

$$\psi = u(\rho, z) H_0^{(1)}(k_0 \rho) \tag{9.4}$$

Inserting this solution into (9.1) and using the definition of Bessel's equation yields the following equation for the factor $u(\rho, z)$:

$$\frac{\partial^2 u}{\partial \rho^2} + \frac{\partial^2 u}{\partial z^2} + \left[ 2\frac{1}{H_0^{(1)}}\frac{d H_0^{(1)}}{d\rho} + \frac{1}{\rho} \right]\frac{\partial u}{\partial \rho} + k_0^2 (n^2 - 1) u = 0 \tag{9.5}$$

The next step is to consider long-range propagation. In the far field limit, the term in square brackets of (9.5) becomes,

$$2\frac{1}{H_0^{(1)}}\frac{d H_0^{(1)}}{d\rho} + \frac{1}{\rho} \approx 2ik_0 + \frac{1}{\rho} \tag{9.6}$$

The PE arises from applying the assumption

$$\left|\frac{\partial^2 u}{\partial \rho^2}\right| \ll 2k_0 \left|\frac{\partial u}{\partial \rho}\right|, \left|\frac{\partial^2 u}{\partial z^2}\right|, \left|k_0^2 (n^2 - 1) u\right| \tag{9.7}$$

to (9.5), which allows the first term to be dropped,

$$\frac{\partial^2 u}{\partial z^2} + 2ik_0 \frac{\partial u}{\partial \rho} + k_0^2 (n^2 - 1)u = 0 \tag{9.8}$$

Rearranging terms, (9.8) can be put in the same form as Schrödinger's time-dependent wave equation from nonrelativistic quantum mechanics.

$$ik_0^{-1} \frac{\partial u}{\partial \rho} = -\frac{k_0^{-2}}{2} \frac{\partial^2 u}{\partial z^2} + \frac{1}{2}(1 - n^2)u \tag{9.9}$$

The range variable in (9.9) takes the place of time in Schrödinger's equation, and the reference wave number can be related to $\hbar$, Plank's constant over $2\pi$. Equation (9.7) is referred to as the paraxial approximation, and is valid for small grazing angles during propagation. A considerable amount of research effort over the past few decades has focused on extending the PE to wide-angle propagation. Modern treatments forego the paraxial approximation in favor of operator factorization.

## 9.3 Operator Factoring

Factorization of the wave equation is commonly used in the presentation of traveling waves in 1-dimension (1-dim). The steps are repeated here to remind the reader of the method and set up some of the terminology used later. The 1-dim wave equation in the time domain is

$$\frac{\partial^2 \psi}{\partial x^2} - \frac{1}{c^2} \frac{\partial^2 \psi}{\partial t^2} = 0 \tag{9.10}$$

For a constant sound speed, this may be written in a form that suggests reduction of order or a change in variables:

$$\left( \frac{\partial}{\partial x} + \frac{1}{c} \frac{\partial}{\partial t} \right) \left( \frac{\partial}{\partial x} - \frac{1}{c} \frac{\partial}{\partial t} \right) \psi = 0 \tag{9.11}$$

The wave equation is satisfied if

$$\left( \frac{\partial}{\partial x} - \frac{1}{c} \frac{\partial}{\partial t} \right) \psi = 0 \tag{9.12}$$

Defining a new variable,

$$\chi = \left( \frac{\partial}{\partial x} - \frac{1}{c} \frac{\partial}{\partial t} \right) \psi \tag{9.13}$$

The second-order wave equation becomes a first-order equation:

$$\left(\frac{\partial}{\partial x} + \frac{1}{c}\frac{\partial}{\partial t}\right)\chi = 0 \tag{9.14}$$

It is fairly straightforward to verify that a solution to either of the first-order equations can be expressed as any function of the single variable $x + ct$ or $x - ct$. Indeed, writing $\theta = x - ct$,

$$\left(\frac{\partial}{\partial x} + \frac{1}{c}\frac{\partial}{\partial t}\right)\chi(\theta) = \frac{\partial\theta}{\partial x}\frac{d\chi}{d\theta} + \frac{1}{c}\frac{\partial\theta}{\partial t}\frac{d\chi}{d\theta} = \frac{d\chi}{d\theta}(1-1) = 0 \tag{9.15}$$

These two types or solutions are referred to as forward and backward solutions. Depending on the sign convention chosen, these can be expressed as forward-time evolution of a shape traveling to the right or left on the $x$ axis. Another example of reduction of order is the reduction of the Klein–Gordon equation describing the relativistic quantum wave function of a massive particle to the first-order Dirac equation.

The PE appears as an approximate form of the full wave equation for long-range propagation in a ducted environment. Early forms of the equation applied to optics and acoustics require the grazing angles, that is, angle of propagation of the energy from a ray theoretic perspective, to be close to the direction of propagation. Other assumptions leading to the PE are weak refractive properties of the medium, all of which are necessary for neglecting undesirable terms in the wave equation. An alternate to the paraxial approximation, which serves as a starting point for operator expansions, is to forego the use of the radial portion of the wave function and simply assume the pressure field can be written as $p(\rho,z) = u(\rho,z)/\sqrt{\rho}$. The radial portion of the wave equation becomes

$$\frac{\partial^2 p}{\partial \rho^2} + \frac{1}{\rho}\frac{\partial p}{\partial \rho} = \frac{1}{\sqrt{\rho}}\frac{\partial^2 u}{\partial \rho^2} + \frac{u}{4\rho^{5/2}} \tag{9.16}$$

Inserting the result in (9.1), ignoring azimuthal dependence, and multiplying by $\sqrt{\rho}$ leads to

$$\frac{\partial^2 u}{\partial \rho^2} + \frac{u}{4\rho^2} + \frac{\partial^2 u}{\partial z^2} + k_0^2 n^2 u = 0 \tag{9.17}$$

The final step involves assuming long-range propagation, justifying the removal of the second term:

$$\frac{\partial^2 u}{\partial \rho^2} + \frac{\partial^2 u}{\partial z^2} + k_0^2 n^2 u = 0 \tag{9.18}$$

The refractive index can be a function of both variables $n(\rho, z)$. There is no need to use cylindrical coordinates in the derivation of (9.18), but it seems natural since many problems to which this method is applied start out as propagation from a point source in a vertically layered medium with weak range-dependent refraction. Leaving the operator in the Cartesian coordinates and choosing one horizontal direction as the preferred direction for modeling propagation lead to a full 3-dim version of (9.18).

$$\frac{\partial^2 u}{\partial x^2} + \nabla_\perp^2 u + k_0^2 n^2 u = 0 \tag{9.19}$$

In (9.19) the $x$ direction has been singled out as the propagation direction, and $\nabla_\perp^2$ is the Laplacian operator relative to the other two, transverse, directions. An immediate consequence of using this approach is that no terms are neglected, and the long-range assumption is not invoked. The propagation direction will be referred to as the "range" variable.

At this point the goal is to find a factorization of the differential operator that produces a first-order equation in the range variable. The approach, and the methods used, appeals to a similarity between (9.19) and the Klein–Gordon equation from the relativistic quantum mechanics. The full relativistic wave equation is a generalization of Schrödinger's equation from nonrelativistic quantum mechanics and is necessary to consider since the nonrelativistic equation does not respect Lorentz symmetry. Dirac proposed an operator factorization of the Klein–Gordon equation that reduced the PDE from a second-order equation in a scalar field to a first-order equation [1]. To accomplish this, the coefficients in the linear operator are required to obey a set of algebraic constraints that prevent them from being simple complex scalars. This process naturally leads to the introduction of spin as a consequence of the reduction of order.

For the Helmholtz equation, range takes the place of time, and the operator being factored is elliptic rather than hyperbolic. An approach to factoring the equation is to define operators:

$$\hat{D} = \frac{\partial}{\partial x} \tag{9.20}$$

$$\hat{Q} = \sqrt{k_0^{-2} \nabla_\perp^2 + n^2} \tag{9.21}$$

The operator in (9.21) is a pseudo-differential operator and is defined formally by its power series expansion. With these definitions (9.19) is expressed as

$$\hat{D}^2 u + k_0^2 \hat{Q}^2 u = 0 \tag{9.22}$$

Some derivations leave the factor of $k_0^2$ in the definition of $\hat{Q}$. At high frequencies, small propagation angles, and weak refraction, one can expect the effect of the differential operator in (9.21) to be small and the refractive index close to 1. This suggests expanding the operator as follows:

$$\hat{Q} = \sqrt{1 + \hat{\Lambda}} = 1 + \frac{1}{2}\hat{\Lambda} - \frac{1}{8}\hat{\Lambda}^2 + \cdots \tag{9.23}$$

where the following operator is defined as

$$\hat{\Lambda} = k_0^{-2} \nabla_\perp^2 + \left[n^2 - 1\right] \tag{9.24}$$

To first order this gives the following for $\hat{Q}$:

$$\hat{Q} \approx 1 + \frac{1}{2}\hat{\Lambda} \tag{9.25}$$

## 9.4  Pauli Spin Matrices

This subsection introduces the Pauli matrices. These are most commonly used in nonrelativistic quantum mechanics to describe electron spin. A formal treatment of relativistic quantum field theory uses the Dirac matrices, which can be expressed in terms of the Pauli spin matrices. While this might seem a little out of place, these matrices naturally occur in the reduction of order and are presented by Levers [2] and used in research by Wurmser [3] and Wurmser et al. [4] and in the presentation of reduction of order applied to the Helmholtz equation in the next section. The Pauli matrices, $\sigma_i$ with $i = 1, 2, 3$, are a set of $2 \times 2$ matrices with complex components. These matrices are a representation of the Lie algebra for the group SU(2), special unitary group in two complex dimensions. The three matrices obey a set of constraints, which define the Lie algebra of the "little group" $su(2)$:

$$\sigma_i^2 = I, \quad \forall i \tag{9.26}$$

$$\sigma_i \sigma_j = i\sigma_k \tag{9.27}$$

for the $(i, j, k)$, an even permutation of $(1, 2, 3)$. The Pauli matrices anti-commute:

$$\sigma_i \sigma_j = -\sigma_j \sigma_i, \quad \forall (i \neq j) \tag{9.28}$$

From these constraints, the commutator of any of the two obeys the following:

$$[\sigma_i, \sigma_j] = 2i\varepsilon_{ijk}\sigma_k \tag{9.29}$$

The Levi-Civita symbol has the following values: $\varepsilon_{ijk} = 0$ if any of the two or indices are the same, $\varepsilon_{ijk} = 1$ if the indices are an even permutation of $(1, 2, 3)$, and $\varepsilon_{ijk} = -1$ if the indices are an odd permutation of $(1, 2, 3)$. Finally, for completeness, the anti-commutator is introduced:

$$\{A, B\} = AB + BA \tag{9.30}$$

As previously stated the Pauli matrices anti-commute:

$$\{\sigma_i, \sigma_j\} = 0, \quad \forall (i \neq j) \tag{9.31}$$

These relations are enough to define the algebra of $\{\sigma_i\}$ in an abstract sense. The common matrix representation of these is provided as follows:

$$\sigma_1 = \begin{bmatrix} 0 & 1 \\ 1 & 0 \end{bmatrix} \tag{9.32}$$

$$\sigma_2 = \begin{bmatrix} 0 & -i \\ i & 0 \end{bmatrix} \tag{9.33}$$

$$\sigma_3 = \begin{bmatrix} 1 & 0 \\ 0 & -1 \end{bmatrix} \tag{9.34}$$

## 9.5   Reduction of Order

There is no unique way to factor the wave equation. A typical starting point is the definition of a two-component vector.

$$\Psi = \begin{bmatrix} u \\ \partial u/\partial x \end{bmatrix} \tag{9.35}$$

This leads to a system of equations:

$$\hat{D}\left(\frac{\partial u}{\partial x}\right) = -\hat{Q}^2 u \tag{9.36}$$

$$\hat{D}u = \frac{\partial u}{\partial x} \tag{9.37}$$

In matrix form these are

$$\hat{D}\begin{bmatrix} u \\ \partial u/\partial x \end{bmatrix} = \begin{bmatrix} 0 & 1 \\ -\hat{Q}^2 & 0 \end{bmatrix}\begin{bmatrix} u \\ \partial u/\partial x \end{bmatrix} \tag{9.38}$$

As it stands this is exact in the sense that nothing has been neglected except for what was needed to arrive at the reduced wave equation, and the original equation with no error or corrections is recovered from this by substitution. The matrix differential operator has the property that it mixes the two components of $\Psi$. A desirable property to have in the factorized equations is a description in terms of forward and backward propagating (also called transmitted and reflected) solutions, $u^+$ and $u^-$, with the complete solution expressed as $u = u^+ + u^-$. To separate the transmitted and reflected modes, a transformation is applied to the vector $\Psi$ that mixes the components:

$$\begin{bmatrix} u^+ \\ u^- \end{bmatrix} = T\begin{bmatrix} u \\ \partial u/\partial x \end{bmatrix} \tag{9.39}$$

The operator $T$ is not yet specified but is assumed to be invertible. The coupled pair of equations are expressed in terms of the new variables:

$$\hat{D}\left(T^{-1}\begin{bmatrix} u^+ \\ u^- \end{bmatrix}\right) = \begin{bmatrix} 0 & 1 \\ -\hat{Q}^2 & 0 \end{bmatrix}T^{-1}\begin{bmatrix} u^+ \\ u^- \end{bmatrix} \tag{9.40}$$

Differentiating the l.h.s. explicitly, moving the derivative of the matrix $T^{-1}$ to the r.h.s., and multiplying by $T$ leads to the equation for $u^+$ and $u^-$:

$$\hat{D}\begin{bmatrix} u^+ \\ u^- \end{bmatrix} = -T\hat{D}(T^{-1})\begin{bmatrix} u^+ \\ u^- \end{bmatrix} + T\begin{bmatrix} 0 & 1 \\ -\hat{Q}^2 & 0 \end{bmatrix}T^{-1}\begin{bmatrix} u^+ \\ u^- \end{bmatrix} \tag{9.41}$$

The goal is to choose $T$ in such a way that (9.41) separates into two decoupled equations for the transmitted and reflected components of the wave. A complete decoupling is not possible but a desired feature is that the resulting equations decouple when $k^2(\vec{x}) = k^2(\vec{x}_\perp)$, that is, when the refractive index is not range dependent. Differentiating the identity $TT^{-1} = T^{-1}T = I$ allows the operator in the first term on the r.h.s. to be replaced by $\hat{D}(T)T^{-1}$. The following transformation matrix diagonalizes the second term in (9.41):

$$T = \begin{bmatrix} 1 & -i/\hat{Q} \\ 1 & i/\hat{Q} \end{bmatrix} \tag{9.42}$$

$$T^{-1} = \frac{1}{2}\begin{bmatrix} 1 & 1 \\ i\hat{Q} & -i\hat{Q} \end{bmatrix} \tag{9.43}$$

$$\frac{1}{2}\begin{bmatrix} 1 & -i/\hat{Q} \\ 1 & i/\hat{Q} \end{bmatrix}\begin{bmatrix} 0 & 1 \\ -\hat{Q}^2 & 0 \end{bmatrix}\begin{bmatrix} 1 & 1 \\ i\hat{Q} & -i\hat{Q} \end{bmatrix} = \frac{1}{2}\begin{bmatrix} 1 & -i/\hat{Q} \\ 1 & i/\hat{Q} \end{bmatrix}\begin{bmatrix} i\hat{Q} & -i\hat{Q} \\ -\hat{Q}^2 & -\hat{Q}^2 \end{bmatrix} = i\hat{Q}\sigma_3 \tag{9.44}$$

Next, differentiating $T^{-1}$ and left multiplying by $T$ gives the first term

$$\hat{D}(T^{-1}) = \frac{1}{2}\begin{bmatrix} 0 & 0 \\ i\hat{D}\hat{Q} & -i\hat{D}\hat{Q} \end{bmatrix} \tag{9.45}$$

$$-T\hat{D}(T^{-1}) = \frac{1}{2}\begin{bmatrix} 1 & -i/\hat{Q} \\ 1 & i/\hat{Q} \end{bmatrix}\begin{bmatrix} 0 & 0 \\ i\hat{D}\hat{Q} & -i\hat{D}\hat{Q} \end{bmatrix} = \frac{1}{2}\frac{1}{\hat{Q}}(\hat{D}\hat{Q})(\sigma_1 - I) \tag{9.46}$$

The final vector form of the wave equation becomes

$$\hat{D}\begin{bmatrix} u^+ \\ u^- \end{bmatrix} = \frac{1}{2}\frac{1}{\hat{Q}}(\hat{D}\hat{Q})(\sigma_1 - I)\begin{bmatrix} u^+ \\ u^- \end{bmatrix} + i\hat{Q}\sigma_3\begin{bmatrix} u^+ \\ u^- \end{bmatrix} \tag{9.47}$$

The explicit form of $(\sigma_1 - I)$ is

$$\begin{bmatrix} -1 & 1 \\ 1 & -1 \end{bmatrix} \tag{9.48}$$

The first term mixes the components of the wave function, whereas the second term leaves them completely decoupled. Since the mixing term is proportional to the range derivative of the operator $\hat{Q}$, which inherits this dependence through $k^2(\vec{x})$, when the refractive effects are independent of range (9.47), it reduces to

$$\hat{D}\begin{bmatrix} u^+ \\ u^- \end{bmatrix} = i\hat{Q}\sigma_3\begin{bmatrix} u^+ \\ u^- \end{bmatrix} \tag{9.49}$$

or in component form

$$\hat{D}u^+ = i\hat{Q}u^+ \tag{9.50}$$

$$\hat{D}u^- = -i\hat{Q}u^- \tag{9.51}$$

The operator $\hat{Q}$ is similar in form to that given in (9.21).

$$\hat{Q} = \sqrt{k^2 + \vec{\nabla}_\perp^2} \tag{9.52}$$

To proceed in developing the PE with this form of the operator, it will need to be Taylor expanded and terms to be neglected beyond some point. While this is not the only possible factoring, it has a desirable feature that the two components are uncoupled when range dependence is absent and are identified with forward (transmitted) and backward (reflected) parts of the full solution. In certain applications, the forward component dominates, which makes this separation useful. The potential downside is clearly that the operator $\hat{Q}$ is not very useful in practice. There are a couple other choices of splitting that appear in the literature worth mentioning; see McDaniel [5] and Levers [2]. For brevity, the generic operator format is defined as follows:

$$T = \begin{bmatrix} 1 & \bar{\beta} \\ 1 & \beta \end{bmatrix} \tag{9.53}$$

The term $\beta$ is a complex valued function or operator. Table 9.1 lists three common choices in the literature.

The first choice leads to the PE developed by Tappert and Hardin [6] and is similar to the starting point in work by Wurmser *et al.* [4]. The second choice was introduced by Corones [7]. The third choice, which introduces a pseudo-differential operator, has been the subject of research by Fishman and McCoy [8] and Saad and Lee [9].

## 9.5.1 The Padé Approximation

Once a factorization has been chosen, effort is focused on obtaining an operator expansion of the square root operator. The obvious low-order Taylor expansion results in the standard PE for

**Table 9.1**  Common choices for operator splitting used to develop the parabolic equation

| Choice of $\beta$ | Effect on reduced order equation |
|---|---|
| $\dfrac{i}{2k_0}$ $\dfrac{i}{2k}$ | Produces a set of coupled equations in the range-independent limit with well-defined transverse differential operator |
| $i\hat{Q}$ | Produces a set of coupled equation that decouple in the range-independent limit but contains a pseudo-differential operator |

small angels. The desire to extend the PE to wide-angle propagation has led to more innovative and accurate representations of the operator. The Padé approximant has been used widely in underwater acoustics [9]. The Padé approximation of a function is formed by the ratio of two polynomials:

$$F(x) = \frac{\sum_{i=0}^{m} a_i x^i}{1 + \sum_{j=1}^{n} b_j x^j} \qquad (9.54)$$

The expansion coefficients $a_i$ and $b_j$ are chosen to match the Taylor expansion of the original function $F(x)$. The finite order of the polynomials in (9.54) defines the Padé approximant, $(m, n)$. For example, a Padé $(1,1)$ approximant would be

$$\frac{a_0 + a_1 x}{1 + b_1 x} \qquad (9.55)$$

The value of this approach is that it naturally leads to a split-step marching algorithm for the original equation. By choosing the coefficients properly, the order of the Padé operator will be lower than that produced by the Taylor expansion for the same accuracy. As an example, consider the Padé $(1,1)$ approximant to $\sqrt{1+\hat{X}}$. The Taylor expansion of the operator function is

$$\sqrt{1+\hat{X}} = 1 + \frac{1}{2}\hat{X} - \frac{1}{8}\hat{X}^2 + \frac{1}{16}\hat{X}^3 - \frac{5}{128}\hat{X}^4 + \cdots \qquad (9.56)$$

A second-order Padé approximation is achieved by a $(1, 1)$ approximant:

$$\frac{1 + a_1 x}{1 + b_1 x} \qquad (9.57)$$

Equating this to (9.56) and collecting like terms up to and including order $\hat{X}^2$ leads to $a_1 = 3/4$ and $b_1 = 1/4$. Accuracy to third order is achieved with a $(2, 1)$ approximant:

$$\frac{1 + x + \frac{1}{8}x^2}{1 + \frac{1}{2}x} \qquad (9.58)$$

And finally, a fourth order Padé $(2, 2)$

$$\frac{1 + \frac{5}{4}x + \frac{5}{16}x^2}{1 + \frac{3}{4}x + \frac{1}{16}x^2} \qquad (9.59)$$

### 9.5.2 Phase Space Representation

Another method of expressing the pseudo-differential operator uses an integral representation which transforms the operator into a phase representation. The application of a phase space representation of the operator and the use of Feynman path $\hat{Q}$ [10] integrals contributed to the development of algorithms for solving the PE in transversely refractive media by Fishman and McCoy [8, 11]. Comparing the Helmholtz factorization with quantum mechanics, the inverse reference number $k_0^{-1}$ is analogous to $\hbar$. In a similar fashion operators are defined for position and momentum variables, $\hat{\boldsymbol{x}} = \vec{x}$ and $\hat{\boldsymbol{p}} = -ik_0^{-1}\vec{\nabla}_{\vec{x}}$. Boldface type is used to express operators so as to not confuse them with unit vectors. The "Hamiltonian" expressed in terms of these operators is the familiar operator $\hat{Q}$. Expressing this as a Weyl pseudo-differential operator introduces a phase space representation of $\hat{Q}$. Following the notation of Fishman and McCoy [8],

$$H\left(-ik_0^{-1}\vec{\nabla}_{\vec{x}}, \vec{x}\right) = \frac{k_0}{2\pi} \int_{R^{2N}} d^N\vec{p}\, d^N\vec{x}\,'\Omega_H\left(\vec{p}, \frac{\vec{x}+\vec{x}'}{2}\right) \exp\left(ik_0\,\vec{p}\cdot\left(\vec{x}+\vec{x}'\right)\right) \tag{9.60}$$

The approach taken by the authors is generic, applying the Helmholtz equation in $N + 1$ dimensions. One of these dimensions is the preferred propagation direction, for example, range, and $N$ dimensions are transverse. The vector $\vec{p}$ is the momentum space equivalent to $\hat{\boldsymbol{p}}$. The square root operator has no meaning until it is expanded in some manner. The function $\Omega_H$ is the Fourier transform of the original operator and its equivalent for the square operator is denoted as $\Omega_{H^2}$; that is, the original transverse component of the Helmholtz operator can be expressed in terms of the transform of $\Omega_H{}^2$ in a formal sense. With this mapping between the operators in place, a procedure emerges for developing an integral representation of the square root operator given the square operator in momentum space. For the factorization presented in the last section, this would be $\Omega_{H^2} = k^2\left(\vec{x}\right) - \vec{p}\cdot\vec{p}$. Inverting the relation between this and the root operator in phase space gives $\Omega_H$, formally an expression for the root operator in coordinate space. The path integral representation is used to develop an algorithm for evaluating the propagator for the one-way PE. Details can be found in Refs. [8, 11, 12].

### 9.5.3 Diagonalizing the Hamiltonian

The methods presented thus far have focused on starting with a representation that is as close to diagonal as possible, exactly diagonal in range-independent environments, and development of approximations of the pseudo-differential operator. Wurmser et al. [4] presented a different approach where the focus is not on starting with a mixture that represents decoupled transmitted and reflected states in range-independent environments but rather on preserving the operator, second order in the derivatives. They start with a mixture using the choice of $\beta$ in the first row of Table 9.1; actually they use $\bar{\beta}$. The components of the vector operator are not initially identified with the standard transmitted/reflected waves. By capitalizing on the analogy of the Klein–Gordon equation pointed out by Levers, the authors succeeded in decoupling the forward and backward modes to increasingly higher orders by application of the Foldy–Wouthuysen transformation used in relativistic quantum mechanics [13]. The process eliminates coupling in steps by diagonalizing the coupling matrix, identified with the Hamiltonian operator, preserving the equation of motion in the process. The effect of the transformation is to add terms to the diagonals. For example, the authors demonstrate that the cumulative effect of the transformation to the third order is a rescaling of the phase speed:

$$c \to c - \frac{1}{8}\frac{\partial^2 c}{k_0^2 \partial x^2} \tag{9.61}$$

Another application of this technique by Wurmser [3] applies the transformation to problems with penetrable rough boundaries in shallow water environments. The approach taken here is different than in the previous work, making use of the pseudo-differential operator $\hat{Q}$. The Foldy–Wouthuysen transformation is then used to obtain new boundary conditions for the higher order terms arising from the operator expansion. Through this process an acoustic analog of the Lamb shift [4] emerges. Lastly, the method is extended to account for density jumps by choosing the component of the transformation matrix; see (9.53), as

$$\beta = -\frac{i\,\rho_0}{k_0\,\rho}\tag{9.62}$$

where $\rho_0$ is a reference density value and $\rho$ is a function of position.

## 9.6 Numerical Approach

Following the analogy of range serving as the equivalent of time, the PE can be solved numerically by methods similar to the FDTD approach. Now the algorithm marches the solution in each transverse slice forward in range. A finite difference evaluation of the operators is applied to the transverse dimension and marched in the range direction. The method employed in the time evolution of the full wave equation was fully explicit. The value of the fields at future times only depended on data at prior times, and no actual matrix inversion was required. The PE has a different structure, second-order derivatives, possibly higher, in space and first order in range. Consider the model equation

$$\frac{\partial f}{\partial r} = a\frac{\partial^2 f}{\partial x^2}\tag{9.63}$$

A second-order explicit finite difference scheme for this is

$$f_i^{(n+1)} - f_i^{(n)} = a\frac{\Delta r}{h^2}\left(f_{i+1}^{(n)} + f_{i-1}^{(n)} - 2f_i^{(n)}\right)\tag{9.64}$$

Applying stability analysis to (9.64) gives the following for the amplitude ratio:

$$1 - 4\frac{a\Delta r}{h^2}\sin^2\left(\frac{kh}{2}\right)\tag{9.65}$$

The stability criterion for this equation is

$$2\frac{a\Delta r}{h^2} \leq 1\tag{9.66}$$

The method is stable but the range is bound by $h^2$ rather than $h$. For practical problems, this can lead to a large number of steps required before any significant change occurs. The method of (9.64) is the FTCS method. Now consider the same form of the equation but with the spatial operator evaluated at the future time value, $n+1$:

$$f_i^{(n+1)} - f_i^{(n)} = \beta \left( f_{i+1}^{(n+1)} + f_{i-1}^{(n+1)} - 2f_i^{(n+1)} \right) \tag{9.67}$$

where $\beta = a\Delta r/h$. This is an implicit scheme; more precisely this is "fully explicit." Rearranging terms to get the future values of the fields on the l.h.s.:

$$-\beta f_{i+1}^{(n+1)} + (1+2\beta)f_i^{(n+1)} - \beta f_{i-1}^{(n+1)} = f_i^{(n)} \tag{9.68}$$

In contrast with the explicit scheme, future values of the fields will be determined by inverting the coefficient matrix on the left, with the addition of suitable boundary conditions. Stability analysis applied to this equation yields the following:

$$\eta = \frac{1}{1 + 4\beta \sin^2\left(\dfrac{kh}{2}\right)} \tag{9.69}$$

This demonstrates that the implicit scheme is unconditionally stable, thus allowing for larger steps. This is an improvement but the equation still has issues with accuracy. Both the FTCS and the implicit method are first-order accurate in time. The aforementioned issues are remedied by averaging the two equations

$$f_i^{(n+1)} - f_i^{(n)} = \frac{\beta}{2} \left( f_{i+1}^{(n+1)} + f_{i-1}^{(n+1)} - 2f_i^{(n+1)} + f_{i+1}^{(n)} + f_{i-1}^{(n)} - 2f_i^{(n)} \right) \tag{9.70}$$

Stability analysis of this method yields

$$\eta = \frac{1 - 2\beta \sin^2\left(\dfrac{kh}{2}\right)}{1 + 2\beta \sin^2\left(\dfrac{kh}{2}\right)} \tag{9.71}$$

which is, again, unconditionally stable. This method is known as the Crank–Nicolson method. Figure 9.1 shows the computational stencil for this scheme.

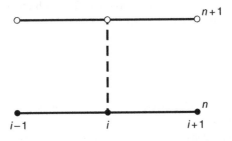

**Figure 9.1**   Computational stencil for the Crank–Nicolson method

The special form of the Crank–Nicolson step is suggestive of another approximation to the operator. Adding a potential term to (9.63) gives the following Crank–Nicolson finite difference:

$$f_i^{(n+1)} - f_i^{(n)} = \frac{\beta}{2}\left(f_{i+1}^{(n+1)} + f_{i-1}^{(n+1)} - 2f_i^{(n+1)} + f_{i+1}^{(n)} + f_{i-1}^{(n)} - 2f_i^{(n)}\right)$$
$$+ \Delta r U_i^{(n+1)} f_i^{(n+1)} + \Delta r U_i^{(n)} f_i^{(n)} \tag{9.72}$$

Rearranging terms with $(n+1)$ on the left and $(n)$ on the right gives

$$f_i^{(n+1)} - \frac{\beta}{2}\left(f_{i+1}^{(n+1)} + f_{i-1}^{(n+1)} - 2f_i^{(n+1)}\right) - \Delta r U_i^{(n+1)} f_i^{(n+1)}$$
$$= f_i^{(n)} + \frac{\beta}{2}\left(f_{i+1}^{(n)} + f_{i-1}^{(n)} - 2f_i^{(n)}\right) + \Delta r U_i^{(n)} f_i^{(n)} \tag{9.73}$$

which can be expressed as

$$\left(I - \Delta r \hat{X}\right) \cdot f_i^{(n+1)} = \left(I + \Delta r \hat{X}\right) \cdot f_i^{(n)} \tag{9.74}$$

where it is understood that the operators generate the second-order finite difference in the variable $i$ when acting on the function and $\hat{X}$ is a scaled version of $\hat{\Lambda}$. The point is that the operator on each side is a first-order approximation to an exponential, for example, $\exp\left(\Delta r \hat{X}\right) \approx I + \Delta r \hat{X}$. This is precisely the form of the propagator in time-dependent quantum mechanics, as pointed out in Numerical Recipes [14]. It is also the foundation of the approach taken by Saad and Lee [9] in their development of a wide-angle PE.

## 9.7   Exercises

1. Develop an implicit finite difference scheme for the PE from the Padè (1, 1) approximant and test its stability.
2. Verify the algebraic relations obeyed by the Pauli spin matrices using their explicit form.
3. Find an exact solution for the paraxial approximation, (9.9), for the cases $n = 1$, $n = a + bz$, and $n = a + bz^2$. How do these compare to the solutions for the full Helmholtz equation?
4. Develop a Crank–Nicolson method for solving (9.74) with one transverse dimension and apply it to a $z$-dependent refractive index.

## References

[1] Dirac, P. A. M., The Principles of Quantum Mechanics, Fourth Edition, Oxford Science Publications, Clarendon Press, Oxford, 1991.

[2] Levers, R. G., Spinning the Helmholtz Equation, in Computational Acoustics: Wave Propagation, Lee, D., Sternberg, R. L., and Schultz, M. H. (Eds.), Elsevier Science Publishers B. V., North-Holland, 1988.

[3] Wurmser, D., A parabolic equation for penetrable rough surfaces: Using the Foldy–Wouthuysen transformation to buffer density jumps, Ann. Phys., Vol. 311, pp. 53–80, 2004.

[4] Wurmser, D., Orris, G. J., and Dashen, R., Application of the Foldy–Wouthuysen transformation to the reduced wave equation in range-dependent environments, J. Acoust. Soc. Am., Vol. 101, No. 3, pp. 1309–1327, 1997.

[5] McDaniel, S. T., Parabolic approximations for underwater sound propagation, J. Acoust. Soc. Am., Vol. 58, No. 6, pp. 1178–1185, 1975.

[6] Tappert, F. D. and Hardin, R. H., in A Synopsis of the AESD Workshop on Acoustic-Propagation Modeling by Non-Ray Tracing Techniques, AESD Tech. Note TN 73-05, November 1973.

[7] Corones, J., Bremmer series that correct parabolic approximations, J. Math. Anal. Appl., Vol. 50, No. 2, pp. 361–372, 1975.

[8] Fishman, L. and McCoy, J. J., Factorization, path integral representations, and the construction of direct and inverse wave propagation theories, IEEE Trans. Geosci. Remote Sens., Vol. GE-22, No. 6, pp. 682–692, 1984.

[9] Saad, Y. and Lee, D., A new algorithm for solving the wide angle wave equation, Research Report YALEU/DCS/RR-485, Yale University, 1986.

[10] Swanson, M., Path Integrals and Quantum Processes, Academic Press, Inc., Boston, 1992.

[11] Fishman, L. and McCoy, J. J., Derivation and application of extended parabolic wave theories II. Path integral representation, J. Math. Phys., Vol. 25, No. 2, pp. 297–308, 1984.

[12] Fishman, L. and Wales, S. C., Phase space methods and path integration: The analysis and computation of scalar wave equations, J. Comput. Appl. Math., Vol. 20, pp. 219–238, 1987.

[13] Bjorken, J. D. and Drell, S. D., Relativistic Quantum Mechanics, McGraw-Hill Book Company, New York, 1964.

[14] Press, W. H., Teukolsky, S. A., Vetterling, W. T., and Flannery, B. P., Numerical Recipes in C++: The Art of Scientific Computing, Second Edition, Cambridge University Press, Cambridge, 2005.

# 10

# Finite Element Method

## 10.1 Introduction

In Chapter 8 the FD and FDTD methods for solving the wave equation were presented. The introduction of the technique started with the discretization of spatial operators then applied these to the Helmholtz equation in the frequency domain. The approaches to numerical modeling presented thus far have relied on techniques for converting derivatives into differences over small intervals. This chapter introduces an entirely new paradigm for generating solutions to PDE, the finite element method (FEM). In the FEM technique, the derivative operators are not approximated to generate a discrete system. Rather, an approximation to the solutions is made. The region of space is discretized, as with the FD technique, but instead of evaluating the equation at the nodes of the discretization, a set of local functions is defined on the subregions, referred to as basis elements, and the PDE projected onto these elements. The unknown field is expressed as a superposition of basis functions. The reader may be wondering how this differs from a mode expansion. The mode expansion used exact or numerical solutions to the 1-dimensional (1-dim) differential equations generated by separation of variables applied to the PDE to model the behavior of the field given a source and boundary conditions. Approximations are made via the truncation of an infinite series, but the function set being used, at least in theory, consists of known solutions to the Helmholtz equation. In the FEM technique, the solution being sought is a set of coefficients that allow the solution to the PDE to be written as a superposition over the basis elements. The process of discretizing the domain, defining elements, and projecting the PDE onto the elements results in a linear matrix equation whose solution vector is the coefficient set.

---

*Computational Acoustics: Theory and Implementation*, First Edition. David R. Bergman.
© 2018 John Wiley & Sons Ltd. Published 2018 by John Wiley & Sons Ltd.

## 10.2    The Finite Element Technique

This section presents the technique at a high level, without focusing on a particular set of equations. The linear PDE being solved is of the following form:

$$L(\varphi(\vec{x})) = S(\vec{x}) \tag{10.1}$$

In (10.1) $L$ is a generic linear operator, containing terms like $\nabla^2$, $\vec{v} \cdot \nabla$, $U(\vec{x})$, and so on, and $S(\vec{x})$ is a source term. The FEM technique assumes a discrete set of functions designed to model the behavior of the solution over small regions of the domain. The basis set models the behavior of the function in the interior of the region. To handle boundary conditions a single function $\tilde{\varphi}$ is chosen that satisfies the boundary conditions of the problem [1]. A set of functions $b_n(\vec{x})$, $n = 1,...,N$ is defined on the interior of the discretization, and the unknown field is expanded in this set:

$$\varphi_N(\vec{x}) = \tilde{\varphi}(\vec{x}) + \sum_{n=1}^{N} a_n b_n(\vec{x}) \tag{10.2}$$

In (10.2) the index $N$ on the function in the l.h.s. is to serve as a reminder that this is not a true solution to (10.1), but an approximation over $N$ basis elements. With the boundary function included as part of the assumed solution, the interior basis functions $b_n(\vec{x})$ are designed to vanish at the boundary. The strategy is to develop an equation to determine the coefficients $\{a_n\}$ from the original differential equation such that the error between the true solution and the approximate solution is minimized. The error is defined by inserting (10.2) in (10.1) and rewriting to yield

$$e_N = L(\varphi_N) - S(\vec{x}) = 0 \tag{10.3}$$

As it stands it should be expected that $e_N \neq 0$. Considering the basis set to be analogous to the independent directions of a vector space and define the inner product,

$$\langle f|g \rangle = \int \bar{f} g \, dV \tag{10.4}$$

While it may not be possible to find a superposition such that $e_N = 0$, the error can be minimized by requiring it to be orthogonal to every element of the space:

$$\langle b_n|e_N \rangle = 0 \tag{10.5}$$

Using the linearity of $L$, (10.5) is used to develop a matrix equation for the determined constants in the basis expansion:

$$\langle b_n | e_N \rangle = \left\langle b_n \left| L(\varphi_N) - S(\vec{x}\,) \right. \right\rangle = \langle b_n | L(\varphi_N) \rangle - \left\langle b_n | S(\vec{x}\,) \right\rangle$$

$$= \left\langle b_n \left| L \left( \sum_{m=1}^{N} a_m b_m \right) \right. \right\rangle + \langle b_n | L(\widetilde{\varphi}) \rangle - \left\langle b_n | S(\vec{x}\,) \right\rangle$$

$$= \sum_{m=1}^{N} \langle b_n | L(b_m) \rangle a_n + \langle b_n | L(\widetilde{\varphi}) \rangle - \left\langle b_n | S(\vec{x}\,) \right\rangle$$

Neglecting the boundary function for the time being, the constraint, (10.5), can now be written as

$$\sum_{m=1}^{N} \langle b_n | L(b_m) \rangle a_n - \left\langle b_n | S(\vec{x}\,) \right\rangle = 0 \tag{10.6}$$

The following arrays are defined:

$$\widetilde{L\varphi}_n = \langle b_n | L(\widetilde{\varphi}) \rangle \tag{10.7}$$

$$S_n = \langle b_n | S \rangle \tag{10.8}$$

$$B_{nm} = \langle b_n | L(b_m) \rangle \tag{10.9}$$

With these definitions the PDE is converted into a linear algebraic equation:

$$\mathbf{B} \cdot \mathbf{a} = \mathbf{S} \tag{10.10}$$

Equation (10.10) is solved for the vector of coefficients $\mathbf{a}$ and a solution is given by (10.2). Before proceeding there is one detail to discuss. Applying the Galerkin method to the wave equation will involve an integral of the following form:

$$\int b_n \nabla^2 b_m dV \tag{10.11}$$

This term imposes the requirement that $b_n$ be differentiable to second order. The integrand is transformed by use of the identity $\vec{\nabla} \cdot (b_n \vec{\nabla} b_m) = \vec{\nabla} b_n \cdot \vec{\nabla} b_m + b_n \nabla^2 b_m$:

$$\int \vec{\nabla} \cdot \left( b_n \vec{\nabla} b_m \right) dV - \int \vec{\nabla} b_n \cdot \vec{\nabla} b_m dV \tag{10.12}$$

Applying the divergence theorem to the first integral changes it to an integral evaluated on the boundary of the integration volume. By choosing the basis elements to vanish at the boundary of their support domain, this term can be made to vanish. With this choice, the differentiability requirement is reduced by one order.

The discretization of the domain is not a trivial endeavor and not unique either. The goal is to formulate a discretization that will allow for easy evaluation the operator matrix, **B**, and source vector, **S**, while providing enough detail to give an accurate estimate of the solution. To this end the spatial continuum is replaced with a set of small discrete regions whose union covers the region. This is done in such a way that the interiors of each region are non-overlapping, but neighboring regions may share boundary points. The size of each region is chosen to be small relative to some length scale of the problem, such as wavelength. In the FD technique, the lattice was chosen to be uniform. This makes it possible to develop FD operators with a minimum number of sample points for a desired order and with the same coefficients at each point in the lattice. In contrast no such issue arises in the application of FEM. Therefore, the patches can be any shape and size, randomly chosen as long as they fill the space without overlapping.

## 10.3   Discretization of the Domain

The discretization of the domain starts with sampling the domain at a collection of points $\vec{x}_i$. For the purpose of defining basis functions over subdomains, attention is focused exclusively on interior points for the time being. The collection of points in the domain can be chosen at random or based on regular intervals defined on coordinate curves. Once the sample points are chosen, a set of non-overlapping domains are defined. These domains can be any geometry, but the most typical are triangular in 2-dim or tetrahedral in 3-dim. One-dimensional problems are particularly easy and will be presented as an example. Without any special circumstances to serve as a guide, regular intervals in Cartesian coordinates are used to generate sample points for the development of FEM in this presentation. Later the use of open-source tools for mesh generation will be discussed. It should be noted that other coordinates may be a better choice for certain problems and that regular spacing of intervals may not be suitable for some problems, for example, if a boundary has an odd shape with tight curves or contours. FEM analysis requires a finite region of space, so there will be a lower-dimensional section to the computational region that requires special treatment. The non-overlapping domains, called patches, can be thought of as intervals or a generalization of an interval in dimensions higher than 1. The individual patches are not the elements used to solve the PDE but are used to define elements. Once a set of patches is created, subsets of patches that have a common point or set of interior points are used to define a basis. The basis function is then defined on the small subset of patchwork.

### 10.3.1   One-Dimensional Domains

Consider a region of a straight line $[a, b]$. The size of the region is $L = b - a$. The region is sampled by a set of points $\{x_i\}$ with $i = 1, \ldots, N$ and $a \le x_i \le b$. The end points are included in the sample set to serve as boundaries. There are $N$ points and $N - 1$ intervals. These need not be the

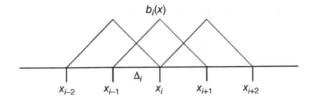

**Figure 10.1**   One-dimensional patches and basis elements defined

same size, but for now they are assumed to be. Hence $\Delta x = x_{i+1} - x_i$, $L = (N-1)\Delta x$, and $x_i = a + (i-1)\Delta x$. This is illustrated in Figure 10.1.

### 10.3.2   Two-Dimensional Domains

For two- and three-dimensional domains, consider regions with Cartesian symmetry and boundaries that coincide with Cartesian coordinate surfaces. Starting with a rectangular region $R = [a_x, b_x] \times [a_y, b_y]$, the length and breadth of the region are defined: $L_x = b_x - a_x$, $L_y = b_y - a_y$. Each axis is sampled at regular intervals as before, generating a set of points $\{x_i, y_j\}$ with $i = 1, \ldots, N$ and $j = 1, \ldots, M$. The boundary in this case is the 1-dim rectangle defined by $\partial R = \{x_i, a_y\} \cup \{x_i, b_y\} \cup \{a_x, y_j\} \cup \{b_x, y_j\}$. Because the sampled points are at constant intervals, the differences, $\Delta x = x_{i+1} - x_i$ and $\Delta y = y_{j+1} - y_j$, are constant and the total length may be expressed in terms of these differences, $L_x = (N-1)\Delta x$ and $L_y = (M-1)\Delta y$. This definition of sampled points does not provide an obvious definition of a patch as it did in one dimension. There are many choices available for defining a patch, or a geometric primitive. One obvious choice are the rectangles, $\{x_i, x_{i+1}\} \times \{y_j, y_{j+1}\}$. It turns out that a preferred primitive geometry is the triangle. Triangles can be defined on the Cartesian grid in several ways. The first is to split each rectangle along a diagonal with positive slope in the $(x, y)$ plane. This results in two triangles per rectangle, $\{x_{i,j}, \vec{x}_{i+1,j}, \vec{x}_{i+1,j+1}\}$ and $\{\vec{x}_{i,j}, \vec{x}_{i,j+1}, \vec{x}_{i+1,j+1}\}$. If the rectangle is split along the other diagonal (negative slope), the following two triangles are obtained, $\{\vec{x}_{i,j}, \vec{x}_{i,j+1}, \vec{x}_{i+1,j}\}$ and $\{\vec{x}_{i+1,j}, \vec{x}_{i,j+1}, \vec{x}_{i+1,j+1}\}$. Each of these choices covers the region with right triangles. Another choice is to add a point to the center of each rectangle and divide it into four isosceles triangles. The center point is $\{x_{i+1/2}, y_{j+1/2}\} \equiv \{x_i + \Delta x/2, y_j + \Delta y/2\}$. Now each rectangle is divided into four triangles:

$$\left\{\vec{x}_{i,j}, \vec{x}_{i+1,j}, \vec{x}_{i+1/2,j+1/2}\right\} \tag{10.13}$$

$$\left\{\vec{x}_{i+1,j}, \vec{x}_{i+1,j+1}, \vec{x}_{i+1/2,j+1/2}\right\} \tag{10.14}$$

$$\left\{\vec{x}_{i,j+1}, \vec{x}_{i+1,j+1}, \vec{x}_{i+1/2,j+1/2}\right\} \tag{10.15}$$

$$\left\{\vec{x}_{i,j}, \vec{x}_{i,j+1}, \vec{x}_{i+1/2,j+1/2}\right\} \tag{10.16}$$

A vector in this 2-dim lattice may be expressed as follows:

$$\vec{x}_{i,j} = x_i\hat{e}_x + y_j\hat{e}_y = \left(x_i, y_j\right)$$

(10.17)

As a final note, the double indexing scheme can be mapped to a single index fairly easily with the following mapping $(i,j) \rightarrow n$:

$$n(i,j) \equiv i + N(j-1), \quad n = 1, \ldots, N$$

(10.18)

A nice feature of sampling points in this manner is that it is easy to develop code that will generate the points and the regions, and one has explicit knowledge of which patches share an edge or vertex. When dealing with problems that have more complicated shaped boundaries and regions, it is necessary to use a variable sampling of the region. This can be done by choosing a set of interior points at random. Once defined, these points can then be grouped together to form non-overlapping triangles. This process is a little more involved since it requires searching for points that are close to a given point and checking to ensure that once a primitive is defined, it does not enclose any of the remaining points. There exist several software packages, many of which are free (open source), that will generate a triangular mesh for a 2-dim (and 3-dim) region and save the results in a variety of formats. One advantage to tiling a region with triangles is that one can fit them into any geometry and define the boundary with the same sample set. Triangular patches make very versatile primitives. Figures 10.2 and 10.3 illustrate the triangular tiling of a Cartesian rectangular region for the two examples described in the previous text. Figure 10.4 illustrates the same for a random tiling generated by Gmsh.

Figure 10.2 shows explicitly the tiling created by cutting a set of identical squares along the diagonal of each square to produce a pair of right triangles. In this case the resulting tiling is not symmetric. The individual patches are drawn in dashed lines. Also, shown in bold solid lines are the perimeters of three sets of connecting patches with a common point. This is the subset

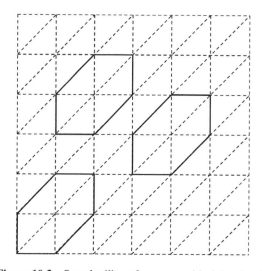

**Figure 10.2**   Sample tiling of a square with right triangle

that will be used to define the basis elements described in the next section. For this particular tiling, every interior point will be shared by exactly six triangles, while boundary points will have fewer triangles. In Figure 10.3 the square lattice is defined the same way as in the previous example, but triangles are made by cutting each square along both diagonals. For this choice of tiling, interior points can have four or eight triangles associated with them; these groupings are illustrated in bold solid lines. The last example of a tiling is provided in Figure 10.4. The tiling was created using the open-source package Gmsh (see Chapter 2). The triangles created by the software is drawn in dashed lines. By inspection one can find interior points with as few as four

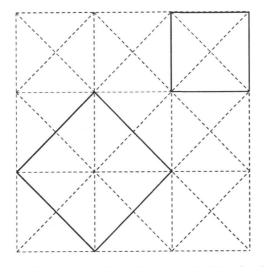

**Figure 10.3**   Sample tiling of a square with right triangles

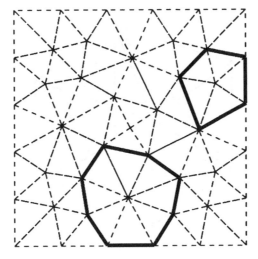

**Figure 10.4**   Example of a random tiling of a square using Gmsh

triangles, and as many as eight, as well as many with five and six triangles. Two groups are singled out by bold solid lines. These will be used to make rooftop basis elements as described in the next section.

### 10.3.3    Three-Dimensional Domains

The final subsection will describe a method for covering a 3-dim volume with small discrete primitives. Following the logical progression of the last two sections, we focus attention on a large cubic volume of 3-dim space $R = [a_x, b_x] \times [a_y, b_y] \times [a_z, b_z]$. The length, breadth, and height are $L_x = b_x - a_x$, $L_y = b_y - a_y$, $L_z = b_z - a_z$, respectively. A set of points is defined, $\{x_i, y_j, z_k\}$ with $i = 1, \ldots, N$, $j = 1, \ldots, M$, and $k = 1, \ldots, L$, and subject to the constraints $a_x \le x_i \le b_x$, $a_y \le y_j \le b_y$, and $a_z \le z_k \le b_z$. These are evenly spaced in each direction, $\Delta x = x_{i+1} - x_i$, $\Delta y = y_{j+1} - y_j$, and $\Delta z = z_{k+1} - z_k$ with $L_x = (N-1)\Delta y$, $L_y = (M-1)\Delta y$, and $L_z = (L-1)\Delta z$. The boundary of the region is given by the union of six coordinate planes:

$$\partial R = \{x_i, y_j, a_z\} \cup \{x_i, y_j, b_z\} \cup \{x_i, a_y, z_k\} \cup \{x_j, b_y, z_k\} \cup \{a_x, y_j, z_k\} \cup \{b_x, y_j, z_k\} \tag{10.19}$$

Following the approach for two-dimensional tiles, the simplest primitive geometry is sought. In 3-dim the generalization of a triangle is the tetrahedron, a four-sided pyramid. In 2-dim rectangles were split into two triangles by dividing along the diagonal. Here the situation is a little more complex. It turns out that a cube can be cut into five regular tetrahedra, illustrated in Figure 10.5.

Four of the five tetrahedra will be identical, which cover the outer boundary of the rectangular solid. The fifth tetrahedron will fill the remaining volume. As in 2-dim this particular division is not unique. Each of these tetrahedral volumes can be identified by the vertices of the rectangular solid that contains them:

$$\{(i,j,k), (i+1,j,k), (i,j+1,k), (i,j,k+1)\} \tag{10.20}$$

$$\{(i+1,j,k+1), (i+1,j,k), (i+1,j+1,k+1), (i,j,k+1)\} \tag{10.21}$$

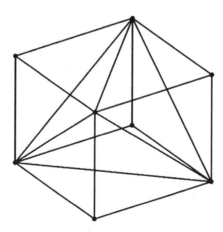

**Figure 10.5**    Cube cut into five tetrahedra

$$\{(i+1,j+1,k),(i+1,j,k),(i+1,j+1,k+1),(i,j+1,k)\} \tag{10.22}$$

$$\{(i,j,k+1),(i,j+1,k+1),(i+1,j+1,k+1),(i,j+1,k)\} \tag{10.23}$$

$$\{(i,j,k+1),(i+1,j,k),(i+1,j+1,k+1),(i,j+1,k)\} \tag{10.24}$$

Regardless of what the rectangular solids look like, they can always be subdivided into five tetrahedra. Again, as in 2-dim this lattice structure is convenient since it provides us with an algorithm for identifying geometric primitives via indices. Hence, a software routine that generates the point set can also generate a complete list of primitives simply by grouping indices according to (10.20)–(10.24). For more complex regions and boundaries, we can generate a random sampling of points and then group 4 neighboring non-coplanar points into tetrahedral primitives. As in 2-dim we will be confronted with the need to identify neighbors, ensure they are not coplanar, and test to ensure they do not enclose other points.

### 10.3.4   Using Gmsh

This section will provide nominal information and Gmsh formats for geometry data. There are several packages available for making surface and volume meshes; this is not an endorsement of Gmsh over other products. Since Gmsh was used to generate many of the tiling of 2-dim and 3-dim regions and has been used by the author for boundary element analysis, it seems appropriate to mention just a few features. Gmsh allows the user to define geometric primitives via a GUI or command line functions. Starting with points, one can define 1-dim entities such as line segments, circular arcs, elliptic arcs, and so on. Surfaces can be defined by translating or rotating curves. This description does not encompass all the possible shapes or ways of making them. The primitives and their relationships are defined in a geo file. An example of the contents of a geo file is provided as follows:

```
Point(1) = {0.25, 0.25, 0, 0.1};
Point(2) = {-0.25, 0.25, 0, 0.1};
Point(3) = {-0.25, -0.25, 0, 0.1};
Point(4) = {0.25, -0.25, 0, 0.1};
Line(1) = {4, 3};
Line(2) = {3, 2};
Line(3) = {2, 1};
Line(4) = {1, 4};
Line Loop(5) = {3, 4, 1, 2};
Plane Surface(6) = {5};
```

The first four lines define points with unique identifiers, Point($k$). The first three entries in the curly brackets are the three coordinates of the point. The fourth entry is a size parameter that is used to determine how small to make the tiles on the surface (or in the volume) defined by this geometry. The next four lines define line segment from pairs of points, for example, Line(2) = {3, 2} is the line with unique id = 2 (the second line drawn) starting at point 3 and ending at point 2. The last two lines define a plane from the four lines. The line loop indicates that the four lines form a closed loop and the last line identifies the plane surface by the boundary defined by the line loop. The mesh created by this file is Figure 10.4. Version 2.12 supports 16 mesh file formats. For the purpose of illustration, the data in two of these formats is presented. The two formats are STL and MSH. Saving in STL format produces a file containing triangular facets. A sample of an STL file for a faceted surface appears as follows:

```
solid Created by Gmsh
facet normal 0 0 1
 outer loop
   vertex -1.0625 1.0625 0
   vertex -1.48032 0.0516304 0
   vertex -0.348278 0.658283 0
 endloop
endfacet
...
endsolid Created by Gmsh
```

Each facet contains the components of the normal vector to a triangular, contained in the second line in the example, and the coordinates of the three vertices of the triangle. The structure between the lines "facet normal" and "endfacet" repeats for every triangle. A second example is illustrated by a sample of the native file format *.msh. A snippet from the file used to create Figure 10.6 is provided as follows:

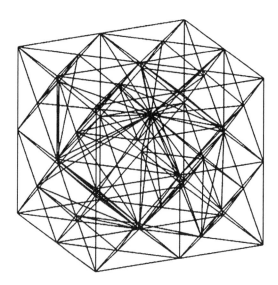

**Figure 10.6**   A cube in 3-dim tiled with tetrahedra using Gmsh

```
$MeshFormat
2.2 0 8
$EndMeshFormat
$Nodes
51
1 0 0 0
2 0 0 1
...

...

51 0.6250000000004479 0.6250000000012492 0.6250000000004479
$EndNodes
$Elements
261
1 15 2 0 1 1
...
13 1 2 0 3 2 11
...
35 2 2 0 14 1 25 15
...
261 4 2 0 26 42 24 3 29
$EndElements
```

The MSH format contains a list of nodes, vertex locations, and geometric elements. Each node and element has a unique id number (column 1). The three Cartesian coordinates of the node follow the id. For the example above, the first node is at the origin, second on the $z$-axis and [0,0,1]. The number of nodes, 51 in this case, is listed before the nodes. Following the nodes are the elements, 261 in this case. A complete description of this format is not presented, but a few numbers are explained in the element list. The first column is the unique element identifier; shown are elements [1,13,35,261]. The second integer identifies the type of element. This example contains 15 = point, 1 = line, 2 = triangle, and 4 = four node tetrahedron. Following this index are a few entries referring to the physical entity that the element belongs to but are not needed for this example. The relevant data for this example are the last set of indices, skipping three after element type. These last indices comprise the list of nodes that are used to create this element. The second element in the example is a line made from nodes 2 and 11. The third element is a triangle made from nodes 1, 25, and 15. Finally the fourth element of this example is a tetrahedron made of nodes 42, 24, 3, and 29. This is by no means a representation of the complete MSH format, but with this information, a parser is easy to write and geometries plotted using any high-level software.

## 10.4   Defining Basis Elements

The last section was devoted to defining a tiling of a spatial region, filling space with small tiles, defined as geometric primitives. This section presents the definition of basis functions defined on these geometric primitives. The choice of functions must be easy to work with so that the

FEM equations can be implemented in software and at the same time obey conditions placed on them related to global boundary conditions, continuity, and differentiability. Simplicity is achieved to some degree by trying to make each basis function as compact as possible, that is, to have the individual basis functions defined on the smallest subregion of space possible. It would be ideal to have each function defined on one, and only one, primitive, and each primitive contain one and only one function. Unfortunately, this makes other restrictions impossible. The goal is to define basis elements in such a way that the local behavior of the field can be described using these computational elements and have their superposition describe the correct large-scale behavior of the field consistent with the field equations. An element will require several geometric primitives to support the basis function. The number of elements required may also depend on the type of function chosen to describe the local behavior of the field. The most common choice of functions is polynomials, and of these the easiest to define and work with are linear polynomials. A complete presentation of linear basis elements for the 1-dim, 2-dim, and 3-dim is provided.

## 10.4.1  One-Dimensional Basis Elements

Consider a 1-dim nonuniform sampling of an interval with $N$ sample points $\{x_i\}$ and $N-1$ intervals $[x_i, x_{i+1}]$, each of size $\Delta_i = x_{i+1} - x_i$. Basis functions are defined on the interior for each pair of consecutive intervals, giving $N-2$ interior basis elements. The basis elements are indexed by the first interval index; hence basis $i$ contains intervals $i$ and $i + 1$ and points $i, i + 1$, and $i + 2$. The basis functions must be at least differentiable to first order. The simplest way to accomplish this is with a rooftop or hat function. An example of this was depicted in Figure 10.1. This is a function that peaks at value 1 at the center point and decays linearly to 0 at the end of the intervals where it is defined. A generic function can be written as a superposition of these for points in the interior:

$$f(x) = \sum_{i=1}^{N-2} a_n b_n(x)$$

(10.25)

Each basis function must be defined in each of the two intervals:

$$b_n^i(x) = c_i + \beta_i x, \, x \in I_i$$

(10.26)

$$b_n^{i+1}(x) = c_{i+1} + \beta_{i+1} x, \, x \in I_{i+1}$$

(10.27)

There are two constraints placed on these functions—that they vanish at the boundary of the basis and they are equal to 1 at the common point:

$$b_n^i(x_i) = c_i + \beta_i x_i = 0$$

(10.28)

$$b_n^{i+1}(x_{i+2}) = c_{i+1} + \beta_{i+1} x_{i+2} = 0$$

(10.29)

Equations (10.28) and (10.29) are used to solve for $c_i$ and $c_{i+1}$:

$$b_n^i(x) = \beta_i(x - x_i)$$

(10.30)

$$b_n^{i+1}(x) = \beta_{i+1}(x - x_{i+2}) \tag{10.31}$$

The requirement that $b_n(x_{1+1}) = 1$ fixes the value of $\beta$:

$$b_i(x) = \begin{cases} \dfrac{(x - x_i)}{\Delta_i} & x \in I_i \\[2mm] -\dfrac{(x - x_{i+2})}{\Delta_{i+1}} & x \in I_{i+1} \end{cases} \tag{10.32}$$

### 10.4.2 Two-Dimensional Basis Elements

Two-dimensional basis elements are defined using the same considerations as in 1-dim. The geometric regions are triangles. This generalizes the interval in 1-dim. A collection of triangles will be used to define the functions $b_n$. The basis functions will peak at a single interior point and vanish on the points that surround the interior basis point, generalizing the rooftop function. A suitable region for the basis is defined by taking all triangles that have a single interior point in common. From an indexing point of view, things get a little complicated, so the interior point is referred to as $\vec{x}_0$ and can be thought of as the origin of a local coordinate system. The search for all triangles containing $\vec{x}_0$ is not guaranteed to yield a fixed number. Regardless, the constraints on each required to define a particular basis are independent of each other. An exact account of how many triangles, and which ones, contribute to a basis element is required to correctly perform the numerical integrations that appear in the definition of the operator matrices and the source terms. For each triangle in a basis, one of its points will be $\vec{x}_0$, and the basis will vanish at the other two. Equation (10.33) gives a generic linear function defined on a triangle:

$$b_n(x) = c_n + \beta_{nx}x + \beta_{ny}y = c_n + \vec{\beta}_n \cdot \vec{x} \tag{10.33}$$

The index $n$ references a particular triangle in the basis region. Figure 10.7 illustrates this situation and defines all quantities. Outer points are labeled similarly. Since each outer point

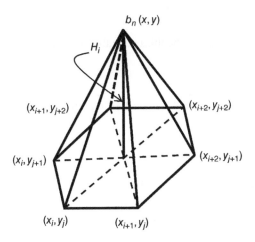

**Figure 10.7**   Two-dimensional rooftop basis defined

is shared by two triangles, the number of outer points is equal to the number of triangles. Finally, an ordering of these points is defined such as to increase in the direction defined by the right-hand rule with the normal being out of the page. For each triangle in the set, the following equations of constraint hold:

$$c_n + \vec{\beta}_n \cdot \vec{x}_0 = 1 \tag{10.34}$$

$$c_n + \vec{\beta}_n \cdot \vec{x}_n = 0 \tag{10.35}$$

$$c_n + \vec{\beta}_n \cdot \vec{x}_{n+1} = 0 \tag{10.36}$$

The following convention is imposed $\vec{x}_{N+1} = \vec{x}_1$. The first constraint allows for the removal of $c_n$:

$$b_n(x) = 1 + \vec{\beta}_n \cdot (\vec{x} - \vec{x}_0) \tag{10.37}$$

The other two boundary conditions are provided as follows:

$$1 + \vec{\beta}_n \cdot (\vec{x}_n - \vec{x}_0) = 0 \tag{10.38}$$

$$1 + \vec{\beta}_n \cdot (\vec{x}_{n+1} - \vec{x}_0) = 0 \tag{10.39}$$

At this point it becomes convenient to define the edges of each triangle. Two edges are defined relative to the common internal point $\vec{x}_0$:

$$\vec{L}_1 = \vec{x}_n - \vec{x}_0 \tag{10.40}$$

$$\vec{L}_2 = \vec{x}_{n+1} - \vec{x}_0 \tag{10.41}$$

The difference of these two gives the third edge that lies on the outer perimeter of the basis function:

$$\vec{L}_3 = \vec{x}_{n+1} - \vec{x}_n = \vec{L}_2 - \vec{L}_1 \tag{10.42}$$

The remaining conditions simplify to

$$1 + \vec{\beta}_n \cdot \vec{L}_1 = 0 \tag{10.43}$$

$$1 + \vec{\beta}_n \cdot \vec{L}_2 = 0 \tag{10.44}$$

Subtracting these equations also yields an alternate equation of constraint:

$$\vec{\beta}_n \cdot \vec{L}_3 = 0 \tag{10.45}$$

There is a very simple geometric interpretation of these constraints. Equation (10.45) states that the parameter vector $\vec{\beta}_n$ is orthogonal to the outer leg of the triangle. In light of this information, it follows that $\vec{\beta}_n \cdot \vec{L}_1 = \vec{\beta}_n \cdot \vec{L}_2 = |\vec{\beta}_n| h$, where $h$ is the height of the triangle. Constraints (10.43) and (10.44) require $\vec{\beta}_n \cdot \vec{L}_{1,2} = -1$. By definition, each of the vectors $\vec{L}_{1,2}$ point in the radial direction with origin, $\vec{x}_0$. The vector $\vec{\beta}_n$, points radially inward, is scaled by $h^{-1}$, and is orthogonal to the vector $\vec{L}_3$. Defining the $z$ unit vector for the normal to the plane $\hat{e}_z$, and the in-plane normal to the edge opposite $\vec{x}_0$ as $\hat{N}_n$,

$$\vec{\beta}_n = -\frac{\hat{N}_n}{h_n} . \tag{10.46}$$

Using some basic geometry the following can be derived:

$$h_n = \frac{2A}{L_3} = \frac{\hat{e}_z \cdot \left(\vec{L}_1 \times \vec{L}_2\right)}{L_3} \tag{10.47}$$

$$\vec{L}_3 \times \hat{e}_z = L_3 \hat{N}_n \tag{10.48}$$

$$f_n(x) = 1 + \frac{\left(\hat{e}_z \times \vec{L}_3\right) \cdot \left(\vec{x} - \vec{x}_0\right)}{\hat{e}_z \cdot \left(\vec{L}_1 \times \vec{L}_2\right)} \tag{10.49}$$

It is a fairly straightforward matter to verify that this function has all the correct behavior. In particular, it vanishes along all of $\vec{L}_3$ for each triangle. The coefficient is defined purely in terms of the geometry of the geometric primitive.

## 10.4.3   Three-Dimensional Basis Elements

Lastly, the linear basis elements defined on tetrahedra in 3-dim are explained. Following the same approach as in the last section, the starting point is a model linear function defined on the tetrahedron:

$$b_n(x) = c_n + \vec{\beta}_n \cdot \vec{x} \tag{10.50}$$

The first condition easily generalizes to

$$b_n(x) = 1 + \vec{\beta}_n \cdot \left(\vec{x} - \vec{x}_0\right) \tag{10.51}$$

Instead of having an outer edge for each triangle, there is an outer triangle for each tetrahedron. The location of the outer points is defined relative to the center point of the basis $\vec{x}_0$, with outer points indexed $\vec{x}_{1,2,3}$:

$$\vec{L}_j = \vec{x}_j - \vec{x}_0, \quad j = 1, 2, 3 \tag{10.52}$$

All three of these vectors point along the radial direction in local spherical coordinates with $\vec{x}_0$ at the origin, leading to the following expression for the other constraints:

$$1 + \vec{\beta}_n \cdot \vec{L}_j = 0 \tag{10.53}$$

The edges of the triangular face opposite the point $\vec{x}_0$ are defined in a way that preserves the sense of traversal of the perimeter relative to the outward facing normal:

$$\Delta \vec{L}_{ij} = \vec{L}_i - \vec{L}_j \tag{10.54}$$

In (10.54) $(i,j) = (2,1), (3,2), (1,3)$. Following the same logic as in two dimensions, it is straight forward to show that the vector $\vec{\beta}_n$ must point in the direction of the normal to the triangle opposite to $\vec{x}_0$. Defining the outward normal $\hat{n}$, we also have $\hat{n} \cdot \vec{L}_j = h$ for all $j = 1,2,3$:

$$\vec{\beta}_n = -\frac{\hat{n}}{h} \tag{10.55}$$

Using geometry, a relation between the height of the tetrahedron and the vertex vectors can be found. This requires the formula for the volume of a tetrahedron:

$$V_t = \frac{1}{3}Ah \tag{10.56}$$

where $A$ is the area of one of the base triangles and $h$ is the height of the apex over the plane of the base, that is, measured along a line orthogonal to the base. The volume can also be expressed in terms of the three vectors $\vec{L}_j$:

$$V_t = \frac{1}{6}\vec{L}_1 \cdot \left(\vec{L}_2 \times \vec{L}_3\right) \tag{10.57}$$

These two definitions for volume are used to rewrite $\vec{\beta}_n$:

$$\vec{\beta}_n = -\frac{\hat{n}2A}{\vec{L}_1 \cdot \left(\vec{L}_2 \times \vec{L}_3\right)} \tag{10.58}$$

The final step involves realizing that the term $\hat{n}2A$ is just the cross product of two of the edges of the outer triangle:

$$\hat{n}2A = \left(\vec{L}_2 - \vec{L}_1\right) \times \left(\vec{L}_3 - \vec{L}_2\right) = \vec{L}_2 \times \vec{L}_3 + \vec{L}_3 \times \vec{L}_1 + \vec{L}_1 \times \vec{L}_2 \tag{10.59}$$

This provides the final expression for the basis element of a tetrahedron:

$$b_n(x) = 1 + \frac{\left(\vec{L}_3 \times \vec{L}_2 + \vec{L}_1 \times \vec{L}_3 + \vec{L}_2 \times \vec{L}_1\right) \cdot \left(\vec{x} - \vec{x}_0\right)}{\vec{L}_1 \cdot \left(\vec{L}_2 \times \vec{L}_3\right)} \tag{10.60}$$

This section concludes with a couple remarks on the formula for the basis elements. It is not difficult to show that the 2-dim basis can be rewritten as

$$f_n(x) = 1 + \frac{\left(\hat{e}_z \times \vec{L}_1 + \vec{L}_2 \times \hat{e}_z\right) \cdot \left(\vec{x} - \vec{x}_0\right)}{\hat{e}_z \cdot \left(\vec{L}_1 \times \vec{L}_2\right)} \tag{10.61}$$

Both expressions have the following form:

$$f_n(x) = 1 + \frac{\left(\sum_{i=1}^{N-1} \partial V_i \hat{n}_i\right) \cdot \left(\vec{x} - \vec{x}_0\right)}{V} \tag{10.62}$$

Where the denominator is the volume, in a rectangular solid with edges $\vec{L}_j$, each term $\partial V_i$ is the area of a face of the rectangular solid containing the special point $\vec{x}_0$, and $\hat{n}_i$ are the outward normal of each of these faces. It is also clear that when evaluated on the face opposite the point $\vec{x}_0$, the numerator evaluates to $-V$, so these functions vanish on the perimeter of the region where the basis function is defined. In lower dimensions this simple geometric interpretation is fairly easy to grasp. The next point is that all of these results come from solving a linear system:

$$-\vec{1} = \begin{pmatrix} \vec{L}_1^T \\ \vec{L}_2^T \\ \vdots \\ \vec{L}_N^T \end{pmatrix} \vec{\beta} \tag{10.63}$$

All vectors exist in an $N$-dimensional vector space $\vec{1} = [1, 1, 1, ..., 1]^T$; each row of the matrix is one of the vertex vectors relative to $\vec{x}_0$. Inverting the matrix will yield

$$\begin{pmatrix} \vec{L}_1^T \\ \vec{L}_2^T \\ \vdots \\ \vec{L}_N^T \end{pmatrix}^{-1} = \frac{1}{\det\left(\vec{L}_j^T\right)} \begin{pmatrix} C_{11} & C_{12} & \cdots & C_{1N} \\ C_{21} & C_{22} & \cdots & C_{2N} \\ \vdots & \vdots & \ddots & \vdots \\ C_{N1} & C_{N2} & \cdots & C_{NN} \end{pmatrix} \tag{10.64}$$

The determinant in the expression is just the total volume of the hyper-parallelepiped with edges $\vec{L}_j$. The entries in the inverse matrix are cofactors $C_{ij}$, developed by taking the determinant of the original matrix with the $i$th row and $j$th column removed. In lower dimensions these are the components of the lower-dimensional hypersurface areas. In three dimensions this is explicit, that is, the components of the inverse are row by row, the cross products of the edges.

## 10.5    Expressing the Helmholtz Equation in the FEM Basis

This section applies the FEM paradigm with the Galerkin approach to the generalized Helmholtz equation. Attention is focused on defining the terms that will contribute to the matrix form of the equation. To illustrate the application of the method in general terms, it will be applied to the form of the wave equation in Chapter 3 with the right-hand side ignored. A generic source term is included. The pressure field is assumed to be the only relevant variable. It is also assumed that the background fields do not contain any time dependence. This can be considered an approximation for small values of fluid motion. The wave operator is presented here for convenience:

$$
-\frac{1}{c^2 \rho_0}\frac{\partial}{\partial t}\left(\frac{\partial}{\partial t}+\vec{v}_0\cdot\vec{\nabla}\right)p-\vec{\nabla}\cdot\left(\frac{\vec{v}_0}{c^2\rho_0}\left(\frac{\partial}{\partial t}+\vec{v}_0\cdot\vec{\nabla}\right)p\right)+\vec{\nabla}\cdot\left(\frac{1}{\rho_0}\vec{\nabla}p\right) \tag{10.65}
$$

Assuming a time-independent environment and passing to the frequency domain via $p=u(\vec{x})\exp(-i\omega t)$,

$$
\frac{\omega^2}{c^2\rho_0}u+i\omega\frac{\vec{v}_0\cdot\vec{\nabla}u}{c^2\rho_0}+\vec{\nabla}\cdot\left(i\omega\frac{\vec{v}_0 u}{c^2\rho_0}-\vec{v}_0\frac{\vec{v}_0\cdot\vec{\nabla}u}{c^2\rho_0}+\frac{1}{\rho_0}\vec{\nabla}u\right) \tag{10.66}
$$

Equation (10.66) is the second-order term in the wave equation derived in Chapter 3 in the frequency domain. In this representation, the coefficients are complex quantities, and the unknown field $u$ is also represented as a complex quantity, $u=u_R+iu_I$. Equation (10.66) is linear and from this, equations for $u_R$ and $u_I$ are derived by separating the real and imaginary parts. The real and imaginary parts of (10.66) are provided separately in the following equations with terms grouped together according to independent field:

$$
\frac{\omega^2 u_R}{c^2\rho_0}-\vec{\nabla}\cdot\left(\frac{\vec{v}_0(\vec{v}_0\cdot\vec{\nabla}u_R)}{c^2\rho_0}-\frac{\vec{\nabla}u_R}{\rho_0}\right)+\omega\left(\frac{\vec{v}_0\cdot\vec{\nabla}u_I}{c^2\rho_0}+\vec{\nabla}\cdot\left(\frac{\vec{v}_0 u_I}{c^2\rho_0}\right)\right) \tag{10.67}
$$

$$
-\omega\left(\frac{\vec{v}_0\cdot\vec{\nabla}u_R}{c^2\rho_0}+\vec{\nabla}\cdot\left(\frac{\vec{v}_0 u_R}{c^2\rho_0}\right)\right)+\frac{\omega^2 u_I}{c^2\rho_0}-\vec{\nabla}\cdot\left(\frac{\vec{v}_0(\vec{v}_0\cdot\vec{\nabla}u_I)}{c^2\rho_0}-\frac{\vec{\nabla}u_I}{\rho_0}\right) \tag{10.68}
$$

The real and imaginary parts are coupled. In the pair of equations, there are two distinct operators:

$$\omega \left( \frac{\vec{v}_0 \cdot \vec{\nabla} \varphi}{c^2 \rho} + \vec{\nabla} \cdot \left( \frac{\vec{v}_0 \varphi}{c^2 \rho} \right) \right)$$

(10.69)

$$\frac{\omega^2 \varphi}{c^2 \rho} - \vec{\nabla} \cdot \left( \frac{\vec{v}_0 \left( \vec{v}_0 \cdot \vec{\nabla} \varphi \right)}{c^2 \rho} - \frac{\vec{\nabla} \varphi}{\rho} \right)$$

(10.70)

The Galerkin method is illustrated by application to these operators using linear basis elements. The following results ignore the boundary function $\widetilde{\varphi}(\vec{x})$ introduced in (10.2). Neglecting the boundary function can be thought of as imposing Dirichlet boundary conditions. The reader can tackle the inclusion of this term as an exercise. The same set of basis functions is used to express both fields ($u_R$, $u_I$), but different coefficients will result for each. The single equation in a complex field is now expressed as an equation for a 2-dim vector field. The divergence in each operator is removed by integration by parts, leading to the following results for each:

$$\omega \int \frac{\vec{v}_0 \cdot \left( \vec{\beta}_m b_n - b_m \vec{\beta}_n \right)}{c^2 \rho} dT_n$$

(10.71)

$$\omega^2 \int \frac{b_n b_m}{c^2 \rho} dT_n - \vec{\beta}_n \cdot \vec{\beta}_m \int \frac{1}{\rho} dT_n + \int \frac{\left( \vec{v}_0 \cdot \vec{\beta}_n \right) \left( \vec{v}_0 \cdot \vec{\beta}_m \right)}{c^2 \rho} dT_n$$

(10.72)

The integration region is indexed to match the projection into the specific basis element. However, the basis elements, $n$ and $m$, must share a common triangle for a nontrivial result so this can be ignored. The different index values refer to the distinct basis element containing the primary cell and hence refer to different vertices of the cell and different direction vectors $\vec{\beta}$. The contribution from (10.71) vanishes for self-coupling since $n = m$. This term is nonzero when evaluated for $n \neq m$. Evaluating these matrix elements will be the primary focus of the remaining part of this chapter. These results can be specialized for the case of no background flow $\vec{v}_0 = 0$. In this case the imaginary term in (10.66) is not present, and the unknown field can be represented by a real function with a single operator:

$$\omega^2 \int \frac{b_n b_m}{c^2 \rho} dT_n - \vec{\beta}_n \cdot \vec{\beta}_m \int \frac{1}{\rho} dT_n$$

(10.73)

In general, these operators will not be diagonal due to the overlap of basis elements by individual triangles or tetrahedra, but they will be sparse since basis elements must share at least one patch to yield a nontrivial result.

Because basis elements overlap, they will be coupled to each other. The rooftop basis functions will couple the nearest neighbors. Consider a single point in the mesh. The boundary points of this basis define the other basis functions that couple to this one. This will always be a finite number much smaller than the total number of points but will not be the same for every mesh. This can be verified by inspection of Figure 10.4. If a regular lattice is chosen, then the number of couplings can be the same for all interior points. In 1-dim there can only be two overlaps for every interior basis, for example, basis $n$ will couple to $n+1$ and $n-1$. Including self-coupling gives a total of three entries for each row of $\mathbf{B}$ corresponding to an interior point. In 2-dim and 3-dim different meshes will produce different size rows. For example, in Figure 10.2 each interior point, not near the boundary, couples to six other basis elements. Interior points near the boundary will have fewer couplings. Inspection of Figure 10.4 shows that in this case some interior elements are coupled to as few as four neighbors and some to six. Despite irregularity, it is fairly straightforward to identify these couplings and keep track of them. It is useful to consider the number of basis elements that share a particular triangle, or tetrahedron. Since each point is a basis and a triangle (tetrahedron) has only three (four) points, this must be the number of elements that share that triangle (tetrahedron). One can evaluate the contributions to the operator matrix elements on a per triangle basis (different use of the word) and store these values. Evaluating the complete coupling amounts to picking and summing the correct pre-calculated values. The boundary conditions are already taken care of with the boundary function. A take-away is that the FEM matrix will be sparse but not necessary tri-diagonal.

## 10.6 Numerical Integration over Triangular and Tetrahedral Domains

The remaining ingredient for completing the operator matrix elements is numerical integration over the basis elements. In the application of the Galerkin method, the original PDE was "projected" onto the same set of functions used to expand the unknown field. The integral over the basis element is a sum of integrals over the triangles in the basis element. Due to the nature of the basis element, this projection will yield a nontrivial result only when the two basis elements in the integrand share common triangles, that is, when the two basis elements overlap:

$$\langle b_n | L(b_m) \rangle = \sum_{T \in I} \int_T b_n L(b_m) dT \tag{10.74}$$

In (10.74) $T$ is a triangle, $dT$ is a differential area element on the triangle, and the notation $T \in I$ means sum over all triangles in the intersection of the two basis elements. For the source term the sum will be over all triangles in the basis $b_n$. The good news is that these integrals are not likely to be poorly behaved. Despite this they are also not likely to be known in closed form in terms of ordinary, or special, functions. A numerical integration technique is needed for evaluating these integrals. Following the theme of this chapter, numerical integration will be introduced for 1-dim intervals. The generalization to integration over rectangular domains in 2-dim and 3-dim is immediate, but for integration over triangular and tetrahedral domains, some additional steps will be needed. The 1-dim treatment will not be comprehensive, defaulting to Gaussian quadrature. This will serve as the foundation to developing higher-dimensional integration, with the exception of a special technique for integration over triangles introduced by Dunavant.

### 10.6.1   Gaussian Quadrature

Gaussian quadrature is introduced here as a foundation for building up integration over triangular and tetrahedral domains. For a complete introduction to numerical integration, the reader is referred to Ref. [2]. Classic numerical integration methods use evenly spaced samples of points on the interval $[a, b]$, $\{x_j\}$. Various methods of approximating the integral to a desired order lead to a choice of weighting factors $\{w_j\}$, associated with each point. The integral of the function $f(x)$, over the interval $[a, b]$, is then approximated by

$$\int_a^b f(x)dx = \sum_{j=1}^N w_j f(x_j) \tag{10.75}$$

Depending on the number of sample points taken, the approximation (10.75) will be exact for polynomials up to some order $N$. It often happens that one encounters integrands with an integrable singularity, expressed as $W(x)$:

$$\int_a^b W(x)f(x)dx \tag{10.76}$$

Gaussian quadrature provides a procedure for choosing not only the weights but also the sample points $\{x_j\}$ using a set of orthogonal polynomial functions, $P_n(x)$ with $n = 0, 1, 2, \ldots$, on $(a, b)$. Rather than equally sampling the interval, the quadrature method takes these points to be the roots of an $N$th-order polynomial from the set to construct an order integration scheme. Using an $N$th-order polynomial yields an integration scheme that is exact for polynomials of degree less than or equal to $2N - 1$. Once the roots of the polynomial $\{x_j\}$ are found to a desired accuracy, the weights are given by

$$w_j = \frac{\langle P_{N-1}|P_{N-1}\rangle}{P_{N-1}(x_j)\, P_N'(x_j)} \tag{10.77}$$

The numerator in (10.77) is defined by the following:

$$\langle P_{N-1}|P_{N-1}\rangle = \int_a^b W(x)P_{N-1}^2(x)dx \tag{10.78}$$

And in the denominator $P_N' = dP_N/dx$. The weights can be chosen to incorporate the function $W(x)$.

## 10.6.2   *Integration over Triangular Domains*

In this section two standard approaches for numerical integration over triangles are presented. Consider numerical integration on a two-dimensional rectangular region, $R = \{(x,y)\,|\,|x| \le a, |y| \le b\}$:

$$I = \int_{-b}^b dy\left\{\int_{-a}^a dx f(x,y)\right\} \tag{10.79}$$

A 2-dim quadrature routine for (10.79) can be built from two sets of sample points and weights for a one-dimensional Gaussian quadrature routine. Let these sample points and weights for the $x$ and $y$ coordinates be $\{x_n, w_n^x\}$, $n = 1, \ldots, N$, and $\{y_m, w_m^y\}$, $m = 1, \ldots, M$, respectively. The value of the integral can be approximated by the following combination of two 1-dim integrations:

$$I \approx \sum_{n=1}^{N} \sum_{m=1}^{M} w_n^x w_m^y f_{nm} \tag{10.80}$$

The function values at the sample points are denoted as $f_{nm}$. Defining a single index for the points on the 2-dim grid, $J(n, m), j = 1, \ldots, NM$, and new weights, $W_j = w_n w_m$, (10.80) is written as a single sum:

$$I \approx \sum_{j=1}^{J} W_j f_j \tag{10.81}$$

Integrating over triangles is a little trickier. One technique is to map the triangle into a square and then perform the integration over the square (see Hussain *et al.* [3]). This is accomplished by mapping the triangle to an isosceles right triangle with unit length legs along the $x$ and $y$ axes, followed by a mapping of this triangle into a square region. Let the vertices of the triangle be given by the vectors $\vec{V}_k$, $k = 1, 2, 3$. A set of parameters $\xi_i$, called simplex coordinates, is defined. An arbitrary point in the triangle is expressed as a weighted sum of the vertices in these coordinates:

$$\vec{x}\left(\vec{\xi}\right) = \xi_1 \vec{V}_1 + \xi_2 \vec{V}_2 + \xi_3 \vec{V}_3 \tag{10.82}$$

Each parameter is constrained: $0 \le \xi_i \le 1$, and $\xi_1 + \xi_2 + \xi_3 = 1$. In this space all triangles are equilateral triangles with vertices, $V_i$, at $\xi_i = 1$. The constraint can be used to eliminate a degree of freedom. For example, solving for $\xi_1$, $\xi_1 = 1 - \xi_2 - \xi_3$, gives the following:

$$\vec{x}\left(\vec{\xi}\right) = \vec{V}_1 + \xi_2\left(\vec{V}_2 - \vec{V}_1\right) + \xi_3\left(\vec{V}_3 - \vec{V}_1\right) \tag{10.83}$$

From Figure 10.8 it is clear that this is a projection of the equilateral triangle onto the $\xi_2, \xi_3$ plane with the vertex $\vec{V}_1$ mapped to the origin. In this representation, the standard triangle is

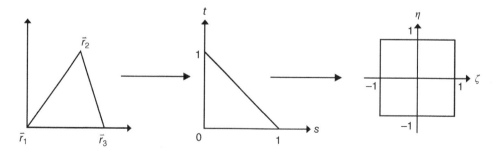

**Figure 10.8**   Mapping of a generic triangle (left) into the standard triangle (center) and to a square (right)

an isosceles right triangle. The procedure can be done for any pair of the three parameters. Without loss of generality, the original triangle is mapped into the $x, y$ plane, and (10.83) maps this into the $\xi_2, \xi_3$ plane. A new set of parameters is defined, $(s, t)$, to represent the two independent simplex coordinates. To express the integral in $(s, t)$ space, the Jacobian of the coordinate transform $(x, y) \rightarrow (s, t)$ is needed. It is straightforward to show that this transform gives

$$\left| \left( \vec{V}_2 - \vec{V}_1 \right) \times \left( \vec{V}_3 - \vec{V}_1 \right) \right| = 2A \tag{10.84}$$

In (10.84) $A$ is the area of the triangle. The next step is to define a map from a square to this right triangle. The square region is described by a new set of parameters $(\zeta, \eta)$, $\{(\zeta, \eta) | |\zeta|, |\eta| \leq 1\}$. These parameter transformations are illustrated in Figure 10.8.

The simplex coordinates $(s, t)$ are functions of these new parameters, $(\zeta, \eta)$:

$$s = \frac{1+\zeta}{2} \tag{10.85}$$

$$t = \left( \frac{1-\zeta}{2} \right) \left( \frac{1+\eta}{2} \right) \tag{10.86}$$

Along with the Jacobian for this transformation

$$J = \frac{1-\zeta}{8} \tag{10.87}$$

the integration over a triangle is transformed as follows:

$$I = \int_T f(x,y) dx dy = 2A \int_0^1 \int_0^{1-s} (s,t) dt ds = 2A \int_{-1}^1 d\eta \int_{-1}^1 d\zeta J(\zeta) f(\zeta, \eta) \tag{10.88}$$

It is understood that the function dependence in each new coordinate comes about through the coordinate functions:

$$f(s,t) = f(x(s,t), y(s,t)) \tag{10.89}$$

$$f(\zeta, \eta) = f(x(s(\zeta, \eta), t(\zeta, \eta)), y(s(\zeta, \eta), t(\zeta, \eta))) \tag{10.90}$$

In this form 1-dim quadrature can be applied to each of the parameters $(\zeta, \eta)$:

$$I = 2A \sum_{j=1}^{NM} W_j J_j f_j \tag{10.91}$$

It is sometimes customary to absorb the Jacobian into the definition of the weight $\hat{W}_j \equiv W_j J_j$. Better still developing a set of quadrature points and weights that incorporates the Jacobian should lead to better precision and less processing time when used in numerical integration routines.

This quadrature technique works well but it tends to cluster points in one corner of the triangle, creating an asymmetric sampling that oversamples one part of the triangle and undersamples other parts. An alternate quadrature is provided by Dunavant [4]. This approach samples the standard triangle in simplex space in a manner that is symmetric with respect to vertex exchange. The resulting quadrature points and weights provide high fidelity numerical integration with much fewer points than the previous technique requires. The theoretical details regarding the technique are beyond the scope of this text. The interested reader is referred to Ref. [5] for details. The original reference provides tables with quadrature points and weights up to order 20, a complete description of the algorithm, and sample code, for evaluating higher orders. Points occur in groups of either 1, 3, or 6, with each member of the group having the same weight and the coordinates of each related by a permutation. The unique parameter values are presented as rows of a matrix of this text, along with the weights as a column vector. An example is provided as follows for $p = 6$ in [5]:

$$P = \begin{bmatrix} 0.2492867 & 0.2492867 & 0.501427 \\ 0.0630890 & 0.0630890 & 0.8738220 \\ 0.3103525 & 0.6365025 & 0.0531450 \end{bmatrix} \tag{10.92}$$

$$w = \begin{bmatrix} 0.1167863 \\ 0.0508449 \\ 0.0828511 \end{bmatrix} \tag{10.93}$$

The results of Dunavant are rounded to seven decimal places. A single row of $P$ is a unique $(\xi_1, \xi_2, \xi_3)$. There are 12 points in all. They are generated by taking all unique permutations of these values. The first two rows each have a repeated value. There are only three unique permutations of these. The last point has three independent values for $(\xi_1, \xi_2, \xi_3)$ and produces a total of six integration points. The weights are ordered to match the rows of $P$. All points generated by permutation have the same weight as the original point. Figure 10.9 provides an

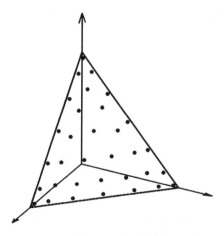

**Figure 10.9**   Example of Dunavant quadrature points in simplex space, order $= 12$

illustration of the Dunavant integration points on the triangle in simplex space for the numerical example given previously.

### 10.6.3  Integration over Tetrahedral Domains

Points in a tetrahedral region, $T$, in 3-dim can be described by three independent coordinates [6], four parameters, $\xi_i$, with $i = 1, 2, 3, 4$, $0 \le \xi_i \le 1$, and the following constraint equation:

$$\sum_{i=1}^{4} \xi_i = 1 \tag{10.94}$$

Points in $T$ are written as a weighted sum of the four vertices:

$$\vec{x}\left(\vec{\xi}\right) = \xi_1 \vec{V}_1 + \xi_2 \vec{V}_2 + \xi_3 \vec{V}_3 + \xi_4 \vec{V}_4 \tag{10.95}$$

Any tetrahedron can be mapped into a standard tetrahedron with vertices at $(0, 0, 0)$, $(1, 0, 0)$, $(0, 1, 0)$, and $(0, 0, 1)$, and the integral of a function over this standard tetrahedron can be expressed in volume coordinates as follows:

$$\int_T f(\vec{r}) d^3 \vec{r} = \int_0^1 d\xi_1 \int_0^{1-\xi_1} d\xi_2 \int_0^{1-\xi_1-\xi_2} d\xi_3 \, Jf(\xi_1, \xi_2, \xi_3) \tag{10.96}$$

The factor of $J$ in (10.96) is the Jacobian of the transformation from $\vec{r}$ to $(\xi_1, \xi_2, \xi_3)$. This coordinate transformation is performed by choosing one vertex of the tetrahedron, say, $\vec{V}_1$, and writing the position vector as

$$\vec{r} = \vec{V}_1 + \left(\vec{V}_2 - \vec{V}_1\right)\xi_2 + \left(\vec{V}_3 - \vec{V}_1\right)\xi_3 + \left(\vec{V}_4 - \vec{V}_1\right)\xi_4$$

The difference vectors define edges of the tetrahedron, $\vec{E}_i = \left(\vec{V}_i - \vec{V}_1\right)$, with common vertex $\vec{V}_1$. Any of the four vertices can be chosen for this. Calculating the Jacobian gives volume of the parallelepiped with edges $\vec{E}_i$ and is equal to or six times the volume of the tetrahedron:

$$J = \left| \vec{E}_2 \cdot \left(\vec{E}_3 \times \vec{E}_4\right) \right| \tag{10.97}$$

Ignoring the Jacobian factor for now, the integral over the simplex parameters can be transformed into a triple integral over a unit cube by a second change of variables. A new set of parameters is defined, $\{(\eta_1, \eta_2, \eta_3) | 0 \le \eta_i \le 1\}$. The transformation $\xi_i \to \eta_j$ is defined as follows:

$$\xi_1 = \eta_1 \eta_2 \eta_3 \tag{10.98}$$
$$\xi_2 = \eta_1 \eta_2 (1 - \eta_3) \tag{10.99}$$

$$\xi_3 = \eta_1(1-\eta_2) \tag{10.100}$$

The Jacobian of this transformation is

$$J = \eta_1^2 \eta_2 \tag{10.101}$$

And with this, the triple integrals in (10.96) are transformed into an integral over a cube:

$$\int_0^1 d\xi_1 \int_0^{1-\xi_1} d\xi_2 \int_0^{1-\xi_1-\xi_2} d\xi_3 f(\xi_1,\xi_2,\xi_3) = \int_0^1 d\eta_1 \int_0^1 d\eta_2 \int_0^1 d\eta_3 \eta_1^2 \eta_2 f(\eta_1,\eta_2,\eta_3) \tag{10.102}$$

Now that the integral is expressed in terms of three independent variables, it can be approximated using a direct product of three 1-dimensional quadrature points and weights:

$$\int_0^1 d\eta_1 \int_0^1 d\eta_2 \int_0^1 d\eta_3 \eta_1^2 \eta_2 f(\eta_1,\eta_2,\eta_3) \approx \sum_{i=1}^{N_1}\sum_{j=1}^{N_2}\sum_{k=1}^{N_3} w_i w_j w_k (\eta_{1i})^2 \eta_{2j} f(\eta_{1i},\eta_{2j},\eta_{3k}) \tag{10.103}$$

## 10.7 Implementation Notes

Like the FD technique the FEM matrices will be sparse. However, if a random tiling of the region is used, the number of nonzero elements of each row will not be constant. Each triangle (tetrahedra) will belong to exactly three (four) elements, but each element can have any number of tiles associated with it. The indexing scheme provided by Gmsh and other meshing tools make it easy to determine which basis elements overlap. Once a model environment is developed, the operator matrix elements can be evaluated, stored, and reused for a variety of sources present in the medium. This chapter presented the acoustic equations for the pressure field only. To account for the full solution, the velocity perturbations may be required. Modeling the velocity field involves the use of additional sample points in the mesh, termed edge or face elements. The pressure field is sampled at the nodes, while velocity vector components are sampled at nodes, or edges, and in the face of the tile. A detailed account of the use of this procedure can be found in Treyssède et al. [7] and Nguyen-Xuan et al. [8]. A comprehensive description of FEM can be found in Bathe [5].

## 10.8 Exercises

1. Include the boundary function $\tilde{\varphi}(\vec{x})$ in the description of the pressure field and work out the additional operator matrix elements that arise from applying FEM to (10.69) and (10.70).
2. With any choice of 1-dim quadrature points and weights, develop a set of points and weights for triangle integration using the first method of Section 10.6, using a mapping from the square to the standard triangle.

(a) Explicitly determine these points and weights for an order $(4, 4)$ quadrature, that is, order 4 in each dimension.

(b) Integrate the function $\exp(s + t)$ over the standard triangular region (1) using the points and weights you developed, (2) using the 12 point Dunavant points, and weight given in this chapter, and (3) analytically.

3. Using the 2-dim tiling in Figure 10.2 and the $O(N)$ method, estimate the size of the FEM operator for an $N \times N$ tiling.

4. Design and develop an FEM software procedure to solve a 2-dim acoustics problem with a position-dependent refractive index and 1-dim flow field with ideal boundary conditions.

# References

[1] Iserles, A., A First Course in the Numerical Analysis of Differential Equations, Cambridge University Press, Cambridge, 1996.

[2] Press, W. H., Teukolsky, S. A., Vetterling, W. T., and Flannery, B. P., Numerical Recipes in C++, Second Edition, Cambridge University Press, Cambridge, 1992.

[3] Hussain, F., Karim, M. S., and Ahmad, R., Appropriate Gaussian quadrature formulae for triangles, Int. J. Appl. Math. Comput., Vol. 4, No. 1, pp. 24–38, 2012.

[4] Dunavant, D. A., High degree efficient symmetrical Gaussian quadrature rules for the triangle, Int. J. Numer. Methods Eng., Vol. 21, pp. 1129–1148, 1985.

[5] Bathe, K., Finite Element Procedures, Second Edition, Klaus-Jürgen Bathe, Watertown, 2014

[6] Rathod, H. T. and Venkatesh, B., Gauss Legendre: Gauss Jacobi quadrature rules over a tetrahedral region, Int. J. Math. Anal., Vol. 5, No. 4, pp. 189–198, 2011.

[7] Treyssède, F., Gabard, G., and Ben Tahar, M., A mixed finite element method for acoustic wave propagation in moving fluids based on an Eulerian–Lagrangian description, J. Acoust. Soc. Am., Vol. 113, No. 2, pp. 705–716, 2003.

[8] Nguyen-Xuan, H., Liu, G. R., Thai-Hoang, C., and Nguyen-Thoi, T., An edge-based smoothed finite element method (ES-FEM) with stabilized discrete shear gap technique for analysis of Reissner-Mindlin plates, Comput. Methods Appl. Mech. Eng., Vol. 199, pp. 471–489, 2010.

# 11

# Boundary Element Method

## 11.1  Introduction

The boundary element method (BEM) is a numerical procedure for solving the boundary integral equations (BIE), an integral version of the Helmholtz equation. In previous chapters, numerical methods were applied to situations describing the propagation of sound through media with nontrivial properties in the frequency and time domains. This chapter focuses on situations where the medium is trivial, and propagation known exactly, but the boundary can be complicated. Boundaries can enclose a source or sources of sound, referred to as the interior problem, or enclose a source-free finite region of space not accessible to measurement, referred to as the exterior problem. The interior problem is well suited for investigating the behavior of sound in an enclosure and can be used in waveguide analysis and room acoustics, while the exterior problem is the foundation of scattering theory and used to determine the scattering cross section of complex shapes. The scattering from nontrivial objects is of particular importance in imaging and remote sensing studies, where an active probe is used to locate an object using monostatic or bistatic methods. This will serve as a motive for examples using the BEM technique.

A few salient points regarding this method, in comparison with previous methods, are mentioned here. The FD, FDTD, and FEM approaches to solving the Helmholtz equation involved discretizing the entire region of space where the acoustic field propagates. These techniques produced sparse matrices. The evaluation of operator matrix elements was fairly simple, trivial in some cases, and involved either function evaluations (as in the FD and FDTD methods) or numerical integrals of well-behaved functions (FEM). In the BEM technique, the propagation is known. The volume of space is replaced with only a description of the closed boundaries present. At first glance this is an improvement in that the size of the system of equations is greatly reduced. However, there are new devils to deal with. First will be the fact that every element will be coupled to every other element by Green's function, resulting in a dense matrix

*Computational Acoustics: Theory and Implementation*, First Edition. David R. Bergman.
© 2018 John Wiley & Sons Ltd. Published 2018 by John Wiley & Sons Ltd.

for the system of equations. Second is that the evaluation of the operator matrix elements requires integrating Green's function and its normal derivatives over two-dimensional surface patches. Because elements will couple to themselves, singular integrals will need to be evaluated. A purely numerical approach will not suffice to produce results for these terms. The technique of singularity extraction is presented for dealing with these integrals. Singular integrals are separated into a well-behaved part that may be evaluated numerically and a singular part that can be evaluated analytically or partially until singular behavior is absent. Three types of singular integrals will be encountered, two of which are known to produce finite results and for which exact closed-form results are known or partial analytic results are finite, making numerical integration tenable.

This chapter presents the theory behind BEM starting from the BIE and their discretization. The techniques for evaluating coupling integrals are explained and examples presented with a focus on scattering from acoustically hard bodies. As a final note this chapter uses a different convention than presented in others for the phase, that is, the engineering convention.

## 11.2   The Boundary Integral Equations

A BIE is developed for the Helmholtz equation. It is assumed that the medium is described by a constant sound speed. Sources are also present in the space, possibly at infinity. For the exterior problem, the source at infinity can be thought of as producing a plane wave for scattering problems. For the interior problem, this will not be present. The total acoustic field is the superposition of all fields present due to the various sources, along with that reflected by the boundaries. An arbitrary point in the free region of space is denoted $\vec{x}$, while sources are labeled by $\vec{x}'$. Lastly, one of the boundaries may be vibrating and serve as a source of sound in the medium. Each contribution to the total field due to source distributions may be written as

$$p_k(\vec{x}) = \int G(\vec{x} - \vec{x}') s_k(\vec{x}') dV' \tag{11.1}$$

The contribution to a plane wave traveling from infinity would be represented by

$$p_{inc}(\vec{x}) = p_0 \exp\left(-i\vec{k}_{inc} \cdot \vec{x}\right) \tag{11.2}$$

Boundaries are closed orientable surfaces in space. The surface may be smooth or it may be piecewise smooth. In the former case, there is a unique tangent plane and normal direction defined at each point, while for the latter case one has curves and points on the boundary where the normal is not unique. An example of the latter case would be the edges and corners on a cube.

There are two classes of boundary problems typically encountered, called interior and exterior. For the interior problem, the volume of space containing sources, and in which the acoustic field is to be determined, is completely enclosed. In the exterior problem, the volume of free space extends to infinity, and the boundary consists of a sphere at infinity and one or more closed surfaces embedded in space. These closed surfaces represent boundaries of other material bodies but do not contain sources embedded within them. One can assume ideal boundary conditions describing hard surfaces at the boundary, but this is not necessary.

The total field contained within free space is expressed as a superposition of that due to various sources and that radiated by the boundary in response to the scattering of acoustic waves from the sources, denoted $p_{sc}$:

$$p(\vec{x}) = \sum p_k(\vec{x}) + p_{inc}(\vec{x}) + p_{sc}(\vec{x}) \tag{11.3}$$

The term $p_{inc}$ represents an incident plane wave propagating through space and acting as an incident field on the other boundaries. For the time being the discussion is restricted to point sources.

The BIE is derived by an application of Green's identity. Consider two arbitrary scalar functions of coordinates, $\varphi_1$ and $\varphi_2$, and form the following vector fields from these functions and their gradients, $\vec{W} = \varphi_1 \vec{\nabla} \varphi_2$ and $\vec{U} = \varphi_2 \vec{\nabla} \varphi_1$. Applying Gauss' divergence theorem to these two vector fields gives the following:

$$\int_V \vec{\nabla} \cdot \left( \varphi_i \vec{\nabla} \varphi_j \right) dV = \int_{\partial V} \varphi_i \left( \hat{n} \cdot \vec{\nabla} \varphi_j \right) dS, \quad i \neq j, \quad i,j = 1,2 \tag{11.4}$$

where $V$ denotes a region of three-dimensional space, $\partial V$ the two-dimensional closed surface boundary to $V$, and $\hat{n}$ the local outward normal of the boundary, that is, $\hat{n}$ points away from $V$. The differential element of the bounding surface is denoted $dS$. Taking the difference of the two equations yields the following by direct substitution:

$$\int_V \left\{ \vec{\nabla} \cdot \left( \varphi_1 \vec{\nabla} \varphi_2 \right) - \vec{\nabla} \cdot \left( \varphi_2 \vec{\nabla} \varphi_1 \right) \right\} dV = \int_{\partial V} \left\{ \varphi_1 \left( \hat{n} \cdot \vec{\nabla} \varphi_2 \right) - \varphi_2 \left( \hat{n} \cdot \vec{\nabla} \varphi_1 \right) \right\} dS \tag{11.5}$$

The left-hand side (l.h.s.) can be reduced further to give

$$\int_V \left\{ \vec{\nabla} \cdot \left( \varphi_1 \vec{\nabla} \varphi_2 \right) - \vec{\nabla} \cdot \left( \varphi_2 \vec{\nabla} \varphi_1 \right) \right\} dV = \int_V \left\{ \varphi_1 \nabla^2 \varphi_2 - \varphi_2 \nabla^2 \varphi_1 \right\} dV \tag{11.6}$$

Equality of the right-hand sides of (11.5) and (11.6) yields Green's identity:

$$\int_V \left\{ \varphi_1 \nabla^2 \varphi_2 - \varphi_2 \nabla^2 \varphi_1 \right\} dV = \int_{\partial V} \left\{ \varphi_1 \left( \hat{n} \cdot \vec{\nabla} \varphi_2 \right) - \varphi_2 \left( \hat{n} \cdot \vec{\nabla} \varphi_1 \right) \right\} dS \tag{11.7}$$

First consider the exterior problem where there is a single closed surface embedded in space. The free space volume is that space outside the surface. To construct a boundary for this volume, a sphere is placed at infinity, that is, a large sphere in the limit as $R \to \infty$, $S_\infty$. Then the boundary of the volume consists of the surface, $S$, and $S_\infty$, $\partial V = S \cup S_\infty$. From the wave equation:

$$G(\nabla^2 + k^2)p - p(\nabla^2 + k^2)G = G\nabla^2 p - p\nabla^2 G = \vec{\nabla} \cdot \left( G\vec{\nabla} p - p\vec{\nabla} G \right) \tag{11.8}$$

Integrating both sides over the free space volume gives the following:

$$\int_V \left(G(\nabla^2 + k^2)p - p(\nabla^2 + k^2)G\right)dV = \int_V \vec{\nabla} \cdot \left(G\vec{\nabla}p - p\vec{\nabla}G\right)dV \qquad (11.9)$$

Applying the divergence theorem to the right-hand side and noting that the normal to the surface (the outward normal) points in the opposite direction as that used in the divergence theorem (*i.e.*, this vector points into the volume bound by $\partial V$) gives

$$\int_V \vec{\nabla} \cdot \left(G\vec{\nabla}p - p\vec{\nabla}G\right)dV = -\int_S \hat{n} \cdot \left(G\vec{\nabla}p - p\vec{\nabla}G\right)dS - \int_{S_\infty} \hat{n} \cdot \left(G\vec{\nabla}p - p\vec{\nabla}G\right)dS_\infty \quad (11.10)$$

The surface $S_\infty$ is a sphere of radius $R$, so the second integral is expressed in spherical coordinates and the limit $R \to \infty$ taken:

$$\lim_{R \to \infty}\left\{\int_{S_R}\left(G\frac{\partial p}{\partial R} - p\frac{\partial G}{\partial R}\right)R^2 d\Omega\right\} \qquad (11.11)$$

In (11.11) the normal is $-\hat{r}$, and $d\Omega$ is the differential solid angle, $\sin\theta d\theta d\varphi$. The integral is rearranged to allow use of the application of the Sommerfeld radiation condition:

$$\lim_{R \to \infty}\left\{\int_{S_R}\left(G\frac{\partial p}{\partial R} + ikGp - p\frac{\partial G}{\partial R} - ikpG\right)R^2 d\Omega\right\}$$

$$= \lim_{R \to \infty}\left\{\int_{S_R}\left((RG)\left(R\frac{\partial p}{\partial R} + ikRp\right) - (Rp)\left(R\frac{\partial G}{\partial R} + ikRG\right)\right)d\Omega\right\}$$

$$= \int_{S_R}\left(\left\{\lim_{R \to \infty}(RG)\right\}\left\{\lim_{R \to \infty}\left(R\frac{\partial p}{\partial R} + ikRp\right)\right\} - \left\{\lim_{R \to \infty}(Rp)\right\}\left\{\lim_{R \to \infty}\left(R\frac{\partial G}{\partial R} + ikRG\right)\right\}\right)d\Omega$$

A factor of $ikpG$ in the integrand has been added and subtracted in the first step. The second equality is merely rearranging factors. The third follows from the fact that the integration is independent of $R$ and that the limit of a product is equal to the product of the limits. Now the Sommerfeld radiation condition is applied. Both functions G and $p$ obey this condition and will go to zero at least as fast as $R^{-1}$ as $R \to \infty$. From this fact, the limits of $RG$ and $Rp$ are, at worst, constant in the limit as $R \to \infty$. This leaves the remaining factor

$$\lim_{R \to \infty}\left(R\frac{\partial f}{\partial R} + ikRf\right) \qquad (11.12)$$

where the function, $f$, may be either G or $p$. This is precisely the Sommerfeld condition. Each of these factors approaches zero as $R \to \infty$. Therefore, this integral vanishes in the limit leaving

$$\int_V \left(G(\nabla^2 + k^2)p - p(\nabla^2 + k^2)G\right)dV = -\int_S \hat{n} \cdot \left(G\vec{\nabla}p - p\vec{\nabla}G\right)dS \qquad (11.13)$$

The remaining work is in reducing the l.h.s. Now that most of the derivation is complete and the integrals in their final form, from this point on, the integration variables will be primed and an arbitrary field point unprimed. Figure 11.1 illustrates the variables used in the exterior problem.

The free Helmholtz equation implies that the first term in the l.h.s. of (11.13) vanishes. The second term evaluates to

$$\int_V p(\vec{x}')\delta(\vec{x} - \vec{x}')dV' \qquad (11.14)$$

The evaluation of this term depends on whether the evaluation point $\vec{x}'$ is inside $S$, outside $S$, or on the surface. For points inside $S$, (11.14) evaluates to zero. For points outside, the result is $p(\vec{x})$. For points on the surface, the result depends on whether the surface is smooth or has corners [1]. The exterior BIE is

$$\alpha(\vec{x})p(\vec{x}) = \int_S \left\{ \hat{n}' \cdot \vec{\nabla}' G(R)p(\vec{x}') - G(R)\hat{n}' \cdot \vec{\nabla}' p(\vec{x}') \right\}dS' + p_{inc}(\vec{x}) \qquad (11.15)$$

The factor $\alpha$ takes different values depending on whether the field point, $\vec{x}$, is outside, inside, or on the closed surface $S$:

$$\alpha = \begin{cases} 1 & \vec{x} \text{ outside } S \\ 1/2 & \vec{x} \text{ on } S, \text{ and } S \text{ smooth} \\ 0 & \vec{x} \text{ inside } S \end{cases} \qquad (11.16)$$

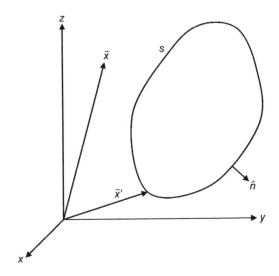

**Figure 11.1**   Definition of terms for the exterior problem

If $S$ has sharp edges, the value of $\alpha$ on $S$ must be determined by the following integral [1]:

$$\alpha(\vec{x}) = 1 + \frac{1}{4\pi} \int_S \hat{n} \cdot \vec{\nabla} \left( \frac{1}{R} \right) dS \tag{11.17}$$

The interior BIE looks identical to the exterior BIE with some changes in convention. For the interior BIE, the unit normal to $S$ is directed into the closed surface, and the factor $\alpha$ takes the following values:

$$\alpha = \begin{cases} 1 & \vec{x} \text{ inside } S \\ 1/2 & \vec{x} \text{ on } S, \text{ and } S \text{smooth} \\ 0 & \vec{x} \text{ outside } S \end{cases} \tag{11.18}$$

For cases where $S$ has sharp edges, the equivalent of (11.17) for the interior problem is

$$\alpha(\vec{x}) = \frac{1}{4\pi} \int_S \hat{n} \cdot \vec{\nabla} \left( \frac{1}{R} \right) dS \tag{11.19}$$

The BIE relates the value of the pressure field at any point in space to the values on the surface. Hence, once the surface data, $p(\vec{x}')$, and $\hat{n}' \cdot \vec{\nabla}' p(\vec{x}')$, is known, the field at any point in space can be calculated using the appropriate BIE. From the equations of fluid dynamics, the particle velocity and the pressure gradient are related:

$$\vec{v} = \frac{i}{\omega \rho_0} \vec{\nabla} p \tag{11.20}$$

In (11.20) the fluid velocity field is that due to acoustic fluctuations and $\rho_0$ is the fluid density. Equation (11.20) can be used to express the BIE in terms of the variables $p$ and $\vec{v}$:

$$\alpha(\vec{x}) p(\vec{x}) = \int_S \left\{ \hat{n}' \cdot \vec{\nabla}' G(R) p(\vec{x}') + G(R) ik Z_0 \hat{n}' \cdot \vec{v}(\vec{x}') \right\} dS' + p_{inc}(\vec{x}) \tag{11.21}$$

In (11.21) the wavenumber, $k = \omega/c$, replaces frequency and the acoustic impedance, $Z_0 = c\rho_0$, has been introduced. Treating $p$, and $\vec{v}$ as independent variables requires a second BIE for determining $\vec{v}$. Treating $p$, and $\vec{v}$ as independent for the time being, an equation for particle velocity is derived by taking the directional derivative of the original BIE with respect to the field point, $\vec{x}$:

$$\vec{\nabla} \left\{ \alpha(\vec{x}) p(\vec{x}) \right\} = \int_S \left\{ \hat{n}' \cdot \vec{\nabla}' \left( \vec{\nabla} G(R) \right) p(\vec{x}') - \vec{\nabla} G(R) \hat{n}' \cdot \vec{\nabla}' p(\vec{x}') \right\} dS' + \vec{\nabla} p_{inc}(\vec{x}) \tag{11.22}$$

The gradient with respect to $\vec{x}$ commutes with $\hat{n}' \cdot \vec{\nabla}'$ since $\vec{x}$ and $\vec{x}'$ are independent and $\hat{n}'$ is a function only of $\vec{x}'$. Equation (11.22) is a vector integral equation. It can be made into a set of scalar equations for each component by projecting onto an orthonormal set of basis vectors at each point in space, $\hat{e}_i$, $i = 1, 2, 3$. Note that the basis need not align with a global Cartesian coordinate system. For smooth surfaces the BIE for the $i$-th component of $\vec{v}$ becomes

$$\hat{e}_i \cdot \vec{\nabla} \, \alpha(\vec{x}) p(\vec{x}) - ikZ_0 \alpha v_i = \int_S \left\{ \hat{n}' \cdot \vec{\nabla}' \left( \hat{e}_i \cdot \vec{\nabla} G \right) p' + \hat{e}_i \cdot \vec{\nabla} \, GikZ_0 v'_n \right\} dS' + \hat{e}_i \cdot \vec{\nabla} \, p_{inc} \quad (11.23)$$

The shorthand notation, $v'_n = \hat{n}' \cdot \vec{v}(\vec{x}')$, is introduced since it is understood at this point which field depend on position, and the dependence of Green's function on $R$ is omitted for brevity. When evaluating (11.23) at a point on $S$, we take $\hat{e}_i = \hat{n}(\vec{x})$.

Equations (11.21) and (11.23) represent four BIE for the four variables $(p, \vec{v})$. For the remainder, it is assumed that $S$ is smooth. Given the appropriate data on $S$, the fields external to the bounded region can be calculated from

$$v_i = \frac{i}{kZ_0} \left\{ \int_S \left\{ \hat{n}' \cdot \vec{\nabla}' \left( \hat{e}_i \cdot \vec{\nabla} G \right) p' + \left( \hat{e}_i \cdot \vec{\nabla} G \right) ikZ_0 v'_n \right\} dS' + \hat{e}_i \cdot \vec{\nabla} \, p_{inc} \right\} \quad (11.24)$$

$$p = \int_S \left\{ \hat{n}' \cdot \vec{\nabla}' G p' + GikZ_0 v'_n \right\} dS' + p_{inc} \quad (11.25)$$

The quantity integrated over $S$ is the scattered field, that portion of the field in the fluid due to the reaction of the surface to the sources or incident field. For scattering problems and interior problems with sources, the primary task of the BEM is to determine the appropriate surface data given the inputs to the problem, that is, the sources or incident field. The BIE are evaluated on the surface by taking $\vec{x}$ to be a point on $S$. The points $\vec{x}$ and $\vec{x}'$ are still independent. From this procedure, a set of BIE for determining the surface data is created:

$$\alpha v_n = \frac{i}{kZ_0} \int_S \left\{ \hat{n}' \cdot \vec{\nabla}' \left( \hat{n} \cdot \vec{\nabla} G \right) p' + \left( \hat{n} \cdot \vec{\nabla} G \right) ikZ_0 v'_n \right\} dS' + \hat{n} \cdot \vec{v}_{inc} \quad (11.26)$$

$$\alpha p = \int_S \left\{ \left( \hat{n}' \cdot \vec{\nabla}' G \right) p' + GikZ_0 v'_n \right\} dS' + p_{inc} \quad (11.27)$$

Similar steps produce the interior equations.

## 11.3 Discretization of the BIE

The primary task at hand is inverting (11.27) and (11.26) to get values for the surface fields. To accomplish this several approximations are made to the continuum version of the problem. The first approximation is to replace the original boundary surface, $S$, with a set of non-overlapping discrete tiles. In theory, these tiles can be any shape, but one of the most popular and convenient is flat triangles. For surfaces with appropriate symmetry, for example, cubes or cylinders, rectangles can be used, but triangles are useful in the most general circumstances. The second approximation is to assume that the surface fields on each tile can be described by fairly simple functions, similar to the approach taken in the development of FEM. Individual tiles may be grouped together to form basis elements, as in the FEM case, or treated as individual basis elements. Basis elements should reflect the local behavior of the surface fields or an assumed reasonable approximation to them. The rooftop basis introduced in the chapter on FEM also works for BEM. In this section the discretization of the equations is independent of the choice of basis. Details regarding the choice of basis and its implementation are handled in a separate section. It is mentioned here that when a basis element contains more than one tile, this relation can be inverted to describe the behavior in one tile by the superposition of all basis elements that contain that tile.

From the first approximation, the integral of any function over the surface can be expressed as a discrete sum over the tiles. The individual tiles are indexed, $j = 1, \ldots, N$, each with a differential area element, $dS'_j$:

$$\int_S f(\vec{x}')dS' = \sum_{j=1}^{N} \int_{S_j} f(\vec{x}')dS'_j \tag{11.28}$$

The individual tiles are chosen to be much smaller than the wavelength of the external source field and small enough to accurately model the local curvature of the boundary. A good rule of thumb is $A < \lambda^2/100$. The surface fields are assumed to have simple behavior within each tile. The set of functions defined on these tiles is referred to as a basis. Let the basis functions on each tile be denoted $b_j(\vec{x})$, and then the surface fields are expressed as a sum over the basis functions:

$$p(\vec{x}) = \sum_{j=1}^{N} P_j b_j(\vec{x}) \tag{11.29}$$

$$v_n(\vec{x}) = \sum_{j=1}^{N} U_{n,j} b_j(\vec{x}) \tag{11.30}$$

When all of these modeling assumptions are used in (11.26) and (11.27), the following discrete versions are the result:

$$\frac{1}{2}\sum_{j=1}^{N} U_{n,j}b_j = \frac{i}{kZ_0}\sum_{j=1}^{N}\left\{P_j\int_{S_j}\frac{\partial^2 G}{\partial n'\partial n}b'_j dS'_j + ikZ_0 U_{n,j}\int_{S_j}\frac{\partial G}{\partial n}b'_j dS'_j\right\} + \hat{n}\cdot\vec{v}_{inc} \tag{11.31}$$

$$\frac{1}{2}\sum_{j=1}^{N} P_j b_j = \sum_{j=1}^{N}\left\{P_j\int_{S_j}\frac{\partial G}{\partial n'}b'_j dS'_j + ikZ_0 U_{n,j}\int_{S_j}G b'_j dS'_j\right\} + p_{inc} \tag{11.32}$$

The development of (11.31) and (11.32) has assumed that the original surface was smooth—hence the factor of $\alpha$ is constant and equal to 1/2 for each BIE. A shorthand notation has also been introduced for the directional derivative operator in the normal direction:

$$\frac{\partial G}{\partial n} = \hat{n} \cdot \vec{\nabla} G \tag{11.33}$$

Each side of (11.31) and (11.32) is a function of $\vec{x}$ constrained to $S$. The final step in developing the fully discrete version of the BIE is to sample the equations on the surface. To do this a set of test functions is defined on each tile, $t_i(\vec{x})$, and each equation integrated over the test functions. As with the basis functions, these test functions are only defined on their respective tile, or set of tiles, and zero everywhere else. The application of the test functions will result in $N$ equations for each of (11.31) and (11.32), one for each tile:

$$\sum_{j=1}^{N} U_{n,j} \mathbf{L}_{ij} = \sum_{j=1}^{N} P_j \mathbf{M}_{ij} + \sum_{j=1}^{N} U_{n,j} \mathbf{N}_{ij} + \mathbf{V}_i^{inc} \tag{11.34}$$

$$\sum_{j=1}^{N} P_j \mathbf{L}_{ij} = \sum_{j=1}^{N} P_j \mathbf{H}_{ij} + \sum_{j=1}^{N} U_{n,j} \mathbf{K}_{ij} + \mathbf{P}_i^{inc} \tag{11.35}$$

The following terms are defined in (11.34) and (11.35):

$$\mathbf{M}_{ij} = \frac{i}{kZ_0} \int_{S_i} \int_{S_j} \frac{\partial^2 G}{\partial n' \partial n} b'_j t_i dS'_j dS_i \tag{11.36}$$

$$\mathbf{N}_{ij} = -\int_{S_i} \int_{S_j} \frac{\partial G}{\partial n} b'_j t_i dS'_j dS_i \tag{11.37}$$

$$\mathbf{K}_{ij} = ikZ_0 \int_{S_i} \int_{S_j} G b'_j t_i dS'_j dS_i \tag{11.38}$$

$$\mathbf{L}_{ij} = \frac{1}{2} \int_{S_i} b'_j t_i dS_i \tag{11.39}$$

$$\mathbf{V}_i^{inc} = \int_{S_i} \hat{n} \cdot \vec{v}_{inc} t_i dS_i \tag{11.40}$$

$$\mathbf{P}_i^{inc} = \int_{S_i} p_{inc} t_i dS_i \tag{11.41}$$

The operator $\mathbf{H}_{ij}$ has a similar form as (11.37). Some authors prefer not to use particle velocity as a variable, opting to keep $\hat{n} \cdot \vec{\nabla} p$ as the other variable. The approach taken here is meant to provide the reader with an understanding of what is involved in converting the BIE to a discrete form, but there are many other approaches to the discretization. Finally, when pressure and normal velocity are treated as separate variables, different basis functions can be used for each. In fact, some authors point out that if a linear basis element is used for pressure, a constant or piecewise constant element should be used for the normal velocity to maintain consistency of (11.20) within the system. Equations (11.34) and (11.35) are recognized as components of a set of vector equations. Defining the vectors **P** and **U** for the unknowns leads to the matrix form of these equations:

$$\{L-N\}\cdot U - M\cdot P = V^{inc} \tag{11.42}$$

$$\{L-H\}\cdot P - K\cdot U = P^{inc} \tag{11.43}$$

Equations (11.42) and (11.43) represent the BEM equations, a discretized version of the BIE. At this point, specific boundary conditions and their effect on the degrees of freedom of the BEM equations are in order. For ideally hard bodies, the Neumann boundary condition $\hat{n}\cdot\overline{\nabla}p=0$ holds on the surface. For ideally hard surfaces, the BEM equations reduce to the following:

$$M\cdot P = -V^{inc} \tag{11.44}$$

$$\{L-H\}\cdot P = P^{inc} \tag{11.45}$$

For scattering from a pressure release surface, $p=0$ and the BEM equations become

$$\{L-N\}\cdot U = V^{inc} \tag{11.46}$$

$$-K\cdot U = P^{inc} \tag{11.47}$$

In more realistic cases the quantities **U** and **P** are related by the Robin boundary condition, relating the pressure and its normal derivative at the surface, $p+\beta v_n = \gamma$. Assuming $\gamma=0$ for illustrating a point, the normal velocity and pressure at the surface are related:

$$U = -\beta^{-1}P \tag{11.48}$$

The factor $\beta^{-1}$ can be a function of the insonification frequency, angle in incidence, and other factors. With (11.48) the particle velocity field can be removed completely:

$$\{\beta^{-1}\{L-N\}+M\}\cdot P = -V^{inc} \tag{11.49}$$

$$\{\{L-H\}+\beta^{-1}K\}\cdot P = P^{inc} \tag{11.50}$$

With this reduction of the degrees of freedom in place, there are now two equations for determining the surface pressure. The incident normal velocity, $V^{inc}$, still needs to be worked out, but this is not an issue as it is completely determined by the model of the incident field. It would seem that (11.50) is sufficient to solve the BEM system and (11.49) superfluous. However, there is a reason to keep (11.49). It turns out that when solving (11.50), the operator will become singular for frequencies that are equal to a natural resonance frequency of the object. This problem is remedied by combining the two sets of equations. Defining the operators,

$$B = \{\beta^{-1}\{L-N\}+M\} \tag{11.51}$$

$$A = \left\{ \{ \mathbf{L} - \mathbf{H} \} + \beta^{-1} \mathbf{K} \right\} \tag{11.52}$$

then multiplying (11.51) by a constant $\eta \in [0,1]$, and adding to (11.52) yields the combined equation

$$\{ \mathbf{A} + \eta \mathbf{B} \} \cdot \mathbf{P} = \mathbf{P}^{inc} - \eta \mathbf{V}^{inc} \tag{11.53}$$

The parameter $\eta$ is tunable; a common choice is 0.5. Discrete versions of the BIE are now in place and the task at hand is the definition of the surface tiles, basis elements, and the evaluation of the coupling integrals. A technical point worth mentioning here is that the evaluation of the matrix elements will require a quadruple integral, or a double double integral, over two patches. In contrast, the FEM only requires integration over a single basis element region. This is due to the FEM being a discretization of a PDE, whereas here in the BEM it is a BIE being discretized.

## 11.4   Basis Elements and Test Functions

The development of the discretized BIE introduced basis elements for expanding the fields on a set of flat tiles and a set of test functions for sampling the equations. Nothing was said about the nature of these functions except that they may cover more than one triangular tile, as in the FEM case. Some comments regarding the test functions are made, followed by a discussion on basis elements. There are two popular choices of test function, or test method. The first is called a point matching. This amounts to sampling the equations at a single point, or weighted average over points, in the interior of the tile. The point match test function can be expressed in the following notation as a Dirac delta:

$$t_i(\vec{x}) = \delta(\vec{x} - \vec{x}_i), \ \vec{x}_i \in S_i \tag{11.54}$$

The other popular choice is to let the test function be equal to the basis function on each tile:

$$t_i(\vec{x}) = b_i(\vec{x}) \tag{11.55}$$

This choice is referred to as Galerkin's method.

The basis element is meant to model the expected behavior of the surface field. When choosing a basis element for evaluating a BEM system, one needs to consider several competing factors. On the one hand, simpler functions lead to easier calculation of the matrix elements. On the other hand, over-trivializing the problem can lead to poor results and slow convergence of the method, requiring a greater number of tiles to improve fidelity. The simplest is the pulse basis, defined in a single patch and constant on that patch, zero everywhere else:

$$b_n(\vec{x}) = \begin{cases} 1 & \vec{x} \in \Delta_n \\ 0 & \text{Otherwise} \end{cases} \tag{11.56}$$

A better choice is to use the rooftop function introduced in the chapter on FEM. This identifies one basis element for each sample point on the surface. The rooftop function previously introduced uses linear functions, but in general higher-order polynomials can be used. One defines a shape function in simplex coordinates on each patch in the basis, $\varphi_n^j(\vec{\xi})$, where $n$ is the basis index and $j$ labels the specific triangle in that basis. The most common shape functions are linear and quadratic. In terms of the shape function operator, matrices can be expressed as

$$\mathbf{T}_{nm} = \sum_{j=1}^{K} \sum_{i=1}^{K'} \int_{T_j} \int_{T_i'} \varphi_n^j(\vec{\xi}) \varphi_m^i(\vec{\xi}') g(\vec{\xi}, \vec{\xi}') dT_i' dT_j \qquad (11.57)$$

In (11.57) $\mathbf{T}_{nm}$ represents any operator matrix and $g(\vec{\xi}, \vec{\xi}')$ is Green's function or one of its derivatives. This generic form represents (11.36)–(11.38).

## 11.5   Coupling Integrals

This section deals with the evaluation of the coupling integrals encountered in the various operator matrices used in BEM. Numerical integration over triangles was presented in the chapter on FEM and the reader is directed there for appropriate formulae. The operator matrix elements involve integration over two triangular patches. The primed integration will be referred to in this section as the inner integration, while the unprimed the outer integration. Some texts refer to the inner integral as a source integral and the outer integration points as test, or receiver points. This is appropriate considering that these matrix elements actually couple pairs of points by a propagator. Time will not be spent on integrations that can be performed readily using triangle quadrature. Of particular interest are the integrations involving Green's function and its derivatives. For pairs of patches at large distances, numerical integration is reliable. However, when two patches are close to each other, the singular behavior of Green's function makes numerical integration unstable. Even worse is the case of self-coupling. Some of the integrals will vanish when the inner and outer integration domains are identical, but others will not. Even when patches do not overlap but share an edge or vertex, the same problem arises. To deal with the singular integrals, a technique called singularity extraction is introduced. Specific closed-form expressions for some of these integrals are presented.

### 11.5.1   Derivation of Coupling Terms

This section focuses on evaluating the quadruple integrals that are encountered in (11.36)–(11.41). First note that the integrals in (11.40) and (11.41) are almost trivial for any choice of basis. For an incident plane wave, these integrals can be completed in closed form. For more complicated source distributions, these are well behaved but require numerical integration to evaluate. The methods introduced for evaluating FEM matrix elements carry over here without modification. The integral (11.39) can be done by hand for almost any reasonable choice of basis and test functions. The hard work comes in evaluating integrals (11.36)–(11.38). These involve Green's function and its first and second derivative. All of these integrands contain some order of singular behavior. When evaluating the coupling between separate non-overlapping basis elements, Green's function will never be evaluated at the singular point $\vec{x} = \vec{x}'$, so these cases are not as much of an issue. The real problem occurs when self-coupling is evaluated. To evaluate these integrals, explicit expressions for the normal derivatives of Green's function are developed. From the chain rule

$$\vec{\nabla} G(R) = \frac{dG}{dR} \vec{\nabla} R = \frac{dG}{dR} \frac{\vec{R}}{R} \tag{11.58}$$

and $\vec{\nabla}'G(R) = -\vec{\nabla} G(R)$. The normal derivatives are

$$\hat{n} \cdot \vec{\nabla} G(R) = \frac{dG}{dR} \frac{\hat{n} \cdot \vec{R}}{R} \tag{11.59}$$

$$\hat{n}' \cdot \vec{\nabla}'G(R) = -\frac{dG}{dR} \frac{\hat{n}' \cdot \vec{R}}{R} \tag{11.60}$$

To evaluate the second derivative of Green's function, the operator $\hat{n} \cdot \vec{\nabla}$ is applied to (11.60). Steps in the derivation are shown as follows:

$$\hat{n} \cdot \vec{\nabla} \left( \hat{n}' \cdot \vec{\nabla}'G(R) \right) = \hat{n} \cdot \vec{\nabla} \left( -G_{,R} \frac{\hat{n}' \cdot \vec{R}}{R} \right) = -\left( \hat{n}' \cdot \vec{R} \right) \hat{n} \cdot \vec{\nabla} \left( \frac{G_{,R}}{R} \right) - \frac{G_{,R}}{R} \hat{n} \cdot \vec{\nabla} \left( \hat{n}' \cdot \vec{R} \right)$$

$$= -\left( \hat{n}' \cdot \vec{R} \right) \frac{d}{dR} \left( \frac{G_{,R}}{R} \right) \frac{\hat{n} \cdot \vec{R}}{R} - \frac{G_{,R}}{R} \hat{n} \cdot \hat{n}'$$

$$= -\left( \hat{n}' \cdot \vec{R} \right) \left( \hat{n} \cdot \vec{R} \right) \frac{1}{R} \left( \frac{G_{,RR}}{R} - \frac{G_{,R}}{R^2} \right) - \frac{G_{,R}}{R} \hat{n} \cdot \hat{n}'$$

$$= -\frac{\left( \hat{n}' \cdot \vec{R} \right)}{R} \frac{\left( \hat{n} \cdot \vec{R} \right)}{R} G_{,RR} + \left( \frac{\left( \hat{n}' \cdot \vec{R} \right)}{R} \frac{\left( \hat{n} \cdot \vec{R} \right)}{R} - \hat{n} \cdot \hat{n}' \right) \frac{G_{,R}}{R}$$

The result is

$$\hat{n} \cdot \vec{\nabla} \left( \hat{n}' \cdot \vec{\nabla}'G(R) \right) = -\frac{\left( \hat{n}' \cdot \vec{R} \right)}{R} \frac{\left( \hat{n} \cdot \vec{R} \right)}{R} G_{,RR} + \left( \frac{\left( \hat{n}' \cdot \vec{R} \right)}{R} \frac{\left( \hat{n} \cdot \vec{R} \right)}{R} - \hat{n} \cdot \hat{n}' \right) \frac{G_{,R}}{R} \tag{11.61}$$

## 11.5.2   Singularity Extraction

An interesting feature of these integrals is the presence of $\hat{n} \cdot \vec{R}$. For self-coupling on flat patches, these terms will vanish. While the singular integrals will not directly contribute to the evaluation of all self-coupling terms, singular behavior will arise when evaluating the coupling between elements that share an edge or vertex. As the integrations are performed, points exactly on a shared edge or vertex will be problematic. However, it turns out that for the integral of Green's function and its first derivative, these results are known to be well behaved. One approach to avoiding the issue is to realize that quadrature rules for integration do not include the boundary points. Therefore, even a pure numerical integration of these types of cases will not evaluate the singularity. This is not really a viable path to numerically integrating these terms as the behavior of the integrals will prevent accurate quadrature evaluation at reasonable orders. Some of these integrals can be performed analytically to yield closed-form results that are functions only of the geometry of the two integration regions and their relative orientation. For the case of integrating Green's function, there is a well-known result for the quadruple integral. For other singular integrals, a popular technique is to perform one of the two integrations analytically and then use numerical quadrature for the other region. The analytic integration removes the singular behavior for all but one of the integrals, leaving a finite well-behaved function defined on the other region. An algorithm for evaluating these integrals will be presented in this section. Lastly, the only real culprit in the evaluation of self-coupling is in the evaluation of the elements of $\mathbf{M}$, (11.36). When both integrations are on the same patch, there is one surviving term:

$$-\frac{G_{,R}}{R} \tag{11.62}$$

This is known as a hypersingular integral with an $R^{-3}$ order singularity. Evaluating the integral of (11.62) has been the subject of research in BEM.

Continuing with the presentation of singularity extraction, the first and second derivatives of $G(R)$ with respect to $R$ are evaluated:

$$G_{,R} = -(ikR+1)\frac{G}{R} \tag{11.63}$$

$$G_{,RR} = (2+2ikR-k^2R^2)\frac{G}{R^2} \tag{11.64}$$

The behavior of the integrand becomes singular when dealing with self-coupling. If a numerical integration is attempted for self-coupling terms, contributions where $\vec{x} = \vec{x}'$ will be encountered. The singular behavior of Green's function in this limit is governed by the real part. Separating Green's function into real and imaginary parts,

$$G(R) = \frac{\cos kR}{4\pi R} - i\frac{\sin kR}{4\pi R} \tag{11.65}$$

The limit of the imaginary part is

$$\lim_{R \to 0} \mathrm{Im}(G(R)) = -\frac{ik}{4\pi} \tag{11.66}$$

while the real part diverges. The integral of $G(R)$ over a two-dimensional region is well behaved, as are many of the more singular integrals coming from derivatives of $G(R)$. Singularity extraction involves splitting Green's function into a part that is well behaved and a part that can be integrated over the patch to yield a closed form for one or both of the double integrations. The well-behaved part will still need to be integrated numerically but will be relatively easy to perform. It is clear that singular behavior occurs in the following functions: $G$, $G_{,R}$, and $G_{,RR}$. The specific singularities can be expressed as

$$\frac{G}{R^n} = \frac{1}{4\pi} \frac{\exp(-ikR)}{R^{n+1}} \tag{11.67}$$

where $n = 0, 1, 2, \dots$ The behavior of these functions near $R = 0$ is determined by looking at the Taylor series of the exponential:

$$\exp(-ikR) = 1 - ikR - \frac{1}{2}k^2 R^2 + \frac{1}{3!}ik^3 R^3 + \frac{1}{4!}k^4 R^4 - \frac{1}{5!}ik^5 R^5 + \cdots$$

$$= \left(1 - \frac{1}{2}k^2 R^2 + \frac{1}{4!}k^4 R^4 + \cdots\right) - i\left(kR - \frac{1}{3!}k^3 R^3 + \frac{1}{5!}k^5 R^5 + \cdots\right)$$

For simplicity, a shorthand notation for the truncated Taylor series out to finite order $n = N$ is introduced:

$$E_N(x) = \sum_{k=0}^{N} \frac{x^k}{k!} \tag{11.68}$$

By definition

$$E_\infty(x) = \exp(x) \tag{11.69}$$

The singularities can be successfully extracted from all of the terms appearing in the BIE integrals by subtracting out terms up to and including order $n$, leaving a leading-order term of order $R^{n+1}$ to cancel out the denominator in (11.67). In general,

$$F_n(R) = \frac{G}{R^n} = \frac{1}{4\pi} \left\{ \frac{\exp(-ikR) - E_n(-ikR)}{R^{n+1}} \right\} + \frac{1}{4\pi} \frac{E_n(-ikR)}{R^{n+1}} \tag{11.70}$$

This is written as the sum of a singular term, denoted, for example, by $F_n^S$, and a remainder, $\tilde{F}_n$, which is well behaved:

$$F_n(R) = \tilde{F}_n(R) + F_n^S(R) \tag{11.71}$$

The definition of each is obvious from the context. This strategy pays off for two reasons. The first is that the quantity in brackets is finite at $R=0$. In fact, the value of this term at $R=0$ for any order $n$ can easily be determined:

$$\tilde{F}_n(0) = \frac{(-i)^{n+1}}{4\pi} \frac{k^{n+1}}{(n+1)!} \tag{11.72}$$

The second payoff is that, for some values of $n$, integrals of the singular term over triangular domains can be done analytically over one triangle, in some cases both triangles. When the integration over one triangle is possible, the result is typically a non-singular function over the second triangular integration domain. Hence the self-coupling and near coupling terms may be performed with good precision and quickly by a combination of numerical and analytical methods.

The singularity extraction is performed on Green's function and its first and second derivatives to illustrate exactly how the procedure works. Based on (11.63) and (11.64), the following are required for this example:

$$G = \frac{1}{4\pi} \left\{ \frac{\exp(-ikR) - 1}{R} \right\} + \frac{1}{4\pi R} \tag{11.73}$$

$$G_{,R} = -\frac{(1 + ikR)}{4\pi} \left\{ \frac{\exp(-ikR) - (1 - ikR)}{R^2} \right\} - \frac{1}{4\pi} \frac{(1 + k^2 R^2)}{R^2} \tag{11.74}$$

$$G_{,RR} = \frac{(1 + ikR - k^2 R^2/2)}{2\pi} \left\{ \frac{\exp(-ikR) - (1 - ikR - k^2 R^2/2)}{R^3} \right\} + \frac{1}{2\pi} \frac{(1 + k^4 R^4/4)}{R^3} \tag{11.75}$$

The remainders of each term need to be evaluated at $R=0$ explicitly to make use of them in a software routine:

$$\tilde{G}(0) = -\frac{1}{4\pi} ik \tag{11.76}$$

$$\tilde{G}_{,R}(0) = \frac{1}{4\pi} \frac{k^2}{2} \tag{11.77}$$

$$\tilde{G}_{,RR}(0) = -\frac{1}{4\pi} i \frac{k^3}{3} \tag{11.78}$$

What remains is combining the following functions with the rest of the integrand and integration over pairs of triangular patches, or the same patch twice:

$$G^S = \frac{1}{4\pi} \frac{1}{R} \tag{11.79}$$

$$G_{,R}^S = -\frac{1}{4\pi}\left(\frac{1}{R^2}+k^2\right)$$

(11.80)

$$G_{,RR}^S = \frac{1}{2\pi}\left(\frac{1}{R^3}+\frac{1}{4}k^4R\right)$$

(11.81)

It is left as an exercise for the reader to work out explicit formulae for the other terms needed to complete the BIE calculations.

It is clear that well-behaved function for numerical quadrature, $\widetilde{F}_n(R)$, has been produced. However, it turns out that this is not always the best choice. In simplex coordinates, there is a cusp in the function $\overline{F}_n(R)$, that is, the derivative is discontinuous along a subset of the integration domain. An alternate approach is to generate singularity extractions that give functions with continuous first derivatives. This can be achieved by removing more terms than required in the extraction process:

$$F_{n,m}(R)=\frac{G}{R^n}=\frac{1}{4\pi}\left\{\frac{\exp(-ikR)-E_m(-ikR)}{R^{n+1}}\right\}+\frac{1}{4\pi}\frac{E_m(-ikR)}{R^{n+1}}$$

(11.82)

with $m>n$. Figure 11.2 shows an example of this applied to Green's function. Shown are the real part of Green's function (solid line), the singular part (dotted line), and the regular part (dashed).

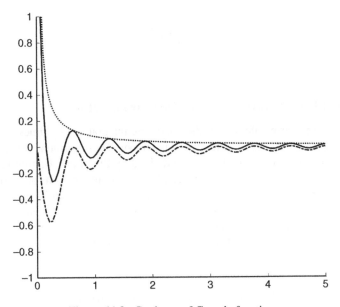

**Figure 11.2**   Real part of Green's function

## 11.5.3    Evaluation of the Singular Part

All the integrals for the operator matrix elements are of the form

$$\int_S f_i(\vec{x}) \left\{ \int_{S'} f_j(\vec{x}') g(\vec{x} - \vec{x}') dS' \right\} dS \tag{11.83}$$

In (11.83) $f_i$ is a non-singular scalar function defined on the patch and $g(\vec{x} - \vec{x}')$ is a singular function, for example, Green's function or any of its normal derivatives. Separating the singular function into its regular and singular parts gives the following:

$$\int_S f_i(\vec{x}) \left\{ \int_{S'} f_j(\vec{x}') \tilde{g} dS' \right\} dS + \int_S f_i(\vec{x}) \left\{ \int_{S'} f_j(\vec{x}') g^S dS' \right\} dS \tag{11.84}$$

The first integral in (11.84) can be evaluated using the triangle quadrature methods introduced for FEM analysis. Depending on the order of the singularity and the nature of the function $f_j$, these integrals may have a closed form. This is the case for integrals of Green's function and linear shape functions integrated over two triangular patches. In more complex cases it may still be possible to extract a closed form for the quadruple integral using Maple or other symbolic math software. The expressions involve multiple occurrences of the logarithm function with complex arguments. Expressing the integrand as the product of a singular and non-singular part allows one to evaluate the singular integrals by evaluating the non-singular part at one point and removing it from the integrand. Thus, attention is focused on integrals of the following form:

$$\int_{S'} \frac{1}{R^n} dS' \tag{11.85}$$

### 11.5.3.1    Closed-Form Expression for the Singular Part of K

A fairly simple closed-form solution for the quadruple integral of $G^S_{,R}$ times two linear basis functions over the same triangular patch twice is known. These formulae are presented here for the reader to use as they see fit. They were derived by Eibert and Hansen [2, 3] and used by the authors in electromagnetic calculations. Closed-form expressions are provided for the integration of linear basis functions over Green's function in Galerkin's method. The evaluation of these integrals makes use of the simplex coordinates defined in triangle patch integration for FEM. The three simplex coordinates are constrained, leaving two independent coordinates, defined here as $(s, t)$. For self-coupling the relative distance can be expressed as a quadratic form:

$$R^2 = a(s - s')^2 + c(t - t')^2 + 2b(s - s')(t - t') \tag{11.86}$$

The coefficients $a$, $c$, and $b$ depend on the geometry of the triangle, $a = L_1^2$, $c = L_2^2$, $a + c - 2b = L_3^2$. The quantities $L_i$ are the lengths of the edges of the triangle. To simplify results the following are defined:

$$\Lambda_1 = \ln \left| \frac{(L_1 + L_2)^2 - L_3^2}{L_2^2 - (L_3 - L_1)^2} \right| \tag{11.87}$$

$$\Lambda_2 = \ln \left| \frac{(L_2 + L_3)^2 - L_1^2}{L_3^2 - (L_1 - L_2)^2} \right| \tag{11.88}$$

$$\Lambda_3 = \ln \left| \frac{(L_3 + L_1)^2 - L_2^2}{L_1^2 - (L_2 - L_3)^2} \right| \tag{11.89}$$

In terms of these quantities, the following integrals are evaluated:

$$\frac{1}{4A^2} \int_T \int_{T'} \frac{1}{R} dS' dS = \frac{\Lambda_1}{3L_1} + \frac{\Lambda_2}{3L_2} + \frac{\Lambda_3}{3L_3} \tag{11.90}$$

$$\frac{1}{4A^2} \int_T \int_{T'} \frac{s'}{R} dS' dS = \frac{\Lambda_1}{8L_1} + \frac{L_1^2 + 5L_2^2 - L_3^2}{48 L_2^3} \Lambda_2 + \frac{L_1^2 - L_2^2 + 5L_3^2}{48 L_3^3} \Lambda_3 + \frac{L_3 - L_1}{24 L_2^2} + \frac{L_2 - L_1}{24 L_3^2} \tag{11.91}$$

$$\frac{1}{4A^2} \int_T \int_{T'} \frac{t's}{R} dS' dS = \frac{\Lambda_3}{40L_3} + \frac{L_1^2 + 3L_2^2 - L_3^2}{80 L_2^3} \Lambda_2 + \frac{3L_1^2 + L_2^2 - L_3^2}{80 L_1^3} \Lambda_1 + \frac{L_3 - L_2}{40 L_1^2} + \frac{L_3 - L_1}{40 L_2^2} \tag{11.92}$$

$$\int_T \int_{T'} \frac{s's}{R} dS' dS = \frac{2}{5} \int_T \int_{T'} \frac{s'}{R} dS' dS \tag{11.93}$$

All other integrals can be worked out using the previous results, redefining the parameters in the integrand, and massaging them into one of the forms shown earlier. Clearly, these are only valuable for evaluating **K** in (11.38).

### 11.5.3.2   Method for Partial Analytic Evaluation

This presentation is based on that given in the appendix of an article by Ylä-Oijala and Taskinen [4]. Their presentation was introduced for electromagnetic BEM but is applicable to some of the integrals encountered here. While this reference was used as a source for the procedure, the presentation in this text uses a slightly different notation that makes explicit the geometric quantities involved in evaluating the integrals using vector notation. Also, integrations presented in the reference that are not relevant to the acoustic BEM are disregarded. Consider the following integrals:

$$K_1^n = \int_T R^n dS' \tag{11.94}$$

$$\vec{K}_2^n = \int_T R^n \left( \vec{x}' - \vec{V} \right) dS' \tag{11.95}$$

$$\vec{K}_3^n = \int_T \vec{\nabla} R^n dS', \quad \vec{x} \notin T \tag{11.96}$$

The index $n$ may take values, $n = -1, 1, 3, 5, \dots$ The integration region $T$ is a triangle and the vector $\vec{V}$ is one of the vertices. The vertices are labeled $\vec{V}_i$, $i = 1, 2, 3$. Points in space relative to this triangle are referenced to a local coordinate frame. This frame is constructed using the normal vector of the tile, $\hat{n}$, and two orthogonal vectors, $\hat{u}$ and $\hat{v}$, in the plane of the triangle. Following the notation of the reference, the $\hat{n}$ coordinate is labeled $w$, and a point in space relative to this frame is labeled $(u, v, w)$. For simplicity $\vec{V}_1$ is placed at the origin and $\vec{V}_2$ along the horizontal axis, $u$, in two-dimensional space. From the vertex vectors three vectors along each edge of the triangle are defined:

$$\vec{L}_3 \equiv \vec{E}_{21} = \vec{V}_2 - \vec{V}_1 \tag{11.97}$$

$$\vec{L}_1 \equiv \vec{E}_{32} = \vec{V}_3 - \vec{V}_2 \tag{11.98}$$

$$\vec{L}_2 \equiv \vec{E}_{13} = \vec{V}_1 - \vec{V}_3 \tag{11.99}$$

Note that the edges of the triangle are labeled with the same index as the vertex opposite each edge, that is, the vertex excluded from the edge. The unit vectors may be written in terms of the triangle vertices:

$$\hat{u} = \frac{\vec{L}_3}{L_3} \tag{11.100}$$

$$\hat{v} = \hat{n} \times \hat{u} \tag{11.101}$$

The coordinate vector, $\vec{x}$, of any point in space, $p$, may be written in this local patch coordinate as follows:

$$u_p = \left( \vec{x} - \vec{V}_1 \right) \cdot \hat{u} \tag{11.102}$$

and similarly for $v_p$ and $w_p$. The projection of $\vec{x}$ into the plane of the triangle is defined, $\vec{\rho} = \vec{x} - w_p \hat{n}$. The coordinates of the third vertex, $\vec{V}_3$, are determined by setting $\vec{x} = \vec{V}_3$. An outward normal vector for each edge of the triangle is defined:

$$\hat{N}_i = \hat{L}_i \times \hat{n}, \quad i = 1, 2, 3 \tag{11.103}$$

Figure 11.3 shows the variables and all geometric quantities defined on the triangle. Capital letters for the normal in the plane are used to avoid confusion with the normal to the surface.

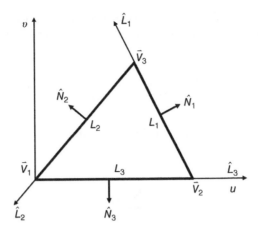

**Figure 11.3**   Variables used for evaluating singular integrals

The solution to the integrals involves a recursive formula, allowing higher-order integrals to be evaluated from the results of lower-order integrals. With these geometric quantities defined, the following parameters are introduced:

$$R_j^+ = R_{j+1}^- = \vec{x} - \vec{V}_{j+2} \tag{11.104}$$

$$s_j^+ = -\left(\vec{x} - \vec{V}_{j-1}\right) \cdot \hat{L}_j \tag{11.105}$$

$$t_j^0 = -\left(\vec{x} - \vec{V}_{j+1}\right) \cdot \hat{N}_j \tag{11.106}$$

$$R_j^0 = \sqrt{\left(t_j^0\right)^2 + w_p^2} \tag{11.107}$$

The indexing in these equations obeys the constraint $j=0 \rightarrow 3$ $j=4 \rightarrow 1$. Using these quantities several well-known results for developing a procedure to evaluate the singular integrals are presented, starting with the following two integrals:

$$I_j^{-1} = \ln\left(\frac{R_j^+ + s_j^+}{R_j^- + s_j^-}\right) \tag{11.108}$$

$$w_p K_1^{-3} = \begin{cases} 0, & w_p = 0 \\ \operatorname{sgn}(w_p) \sum_{j=1}^{3} \beta_j, & \text{otherwise} \end{cases} \tag{11.109}$$

$$\beta_j = \arctan\left(\frac{t_j^0\, s_j^+}{\left(R_j^0\right)^2 + |w_p|\, R_j^+}\right) - \arctan\left(\frac{t_j^0\, s_j^-}{\left(R_j^0\right)^2 + |w_p|\, R_j^-}\right) \tag{11.110}$$

The terms $\beta_j$ are only defined when $w_p \neq 0$. These two integrals are the starting point for a set of recursion relations that are used to evaluate any of the aforementioned integrals for arbitrary $n$:

$$I_j^n = \int_{\partial_j T} R^n dl' = \frac{1}{n+1} \left( s_j^+ \left( R_j^+ \right)^n - s_j^- \left( R_j^- \right)^n + n \left( R_j^0 \right)^2 I_j^{n-2} \right) \tag{11.111}$$

where $n = 1, 3, 5, \ldots$. Vector and scalar quantities are defined by taking a weighted sum of the previous integral:

$$\vec{I}^n = \sum_{j=1}^{3} \hat{N}_j I_j^n \tag{11.112}$$

$$I_0^n = \sum_{j=1}^{3} t_j^0 I_j^n \tag{11.113}$$

These combine to give the following for (11.94)–(11.96):

$$K_1^n = \frac{1}{n+2} \begin{cases} I_0^n, & w_p = 0 \\ n w_p^2 K_1^{n-2} + I_0^n, & w_p \neq 0 \end{cases} \tag{11.114}$$

$$\vec{K}_2^n = \frac{1}{n+2} \vec{I}^{n+2} + \left( \vec{\rho} - \vec{V} \right) K_1^n \tag{11.115}$$

$$\vec{K}_3^n = n w_p^2 \hat{n} K_1^{n-2} - \vec{I}^n \tag{11.116}$$

When using these formulae for near terms, one will likely have to deal with the $w_p \neq 0$ case. For self-coupling the position vector, $\vec{x}$, will be inside the same triangle as the one integrated over in developing these functions. The formulae simplify now since $w_p = 0$. There is only one case where this will be necessary and that is the integral (11.38). These formulae can also be used for the $R^{-3}$ singularity in cases except the self-coupling case due to the requirement of $w_0 = 0$. Regardless of the test function employed, if numerical integration is performed with respect to the unprimed variables, these functions will be evaluated at a discrete set of points and summed with appropriate weights.

Evaluation of some of these expressions involves functions that can become ill behaved for certain configurations. This needs to be checked in the process and separate computational paths formed for each possible case where a direct evaluation fails. Dealing with the logarithms in the closed-form expressions require care. There are configurations where the numerator or denominator is close to zero, leading to infinities. One also has to be careful regarding cases where either the numerator or denominator becomes negative due to roundoff error. For deciding these cases the vector form of the expressions is helpful. By inspection one can verify that factors like $R_j^\pm + s_j^\pm$ are positive definite. However, computers don't know this and it is likely that these terms will occasionally generate small negative numbers. Focusing on the logarithm function is helpful. Rather, the complete integral expression needs to be investigated. From (11.94),

$$K_1^{-1} = \int_T R^{-1} dS' \tag{11.117}$$

with $w_p = 0$. Now from (11.108) and (11.113),

$$K_1^{-1} = t_1^0 I_1^{-1} + t_2^0 I_2^{-1} + t_3^0 I_3^{-1} \tag{11.118}$$

Before getting too involved in trying to regulate the behavior of $I_j^{-1}$, take a deeper look at the combination, $t_j^0 I_j^{-1}$, for one value of the index:

$$t_1^0 I_1^{-1} = -\left(\vec{x} - \vec{V}_2\right) \cdot \hat{N}_1 \ln\left(\frac{\left\|\vec{x} - \vec{V}_3\right\| - \left(\vec{x} - \vec{V}_3\right) \cdot \hat{L}_1}{\left\|\vec{x} - \vec{V}_2\right\| - \left(\vec{x} - \vec{V}_2\right) \cdot \hat{L}_1}\right) \tag{11.119}$$

Referring to Figure 11.3, the singular behavior of the logarithm will occur inside the triangle for points along the line defined by $\hat{L}_1$. For any point, $\vec{x}$, on this line, and between the points $\vec{V}_2$ and $\vec{V}_3$, the vector $\left(\vec{x} - \vec{V}_2\right)$ will be parallel to $\hat{L}_1$. This is really just the point $\vec{x}$ relative to $\vec{V}_2$ as the origin of a new coordinate system. The vector $\left(\vec{x} - \vec{V}_3\right)$ will be antiparallel to $\hat{L}_1$. Hence the numerator of the logarithm argument will be $2\|\vec{x} - \vec{V}_3\|$, while the denominator will be zero. This is the culprit preventing an easy evaluation of the integral. Now consider the factor $t_1^0$. Based on previous comments it is clear that this term vanishes identically for any point on the $\hat{L}_1$ axis. In reality the term evaluates to $0(\infty)$ along the edge. For proper evaluation, a small amount, $\varepsilon$, is added in the $-\hat{N}_1$ to a point $\vec{x}$ on the edge, but not a vertex, and the behavior as $\varepsilon \to 0$ investigated. Applying L'Hôpital's rule twice gives a term that vanishes in the limit $\varepsilon \to 0$. This term is identically zero for any point on the $\hat{L}_1$ edge. The same results are obtained for the second and third terms in (11.119) for points on the $\hat{L}_2$ and $\hat{L}_3$ edges, respectively.

It is worth mentioning that the analysis of the behavior was made easier by the vector notation of the terms and their geometric interpretation. It is rare, if not impossible, to have a true edge point present in a numerical integration procedure, but it is wise to plan for them. Figure 11.4 illustrates the removal of the singularity by showing a plot of $K_1^{-1}$ for a test triangle with vertices $\vec{V}_1 = [0,0,0]^T$, $\vec{V}_2 = [0.1,0,0]^T$, and $\vec{V}_3 = [0.7,0.5,0]^T$. The horizontal coordinates are $(u, v)$, which are potential locations of outer triangle integration points. Figure 11.5 is a contour plot version of Figure 11.4.

The equation for $\vec{K}_3^{-1}$ can be used to evaluate the terms containing the first-order normal derivatives of Green's function, that is, $\mathbf{N}$. The method presented in this subsection also applies to near coupling that can regulate integration when the two patches share an edge or vertex, whereas the closed form cannot. As for equation (11.96) only being defined for points outside the plane in the inner triangle, this is not a deficit since $\mathbf{N}$ will vanish for these configurations.

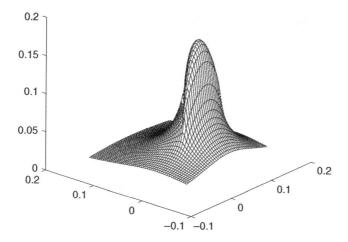

**Figure 11.4**   Evaluation of $K_1^{-1}$

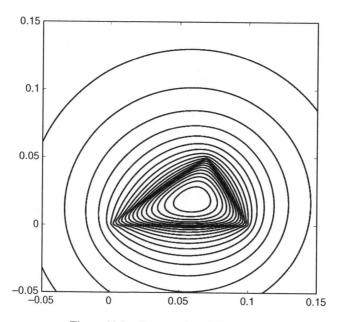

**Figure 11.5**   Contour plot of Figure 11.4

### 11.5.3.3   The Hypersingular Integral

So far details on the regularization of singular integrals have been discussed along with specific formulae form the literature that are useful for most of the integrals encountered. The remaining integral that occurs in self-coupling is the hypersingular integral. This is the remaining term in the second normal derivative of Green's function in the operator **M**. Evaluating (11.61) for self-coupling yields

$$-\frac{G_{,R}}{R} = (ikR + 1)\frac{\exp(-ikR)}{4\pi R^3}$$  (11.120)

The hypersingular term is given by

$$\frac{1}{4\pi R^3}$$

This type of function integrated over a 2-dimensional (2-dim) patch by standard methods does not completely remove the singular behavior, in contrast to the other singular integrals encountered. This is termed a hypersingular integral and analysis of integration of hypersingular functions requires a bit more care. An exact formulation of this integral will not be presented here, the reader is referred to numerous treatments in the literature [5–12]. There is more than one approach to evaluating this type of integral. An analysis of the Cauchy principal value (CPV) of the integral illustrates that a logarithmic singularity survives. Salvation comes through the discovery that the same order singularity occurs on neighboring triangular patches in the basis. Dropping the singularity and keeping the finite part of each patch integration is referred to as the Hadamard regularization. Other approaches apply integration by parts to reduce the singularity of the integrand. In a paper by Aimi and Diligenti [5], explicit expression for the inner integration and a proof of cancelation of the singularities is provided. The approach taken in Ref. [5] is clear and illustrates where the issue arises for dealing with hypersingular integrals proving examples.

## 11.6   Scattering from Closed Surfaces

The final equations of the last section are the starting point for the development of the numerical technique known and BEM. In a nut shell the method seeks to discretize the BIE so that an estimate of the surface fields may be determined. Then from these surface fields, the acoustic field at any point in the volume may be determined. An interesting and important point is that the same BIE is used for both calculations. To determine the surface fields, the BIE is evaluated for points, $x$, on the surface, $S$. Once known, these values are used as sources and the BIE is evaluated for points, $\vec{x}$, in the volume, $V$. Exterior problems are often used to model the scattering cross section of an extended object in space. The concept of scattering cross section is generic and appears in electromagnetism, acoustics, and particle physics. A specific application is the detection, tracking, and even imaging of objects using active remote sensors, typically arrays of sensors. This is the principle behind active radar and sonar systems. In such cases space is probed by sending out pulses of sound or electromagnetic waves of a particular dominant frequency. To generate a pulse of finite time duration requires some bandwidth in the signal. A truly monochromatic wave would have an infinite duration and is referred to as a c.w. (for continuous wave). If the pulse width is larger than the object, one can consider the interaction of the pulse and the object to be close to that of a c.w. signal. The emitted pulses may be omnidirectional or focused in narrow beam and aimed at a particular direction in space. Emitted signals propagate out, interact with a variety of obstacles, and return to detectors. The surfaces of the reflecting bodies act as sources, generating an acoustic field that propagates back to the receivers. When the detector is at the same position as the emitter, the process is referred to as monostatic detection; otherwise it is bistatic detection. In either case both the transmitter and receiver are considered to be very far away from the detected objects; being somewhat subjective, "very far" is defined relative to the object size, wavelength, and aperture of the array. In developing a model of detection for a remote sensor, one typically starts with a far-field assumption and an ideally hard body. Thus, it is beneficial to work out an expression for the BIE in the limit where $\| \vec{x} \| \gg \| \vec{x}' \|$. The scattered pressure wave is determined by

$$p_{sc}(\vec{x}) = \int_S \left( p(\vec{x}') \hat{n}(\vec{x}') \cdot \vec{\nabla}' G(R) - G(R) \hat{n}(\vec{x}') \cdot \vec{\nabla}' p(\vec{x}') \right) dS' \tag{11.121}$$

The only thing in the integrand that depends on $\vec{x}$ is Green's function. Therefore, a far-field version of this equation will require expanding $G(R)$ in this limit. The relative distance, $R = | \vec{x} - \vec{x}' |$, is Taylor expanded under the assumption $| \vec{x} | \gg | \vec{x}' |$:

$$R = \sqrt{x^2 + x'^2 - 2 \vec{x} \cdot \vec{x}'} = x \sqrt{1 + \frac{x'^2 - 2 \vec{x} \cdot \vec{x}'}{x^2}} = x \left( 1 + \frac{1}{2} \frac{x'^2 - 2 \vec{x} \cdot \vec{x}'}{x^2} + \cdots \right)$$

Keeping only terms linear in $\vec{x}'$,

$$R \approx x \left( 1 - \frac{\vec{x} \cdot \vec{x}'}{x^2} \right) = x - \hat{x} \cdot \vec{x}' \tag{11.122}$$

Both terms are retained when approximating the phase in the exponential but only the first term is used for the amplitude. With these approximations in place, the far-field Green's function becomes

$$G(R) \approx \frac{1}{4\pi} \frac{\exp(-ikx)}{x} \exp(ik\hat{x} \cdot \vec{x}') = G(x) \exp(ik\hat{x} \cdot \vec{x}') \tag{11.123}$$

The gradient of $G(R)$ in this approximation is

$$\vec{\nabla}' G(R) \approx G(x)(ik\hat{x}) \exp(ik\hat{x} \cdot \vec{x}') \tag{11.124}$$

Inserting these expressions into (11.121) gives the far-field version of the scattered field in the far-field limit:

$$p_{sc}(\vec{x}) \approx G(x) \int_S \left( ikp' \hat{n}' \cdot \hat{x} - \left( \frac{\partial p}{\partial n} \right)' \right) \exp(ik\hat{x} \cdot \vec{x}') dS' \tag{11.125}$$

Combining this with the output from the full BEM matrix equation provides a model of far-field scattering from complex objects. This is illustrated in Figure 11.6 for the case of bistatic

**Figure 11.6** Example of far field bi-static scattering from a sphere

scattering of an incident pressure wave from a hard sphere. For this example, the radius of the sphere is $r = 0.2$m and $kr = 12.6$. The sphere was covered with 11,860 triangular tiles.

## 11.7 Implementation Notes

Some items relevant to implementation of BEM are discussed here in reference to the exterior problem and scattering but are applicable in general. The BEM produces a dense matrix for the reaction of the surface to an incident field. This operator matrix is frequency dependent but otherwise independent of the incident field for ideal boundary conditions. There is no way around inverting this matrix, but for scattering studies where an object will be probed at multiple, perhaps thousands, of look directions, it is useful to perform an LU decomposition of the operators and save the results for multiple uses. Studies involving pulses rather than c.w. signal, that is, broadband, will require multiple BEM instantiations in the frequency domain. Depending on the type of studies being considered, it will be worth the investment to solve and store LU for a band of frequencies for a given scattering body. Once this information is known, the same LU results can be invoked for countless studies involving pulse envelopes and look directions. Such a collection constitutes an acoustic model of the body for that band. Storing LU reduces the amount of work in solving problems for multiple inputs. One of the costliest parts of BEM calculations in the author's experience is the evaluation of the matrix elements. This entire process can, and should, be parallelized to the best of one's ability. The cost of individual coupling evaluation depends on the approach taken and the details of the implementation. The techniques presented here, and in the references, require some experimentation to find optimal combinations of analytic and numerical techniques. When using rooftop basis or any that are defined on multiple tiles, it is better to evaluate tile to tile coupling and store these results, then full basis coupling evaluations are a matter of reading and combining data from an array of values. Additionally, the singular parts are independent of frequency, allowing one to reuse evaluation of these results for a single body over multiple frequencies. Related to this last comment is the choice of tile size. For a single frequency problem, a rule of thumb is to choose tiles with a characteristic length that is smaller than $\lambda/10$ where $\lambda$ is the wavelength of the incident plane wave. From a computational standpoint, it is desirable to push the envelope and try a larger tile size, for example, $\lambda/5$, to speed up processing and reduce size at the cost of fidelity. This is a personal choice that sometimes works. In the author's experience, when objects are flat or smooth with large radius of curvature compared with the wavelength, results can be "good," meaning they exhibit the correct behavior and provide reasonable peak strength for the scattered field. However, one has to be cognizant of whether the result is converging. This requires testing the scattered field again on a finer tessellation and comparing results. When dealing with problems that require a band of frequencies close to a center frequency, it is best to tile the body surface to satisfy the requirement imposed by the smallest wavelength and use this model for all wavelengths in the band. For new studies of the same body at a new set of frequencies, a new tessellation should be made. Consider a rooftop basis defined on a surface with $N$ vertices. To get a rough estimate the size of the BEM system of equation, the area of a tile is chosen to be $(\lambda/n)^2$, where $n$ is some divisor (typically 10). The number of tiles needed to cover the surface is roughly

$$N_T \sim n^2 \frac{A}{\lambda^2} \tag{11.126}$$

where $A$ is the surface area of the body. For a rooftop basis, the number of vertices is needed. For a closed surface the vertices, faces, and edges of a tessellation are related by the Euler's formula. When the number of tiles is very large, the number of tiles and number of vertices are related by the approximate formula $N_T \approx 2N_V$. If one considers as a model for interpreting (11.126) a large rectangle, then the formula essentially divides this into small rectangle of edge length $L \sim n\sqrt{A}/\lambda$. For triangle patches the number should be increased by a factor of 2. Hence it is reasonable to replace $N_T$ by $N_V$ in (11.126). As a concrete example, consider an underwater sonar with center frequency of 10 kHz operating in an environment with $c_0 \sim 1500$ m/s and scattering off a body modeled as a closed cylinder with length $L = 3.5$ m and radius $r = 0.5$ m. The wavelength is $\lambda = 0.15$ m. From the dimensions given the area of the cylinder is $A \sim 11$ m$^2$. To achieve a sampling that is approximately $\lambda/10$ requires $N_V \sim 49,000$, rounded up. The BEM matrix will be $N_V \times N_V$, containing 2.4 billion entries. If these are calculated and stored in double precision, the matrix will require $\sim$38GB of memory (why?). This is just for one matrix. It is possible to take advantage of symmetry properties of the matrices to reduce the required memory.

On the topic of choosing tile size, there is one more thing to consider. What is actually being solved is BIE for a piecewise flat surface, not a smooth continuous surface. The model was chosen to make the system solvable. Ideally the results should reveal information about the large-scale structure of the surface and ignore the small-scale structure of the tiles. This is why such a small tile was chosen, to act as an atom relative to this macro analysis. In practice the discrete facetization of the surface will inject some artifacts in the scattered field. This is sometimes referred to as facet noise. For appropriately chosen facet size, the artifacts are tolerable. The reason this is pointed out is to set a reasonable expectation for the reader who is building and testing their own BEM routines for the first time. The effects of the discrete facets can be seen by testing on symmetric bodies such as cylinders and spheres. Where one should expect a constant value for all incident field directions for the monostatic backscatter of a sphere, closer examination of a BEM calculation will reveal deviations in the estimated value as a function of the incident angles, $(\theta, \varphi)$. The mean value over all directions should be close to the theoretical value for the cross section of the sphere, but a true constant will not occur. For the example of Figure 11.6, the monostatic scattering cross section contains deviations when sampled at a variety of input directions, $(\theta, \varphi)$. The ratio of the mean value of $|p_{sc}|$ to the standard deviation is $\sim 8.1 \times 10^{-4}$. This is dependent on the size of the tiles; finer tiling will lead to a smaller deviation in monostatic cross section.

## 11.8   Comments on Additional Techniques

### 11.8.1   Higher-Order Methods

From the presentation of the BEM, one can see that it is an elegant approach to scattering problems in theory, but the details get involved fairly quickly. The fidelity of a particular implementation depends on several factors. Among these are the particular discretization of the surface, the choice of basis elements and test function method, the particular method for handling singularities in the self-coupling integrals, and the order of numerical integration for non-singular coupling between elements. The reader can plainly see that many parameters can be changed in the BEM algorithms, each affecting the numerical result and run time.

Demonstration of convergence is prudent to ensure good results are being obtained at each step. This puts one in a position to weigh the need for fidelity against speed. A poor fidelity operator matrix, or undersampled discretization of the surface, can lead to erroneous results. Two criteria for choosing patch size are wavelength and radius of curvature of the body. Given an incident field of wavelength $\lambda$, one is advised to choose triangle tiles subject to the constraint $A \lesssim (\lambda/10)^2$. This will ensure that the artificial surface details won't be probed by the incident field. But it must be noted that for some shapes one can get away with slightly bigger tiles, for example, $A \lesssim (\lambda/5)^2$. The other criterion is local radius of curvature $R_c$. If the surface has a variety of characteristic length scales, then smaller tiles should be used in regions where the surface curvature is large. In theory, if the smallest patch necessary is used to cover the entire surface, everything should work out. However, one will have to deal with unnecessarily high computation time. A better approach is to use triangles of different sizes on different regions of the surface. While none will be larger than $A \lesssim (\lambda/10)^2$, some will be chosen such that $A \lesssim (R_c/10)^2$, or some other sensible criterion. As tiles are made smaller, $N$ grows and with it the amount of memory required to hold the full matrix. The inversion time will also grow for larger matrices. This situation can be remedied to some degree with higher-order methods.

Higher-order methods employ two generalizations of the procedure presented here. The first is the use of basis functions with more detail, for example, quadratic instead of linear. The second is the use of curvilinear tiles to better account for the local curvature. The need to keep triangles smaller than $(\lambda/10)^2$ was in part driven by the fact that they contained no geometric information. From the general theory of curved surfaces embedded in $\mathbf{R}^3$, a point on a surface can be described by two parameters, $\vec{x}(u,v)$. This is a generalization of the theory of parameterized curves in $\mathbf{R}^3$, $\vec{x}(t)$. A collection of patches is defined that cover the surface. Given the description of a constrained vector on a patch of the surface, we can determine local tangent vectors at every point within the patch:

$$\vec{u} = \frac{\partial \vec{x}}{\partial u} \tag{11.127}$$

$$\vec{v} = \frac{\partial \vec{x}}{\partial v} \tag{11.128}$$

The parameters can always be scaled to ensure these are unit vectors. The local normal vector is then determined at each point by the cross product of these two tangent vectors:

$$\hat{n} = \frac{\vec{u} \times \vec{v}}{|\vec{u} \times \vec{v}|} \tag{11.129}$$

The metric of the surface is defined by the inner products of the tangent vectors:

$$g_{uu} = \frac{\partial \vec{x}}{\partial u} \cdot \frac{\partial \vec{x}}{\partial u} \tag{11.130}$$

$$g_{vv} = \frac{\partial \vec{x}}{\partial v} \cdot \frac{\partial \vec{x}}{\partial v} \tag{11.131}$$

$$g_{uv} = g_{vu} = \frac{\partial \vec{x}}{\partial u} \cdot \frac{\partial \vec{x}}{\partial v} \tag{11.132}$$

With this in place all of the boundary integrals are expressed in curvilinear coordinates. With curved patches, it is customary to define higher-order basis functions:

$$b(u,v) = \sum_{m,n=0}^{M,N} b_{mn} u^m v^n \tag{11.133}$$

Equation (11.133) generically expresses the behavior of a basis element that has higher-order than linear behavior in the interior parameters $u^m v^n$ using the notation of the section on singular integrals. These parameters are related to the simplex coordinates introduced in numerical integration over triangle and tetrahedra. The coefficients $b_{mn}$ are chosen to match a set of points on the patch. This can also be expressed in terms of a set of polynomials, an example of which are the Lagrange–Sylvester interpolation polynomials appearing in Wilton et al. [13].

## 11.8.2 Body of Revolution

This chapter closes with a brief mention of BEM modeling for bodies of revolution (BoR). This chapter has presented an introduction to the BEM technique applied to general bodies in 3-dim using a tiling of the object's surface with small flat triangles. When modeling bodies with rotational symmetry about one body axis, it is possible to formulate a complete description of the BEM in which one part of the solution involves the uses of a function expansion to account for the angular behavior of the surface fields, leaving only a 1-dim profile requiring discretization. Furthermore, if the boundary conditions have the same symmetry as the body shape, then the angular dependence vanishes completely, reducing the problem to a purely 1-dim problem.

## 11.9　Exercises

1. Perform a complete singularity extraction on the first and second derivatives of Green's function presented in this chapter.
2. Following the analysis of $K_1^{-1}$, determine what special cases need to be considered for evaluating $K_1^{-3}$, when $w_p \neq 0$ numerically. With MATLAB or other equivalent software, plot the function $K_1^{-3}$ over the parameter space of the outer integral.
3. Derive a boundary integral equation and BEM discretization for purely 2-dim scattering using $G = iH_0^{(1,2)}/4$ for Green's function (choice of super script depends on convention).
4. Estimate the number of patches and memory required for a flat tile patch BEM model of a submarine with approximate dimensions, length $L = 170\,\text{m}$ and mean radius $R \sim 7\,\text{m}$, insonified by a sonar with frequency, $f = 5\,\text{kHz}$, $f = 10\,\text{kHz}$, and $f = 25\,\text{kHz}$. Assuming that a curved tile patch can be chosen to be $\sim 1.5$ times the size for similar fidelity, what are the size and memory estimates if curved tiles are used?

## References

[1] Seybert, A. F., Soenarko, B., Rizzo, F. J., and Shippy, D. J., An advanced computational method for radiation and scattering of acoustic waves in three dimensions, J. Acoust. Soc. Am., Vol. 77, No. 2, pp. 362–368, 1985.

[2] Eibert, T. F. and Hansen, V., On the calculation of potential integrals for linear source distributions on triangular domains, IEEE Trans. Antennas Propag., Vol. 43, No. 12, pp. 1499–1502, 1995.

[3] Sievers, D., Eibert, T. F., and Hansen, V., Correction to "on the calculation of potential integrals for linear source distributions on triangular domains," IEEE Trans. Antennas Propag., Vol. 53, No. 9, pp. 3113, 2005.

[4] Ylä-Oijala, P. and Taskinen, M., Calculation of CFIE impedance matrix elements with RWG and n × RWG functions, IEEE Trans. Antennas Propag., Vol. 51, No. 8, pp. 1837–1846, 2003.

[5] Aimi, A. and Diligenti, M., Hypersingular kernel integration in 3D Galerkin boundary element method, J. Comput. Appl. Math., Vol. 138, pp. 51–72, 2002.

[6] Chien, C. C., Rajiyah, H., and Atluri, N., An effective method for solving the hypersingular integral equation in 3-S acoustics, J. Acoust. Soc. Am., Vol. 88, No. 2, pp. 918–937, 1990.

[7] Ervin, V. J. and Stephan, E. P., A boundary element Galerkin method for a hypersingular integral equation on open surfaces, Math. Methods Appl. Sci., Vol. 13, pp. 281–289, 1990.

[8] Ervin, V. J. and Stephan, E. P., Collocation with Chebyshev polynomials for a hypersingular integral equation on an interval, J. Comput. Appl. Math., Vol. 43, pp. 221–229, 1992.

[9] Gray, L. J., Glaeser, J. M., and Kaplan, T., Direct evaluation of hypersingular Galerkin surface integrals, SIAM J. Sci. Comput., Vol. 25, No. 5, pp. 1534–1556, 2004.

[10] Guiggiani, M., Krishnasamy, G., and Rudolphi, T. J., Numerical solution of hypersingular boundary integral equations, Trans. ASME, Vol. 59, pp. 604–614, 1992.

[11] Karafiat, A., Adaptive integration techniques for almost singular functions in the boundary element method, Comput. Math. Appl., Vol. 32, No. 5, pp. 11–30, 1996.

[12] Monegato, G., Numerical evaluation of hypersingular integrals, J. Comput. Appl. Math., Vol. 50, pp. 9–31, 1994.

[13] Wilton, D. R., Rao, S. M., Glisson, A. W., Schaubert, D. H., Al-Bundak, O. M., and Butler, C. M., Potential integrals for uniform and linear source distributions on polygonal and polyhedral domains, IEEE Trans. Antennas Propag., Vol. AP-32, No. 3, pp. 276–281, 1984.

# Index

*Computational Acoustics: Theory and Implementation*, First Edition. David R. Bergman.
© 2018 John Wiley & Sons Ltd. Published 2018 by John Wiley & Sons Ltd.

Printed and bound by CPI Group (UK) Ltd, Croydon, CR0 4YY

16/04/2025

14658504-0003